Waste Management in the Circular Economy

Suhaib A. Bandh · Fayaz A. Malla
Editors

Waste Management in the Circular Economy

Springer

Editors
Suhaib A. Bandh
Department of Higher Education
Government of Jammu and Kashmir
Srinagar, Jammu and Kashmir, India

Fayaz A. Malla
Department of Environmental Science
Government Degree College
Tral, Jammu and Kashmir, India

ISBN 978-3-031-42425-0 ISBN 978-3-031-42426-7 (eBook)
https://doi.org/10.1007/978-3-031-42426-7

This Springer imprint is published by the registered company Springer Nature Switzerland AG
The registered company address is: Gewerbestrasse 11, 6330 Cham, Switzerland

Paper in this product is recyclable.

Contents

List of Figures

List of Tables

Chapter 1
Waste Management and Circular Economy

Suhaib A. Bandh, Fayaz A. Malla, Shahid A. Wani, and Anh Tuan Hoang

1.1 Introduction

With the growing quantity of waste generation, the waste management has become an essential issue that needs urgent attention and action. There is a need for a more sustainable approach to waste management, and the solution could be the circular economy concept (Ribić et al., 2017).

Waste management and the circular economy are closely related and mutually dependent concepts that have gained increased attention from policymakers and eco-conscious companies in recent years. Waste management involves the collection, transportation, and disposal of waste materials through various methods, such as landfilling, incineration, or recycling (Nanda & Berruti, 2021). On the other hand, the circular economy is an economic model that aims to keep resources in use instead of throwing them away by designing out waste and pollution, keeping products and materials in use, and regenerating natural systems.

The traditional linear economy model, which is based on a take-make-use-throwaway approach, is no longer sustainable given the growing population, the limited availability of resources, and the pressing environmental issues we face today, such as climate change, air, and water pollution (Dissanayake, 2022). Therefore, the

S. A. Bandh (✉)
Department of Higher Education, Government of Jammu & Kashmir, Srinagar 190001, India
e-mail: suhaibbandh@gmail.com

F. A. Malla
Department of Environmental Science, Government Degree College Tral (Higher Education Department Government of Jammu & Kashmir), Srinagar, India

S. A. Wani
Sri Pratap College Campus, Cluster University Srinagar, Srinagar, Jammu & Kashmir 190001, India

A. T. Hoang
Faculty of Automotive Engineering, Dong A University, Danang, Vietnam

S. A. Bandh and F. A. Malla (eds.), *Waste Management in the Circular Economy*,
https://doi.org/10.1007/978-3-031-42426-7_1

circular economy offers a more sustainable alternative that benefits the economy, society, and the environment. It promotes a closed-loop system that emphasizes waste reduction, reuse, refurbishment, and recycling instead of the linear model's end-of-life disposal (Rajak et al., 2022).

Waste management plays a critical role in supporting the circular economy by collecting and separating waste materials that can be reused in the production process (Arena et al., 2021). For example, recycling plastic bottles can go a long way in reducing the demand for virgin materials and cutting down on greenhouse gas emissions associated with extraction and transportation. Waste-to-energy technologies, such as incineration or gasification, can also be used to convert waste into energy, reducing reliance on fossil fuels and mitigating climate change.

One prominent example of a company that has embraced the circular economy is the multinational IT giant, Dell (Zhang et al., 2022). The company has implemented a closed-loop recycling process that collects used computers, recycles them into new ones, and sells them back to customers. Dell's approach not only reduces waste sent to landfills but also delivers cost savings and energy efficiency benefits.

Another example of a circular waste management system is the city of Amsterdam's approach to managing biosolids, which are the solid by-products from wastewater treatment. Amsterdam has installed an innovative facility that processes biosolids into phosphorus, a vital ingredient in fertilizer (Gianico et al., 2021). By doing so, the city has reduced the environmental impact of waste disposal and prevented the loss of valuable resources.

The relationship between waste management and the circular economy is symbiotic, with each supporting and reinforcing the other's goals and objectives. Waste management helps to ensure the efficient collection, separation, and processing of waste materials that contribute to the circular economy's closed-loop system. The benefits of adopting a circular economy model are significant, with reduced waste and pollution, new business opportunities, cost savings, and more sustainable economic growth (Androniceanu et al., 2021). While implementing a circular economy requires significant investment from governments and companies, the potential rewards are well worth the effort.

1.2 Waste Management

Waste management has become an increasingly important issue in our world. As economies continue to develop and populations grow, the amount of waste generated by us has also increased rapidly. Therefore, it is important to tackle this issue as the amount of waste produced can have serious consequences for the environment and our health (Ray et al., 2022). Waste management, the process of treating, storing, or disposing of waste involves the following steps:

1. The first step in waste management is to reduce the amount of waste we produce. This can be achieved by adopting a recycling and reuse strategy as well as composting of organic waste to produce fertilizer. Recycling reduces the amount of raw materials extracted from the environment, slows the depletion of non-renewable resources, and saves landfill space (Musarat et al., 2022). When we recycle, we reduce greenhouse gas emissions by decreasing the energy used to manufacture new products. Promoting reuse culture reduces the need for manufacturing new products, reducing waste and the energy used to manufacture such products.
2. The second step is to dispose of the waste properly. Ineffective disposal of waste can lead to air and water pollution, the spread of harmful diseases, and soil contamination (Singh et al., 2020). To ensure proper disposal of waste, there are several methods such as landfilling, incineration, and recycling. Landfills involve burying the waste in the ground in a way that prevents it from contaminating the surrounding environment. Incineration involves burning the waste to reduce its volume, toxicity, and weight. Recycling entails repurposing waste into new products.
3. The third step is to ensure that hazardous waste is disposed of in an environmentally friendly manner. Hazardous waste is a significant risk to human and environmental health. It requires specialist treatment. These wastes include chemicals, batteries, medical waste, and electronic waste known as e-waste. Hazardous waste must be treated and disposed according to the country's regulations (Chisholm et al., 2021).
4. The fourth step is to adopt a circular economy model that focuses on maintaining the value of products as high as possible for as long as possible. Products are kept in use and are repurposed, upgraded, reused, or recycled where possible (Han et al., 2020). This model aims to minimize waste, support environmental sustainability, and stimulate economic growth.

Waste management is an essential strategy in minimizing the risks associated with waste to the environment and human health. Together, these actions will significantly contribute towards the building of a better world.

1.3 Circular Economy

In today's world, there is an increasing concern about the rate at which we are consuming the earth's resources. The traditional linear economy, based on a "take-make-dispose" model, is no longer sustainable (Elisha, 2020). It is now widely acknowledged that we need to shift to a more circular economy that aims to reduce waste and create more sustainable products and processes. A circular economy is a system that aims to keep products, components, and materials in use for as long as possible, maximizing their value and reducing the amount of waste generated

(Morseletto, 2020). It is based on three principles: designing out waste and pollution, keeping products and materials in use, and regenerating natural systems. The circular economy aims to replace the current linear economy, where products are created, used, and then discarded. The circular economy has numerous potential benefits. For example, environmentally, a circular economy would help reduce the depletion of natural resources, reduce waste, and reduce pollution. Economically, a circular economy would create new job opportunities and help increase the efficiency of the economy. Socially, it would lead to a more equitable distribution of wealth and resources, as well as greater consumer choice and improved access to goods and services (Schröder et al., 2020).

1.4 Principles of Circular Economy

The concept of a circular economy is gaining momentum as a way to address the environmental, economic, and social challenges we face today. By adopting the principles of designing out waste and pollution, keeping products and materials in use, and regenerating natural systems, a circular economy can create a more sustainable future for us all.

(i) **Designing out waste and pollution**: It involves creating products and processes that generate the least amount of waste and pollution possible. It can be achieved through the use of sustainable materials, designing products for longevity and recyclability, and ensuring that resources are used in the most efficient way possible. The circular economy also puts a focus on companies taking responsibility for the entire life cycle of their products, from the initial design to the end of its useful life (Maitre-Ekern, 2021).

(ii) **Keeping products and materials in use**: It means extending the life of products through reuse, repair, and refurbishment. Instead of discarding a product, it is resold or upgraded. Recycling is also an important consideration in a circular economy, and the aim is to create a closed-loop system where waste is transformed into new products, minimizing the need for virgin materials (Ribul et al., 2021).

(iii) **Regenerating natural systems**: It means restoring natural systems that have been depleted or damaged by human activities, for example, protecting and regenerating forests, oceans, and other ecosystems. In a circular economy, a systemic approach is taken, in which natural processes are integrated into production processes (Desing et al., 2020).

1.5 The Importance of Waste Management in Circular Economy

Waste management and circular economy are two concepts that are interconnected and have significant importance in the contemporary world. The circular economy is an alternative to the traditional linear economy, which follows the take-make-use-dispose model and promotes the consumption of natural resources (Barros et al., 2020). Differently, the circular economy aims to minimize the consumption of resources, waste production, and environmental impact. Waste management is an essential component of the circular economy and is critical in the transition from a linear to a circular model. Efficient waste management is necessary to ensure that waste materials are collected, sorted, and treated, following proper disposal protocols (Fatimah et al., 2020). This process contributes to the recovery and recycling of materials that can be used for other products or services, thus minimizing the production of waste. Waste management is crucial in reducing environmental pollution, preserving natural resources, and mitigating climate change. Proper management of waste reduces the volume of waste that ends up in landfills or dumped illegally in open spaces. It also helps to prevent the release of greenhouse gases, i.e. carbon dioxide, which contributes to global warming. By reducing the amount of waste that ends up in landfills, waste management plays a vital role in the conservation of valuable natural resources such as land, water, air, and minerals (Ramanayaka et al., 2020).

The importance of waste management in the circular economy lies in the fact that it creates a closed-loop system, where waste materials are collected, processed, and transformed into new products, reducing the need for virgin materials. The process of recycling and recovering resources helps to reduce the environmental footprint of production and consumption activities. Additionally, it presents economic opportunities through job creation and the development of new industries that use recycled materials to create eco-friendly products. Additionally, the proper waste management in the circular economy helps in the reduction of energy consumption. By recycling or recovering materials, it consumes less energy than producing products from virgin materials meaning the emission of fewer greenhouse gases and overall reduction of carbon footprint (Nguyen et al., 2021). In addition, waste can be used as a source of energy through landfill gas recovery, anaerobic digestion, and incineration, contributing to the production of clean energy.

Waste management is critical in the transition towards a circular economy, where resources are conserved, waste is minimized, and the environment is protected. It plays a crucial role in reducing environmental pollution, preserving natural resources such as land, water, air, and minerals, and mitigating climate change. By recycling and recovering materials, it creates job opportunities and economic benefits and promotes the development of eco-friendly industries (Kurniawan et al., 2022). Waste management also contributes to the reduction of energy consumption and the production of clean energy, making it an essential component of the circular economy. It is important to promote awareness and encourage participation in waste management initiatives to achieve the goals of a sustainable circular economy.

1.6 Types of Waste

Waste is a by-product or unwanted material that results from human activities, household, industrial, agricultural, or other sources. It can be managed and disposed of properly to avoid environmental degradation and to promote sustainable development. The types of wastes are diverse and can be classified based on their origin, composition, and physical properties. Following are a few important categories of wastes:

(i) **Municipal solid waste (MSW)**: The municipal solid waste includes garbage and other household products such as newspapers, plastic packaging, food wastes, and other types of materials (Singh et al., 2014). These wastes are typically collected and transported by local authorities to landfills for disposal. However, landfills are not always environmentally friendly. MSW can also be incinerated to generate power or recycled.

(ii) **Hazardous waste**: These wastes are produced by industrial and medical practices. Hazardous wastes include toxic chemicals, radioactive materials, and bio-hazardous waste, which pose a significant risk to both human health and the environment (Ghasemi et al., 2016). Hazardous wastes require specialized disposal methods to prevent their spread and potential damage.

(iii) **Construction and demolition waste**: These are the wastes that result from building and demolishing structures. These materials include steel, concrete, wood, and other debris and can have a significant impact on the environment (Kucukvar et al., 2014). While some waste is often reused in new construction projects, most of it is hauled to landfills for disposal.

(iv) **Electronic waste, or e-waste**: these wastes include computers, televisions, phones, and other electronic devices that contain hazardous materials such as lead, mercury, and cadmium. Improper disposal of e-waste can lead to the release of toxic chemicals that can contaminate the soil and waterways (Perkins et al., 2014).

(v) **Organic waste**: These are the wastes that comprise biodegradable materials such as food waste, livestock manure, yard waste, and other decomposable material. Organic waste can be recycled into fertilizer through composting and used to enrich soil and gardens. Composting not only reduces the amount of waste sent to landfills but also plays an essential role in reducing environmental pollution (Gonawala & Jardosh, 2018).

(vi) **Agricultural waste**: It is a type of waste associated with agricultural activities and includes the remains of crops and animal waste. Livestock manure can contaminate rivers and other water bodies if not disposed of appropriately, while crop residues can be used to produce biofuels and other commodities.

1.7 The Three Rs of Circular Economy

The concept of circular economy has been gaining traction worldwide as a potential solution to the growing problem of waste management. Circular economy aims to reduce waste by designing systems that are regenerative and sustainable (Suárez-Eiroa et al., 2019). One of the most important aspects of a circular economy is the "three Rs" principle: Reduce, Reuse, and Recycle. This principle is a simple but effective way to guide individuals and organizations towards a more sustainable and responsible way of living.

The first "R" of the principle is "reduce". It is the most effective way of preventing waste from being generated, thus reducing the overall use of resources to the minimum necessary level. There are numerous ways to reduce the amount of waste that we generate. Some of these methods include buying products with minimal packaging, consuming less meat, using public transport instead of cars, and using energy-efficient appliances. By reducing the amount of waste generated, we can reduce the stress on our environment by conserving natural resources (Ragheb et al., 2016).

The second "R" is the "reuse". It encourages people to reuse products in new ways instead of buying new ones. Reusing items helps to extend their lifespan and reduces the overall demand for resources. Reusing could be as simple as using a shopping bag more than once or as complex as upcycling furniture instead of throwing it away. Reusing items also saves money and helps to reduce the environmental damage that comes with the production of new products (Castellani et al., 2015).

The third "R" is the "recycle". It is the process of converting waste materials into new materials and objects. Recycling helps to conserve natural resources, reduce greenhouse gases, and reduce the amount of waste that ends up in landfills. Recycling has numerous benefits beyond environmental protection, including energy savings, creating jobs, and reducing the cost of raw materials.

The three Rs of circular economy are essential to achieving a more sustainable, thriving, and circular system. Though there are challenges to implementing these principles, the benefits far outweigh the costs. By reducing, reusing, and recycling, we can achieve a future where our resources are conserved, waste is minimized, and the environment is protected. It is our responsibility as individuals and organizations to develop the necessary habits and infrastructures to move towards a circular economy and make our world a better place for future generations (Gazzola et al., 2020).

1.8 Challenges in the Implementation of 3Rs

There are many challenges when it comes to implementing the three Rs of circular economy.

(i) ***Getting people to change their behaviour***: It takes time and effort to switch from a consumptive model of behaviour to one that is more mindful of the use and disposal of resources (Terlau & Hirsch, 2015). However, with growing environmental concerns, there has been an increase in public awareness about the need for sustainable living, leading to an increased adoption of the three Rs.

(ii) ***Lack of infrastructure***: Lack of infrastructure is another challenge in the implementation of a circular economy model (Sharma et al., 2019). Building the necessary infrastructure, such as recycling and composting facilities, can be expensive. However, the demand for e-waste recycling, organic waste management, and plastic recycling, for example, has stimulated various initiatives to build facilities and create systems that allow for effective waste management.

1.9 Key Elements of Circular Economy Model in Waste Management

Implementing the key elements of the circular economy model in waste management is essential for achieving a sustainable and regenerative system.

(i) **Extended producer responsibility (EPR)**

Extended producer responsibility is a policy approach that holds manufacturers responsible for the entire life cycle of the products they produce. This includes ensuring that their products are designed for easy disassembly and recycling at the end of their life cycle.

(ii) **Product life extension and maintenance**

Extending the life of products through maintenance and repair is another key element of the circular economy model. This can be achieved through designing products that are easy to repair and providing access to repair services.

(iii) **Waste-to-energy and biomimicry**

Waste-to-energy and biomimicry are two strategies that can be used to transform waste into valuable resources. Waste-to-energy involves converting waste into energy through incineration or other methods, while biomimicry involves emulating natural systems to create sustainable solutions to waste management.

1.10 Strategies for Effective Waste Management in Circular Economy

According to a report released by the World Bank, global waste generation is expected to increase by 70% in the next few years (Ghosh, 2020). As such, waste management is becoming increasingly critical to environmental sustainability and mitigating resulting adverse effects. The circular economy model provides a framework to reduce waste generation and promote a circular flow of materials, products, and resources (Perey et al., 2018). Here are some strategies for effective waste management in the circular economy.

(i) ***Prioritizing waste prevention over waste recycling***: This approach seeks to reduce the volume of waste produced by promoting sustainable consumption and production. For instance, firms can employ eco-design principles to incorporate the remanufacturing or recycling opportunities into the initial product design. By extending the product's life cycle, the amount of waste produced will reduce significantly (Yeheyis et al., 2013). This strategy also includes a focus on reuse options and alternative solutions to single-use items.

(ii) ***Creating a predetermined plan for waste management at each stage of the supply chain***: This can be achieved by deploying reverse logistics or the 5R principles (Reuse, Reduce, Recycle, Retain, and Recover). Firms can work together to develop channels that support the integration of secondary materials into the product value chain and direct product flow back to manufacturers for reuse (Frei et al., 2020).

(iii) ***Deploying innovative technologies and processes to minimize waste and increase sustainable practices***: These can include the development of closed-loop processes that support waste minimization through the sharing of resources. For example, companies can implement advanced sorting and processing technologies for remanufacturing. This will enable the separation of different waste streams for reprocessing at the source to create a cycle of production and consumption.

(iv) ***Establishing a regulatory framework that promotes waste management, emphasizing economic and social responsibilities***: Policy-makers can advance initiatives that incentivize companies to adopt circular economy practices. Monetary incentives on the one hand and disincentives such as fines in case of non-compliance on the other hand should be set in place to drive forward sustainability.

(v) ***Implementing circular economy waste management strategies through cross-sectoral collaborations and partnerships***: Collaboration with non-governmental organizations has the potential to increase awareness of the importance of waste management, while partnerships with communities can increase their engagement and participation in the implementation.

(vi) ***Designing for a circular economy: cradle to cradle***: Designing products and materials with a cradle-to-cradle approach is an essential strategy for creating a closed-loop system. This means designing products and materials that can be easily disassembled, recycled, or composted at the end of their life cycle.

(vii) ***Closed-loop systems and sustainable resource management***: Creating closed-loop systems that minimize waste and maximize resource efficiency is another key strategy for waste management in a circular economy. This includes implementing sustainable resource management practices, such as using renewable energy sources and reducing the use of virgin resources.

1.11 Waste Management and Corporate Social Responsibility

As our society continues to grow and expand, the issue of waste management becomes increasingly important. We produce more waste than ever before, and it is essential that we handle it responsibly. Waste management is not just an environmental issue, but also an issue of corporate social responsibility. Businesses have a responsibility to manage their waste in an environmentally friendly manner and to minimize their impact on the environment (Chuang and Huang, 2018). One of the most significant impacts of waste is its effect on the environment. Landfills, where most waste ends up, emit large amounts of greenhouse gases such as methane, which contributes to climate change. In addition, waste can contaminate air, water, and soil, which can have serious health effects for humans and wildlife. By properly managing their waste, companies can help to mitigate these impacts and promote sustainability.

Corporate social responsibility requires businesses to be accountable for their actions and recognize the impact they have on society and the environment (Ashrafi et al., 2020). Managing waste responsibly is part of this responsibility. Companies can improve their waste management practices by evaluating their waste production and finding ways to reduce it. They can also implement recycling and composting programmes, which can reduce the amount of waste sent to landfills. By being conscious of their waste management practices, businesses can prove to consumers that they are committed to being sustainable and environmentally responsible. In addition to reducing their own waste, companies can also contribute to waste reduction on a larger scale. They can work with suppliers to reduce packaging and find ways to use more sustainable materials. They can also partner with organizations that promote recycling and waste reduction and support legislation aimed at reducing waste. By taking these steps, businesses can help to reduce waste throughout the entire supply chain and promote sustainability on a larger scale.

Effective waste management requires collaboration between businesses, government, and consumers (Kazancoglu et al., 2021). Governments can create and enforce policies to promote waste reduction and provide incentives for companies to implement sustainable practices. Consumers can also play a role by choosing to buy from companies that have sustainable policies and by properly disposing of their own waste. By working together, we can create a more sustainable future.

1.12 Innovations in Waste Management

Waste management is a growing concern all across the world. The increasing population, rapid urbanization, and industrialization have led to the rise in the amount of waste generated and the consequent problems associated with their disposal. However, with the innovations in waste management, it is now possible to overcome some of these challenges. The new technologies and approaches have ensured that waste is no longer seen as just a problem but also as a resource that can be transformed into useful products. The circular economy concept, waste-to-energy technologies, automatic segregation, bioreactor landfills, the Internet of things, and plastic waste-to-fuel technologies are some of the innovations that are transforming waste management (Sonu et al., 2023). These innovations are essential in ensuring that waste is no longer seen as just a problem but also as a resource that can be transformed into useful products. With these solutions in place, it is possible to overcome some of the challenges associated with waste management and create a sustainable future where waste is not just disposed of but is utilized as a valuable resource.

One of the most significant innovations in waste management is the introduction of the circular economy concept, based on the principle of reduce, reuse, and recycle. It emphasizes the importance of reducing waste generation by changing consumption patterns, reusing materials, and recycling what cannot be reused. The circular economy approach aims to create a closed-loop system where waste is seen as a resource and its value is preserved. It envisions a future where waste does not exist at all, and everything is used and reused in a loop.

Another innovation is the use of waste-to-energy (WTE) technologies that transform waste into electricity or heat (Mayer et al., 2019). The WTE technologies have a two-fold benefit for waste management. On the one hand, they help to reduce the volume and mass of waste that needs to be disposed of in landfills. On the other hand, they provide an alternative source of energy that reduces dependence on fossil fuel. WTE technologies include incineration, gasification, and pyrolysis, which are used to convert waste into useful energy products.

One of the most critical aspects of waste management is segregation into different categories before disposal. The segregation enables efficient recycling of materials and processing of waste. However, manual segregation of waste is time consuming and inefficient. Innovations in waste management have introduced automatic waste segregation technologies that can segregate waste based on its composition. These machines use sensors and artificial intelligence to identify and sort waste based on its characteristics. Automatic segregation of waste improves the efficiency of recycling and reduces the cost of waste management.

Another novel method of waste management is the use of bioreactor landfills. Unlike traditional landfills that are passive, bioreactor landfills aim to accelerate the decomposition process by controlling the moisture, temperature, and microbial activity. The faster rate of decomposition results in the faster production of landfill gas, which can be collected and used as an energy source. The bioreactor landfill is a sustainable waste management option that maximizes the use of waste as a resource.

The Internet of things (IoT) is another technological innovation that is transforming waste management (Nižetić et al., 2020). IoT devices are used to collect data on waste generation, collection, and transport. The device sensors can detect the fill level of waste bins and alert garbage collectors for timely collection. The data collected through the IoT devices can be used to optimize waste management processes, reduce the amount of waste generated, and improve the efficiency of collection and transport. Innovations in waste management have introduced plastic waste-to-fuel technologies that transform plastic waste into fuels, like diesel, gasoline, and jet fuel. These technologies have a two-fold benefit; they help reduce the volume of plastic waste and provide an alternative source of energy.

1.13 Circular Economy Success Stories in Waste Management

(a) ***The Zero Waste Communities***

In the pursuit of creating sustainable communities, several cities around the world have adopted a "Zero Waste" policy. These communities prioritize reducing waste by promoting the use of composting, recycling, and reusing materials. For example, the city of San Francisco set a target of achieving Zero Waste by 2020 and their efforts led to an impressive diversion rate of 80% of waste from landfills, achieved through innovative recycling programmes and the implementation of composting.

(b) ***The Ellen MacArthur Foundation's Circular Economy Initiatives***

The Ellen MacArthur Foundation is a pioneer in promoting the circular economy and has been responsible for several successful initiatives in waste management. One such initiative is "The New Plastics Economy", which aims to address the plastic waste issue through the application of circular economy principles. The foundation has also launched "Circular Fibres Initiative", which promotes the adoption of circular economy principles in the fashion industry, to reduce waste in textile production and consumption.

1.14 Future of Waste Management in Circular Economy

The circular economy is a new paradigm aimed at reducing waste by reusing and recycling resources. In this context, waste is no longer considered just trash, but rather a resource that can be used to benefit mankind. Therefore, the future of waste management promises to be a crucial aspect of the circular economy. The goal of waste management is to minimize the negative impact of waste on the environment, human health, and the economy. It is a complex process that requires coordination between different stakeholders, including citizens, waste management companies, governments, and businesses. In the circular economy, waste management takes on an even greater significance because the focus is on regenerating wasted resources to create a sustainable economy that can reduce the exploitation of natural resources (Smol et al., 2020).

One of the most significant challenges in waste management is the proper collection and segregation of waste. Often, contaminated waste streams lead to less efficient recycling and reuse of resources. In a circular economy, waste management requires a more standardized collection and sorting process to ensure that the waste can be quickly sorted by type and quality (Hossain et al., 2022). This is important to standardize the waste management process and make it easier to generate revenue from resources that were once considered garbage.

Another significant challenge in waste management is the limited availability of resources required for recycling. Most recycling facilities worldwide lack these resources, creating a bottleneck. In a circular economy, the focus on reducing waste creation reduces the demand for these resources, making the recycling process easier (Xavier et al., 2021). Also, recycling facilities required an update and have necessary equipment to improve the efficiency of resource recycling, considering that the facility and equipment may help to automate the process and reduce the need for hand-sorting.

Circular economy expects that waste be returned as a valuable resource to the economy through different means, including recycling, reuse, refurbishment, and remanufacturing. The recycling process of waste management is crucial for supporting the circular economy. A new approach introduced in waste management, called Zero Waste, aims to eliminate waste by-products entirely. The idea behind Zero Waste is that waste is a misused resource and that it can be reused for a more significant productive economy and that nothing is wasted.

The future of waste management in a circular economy is a combination of using cleaner production techniques, improving collection and segregation processes, increasing the efficiency of recycling facilities, and establishing Zero Waste policies. Moreover, more efficient and greener manufacturing techniques would generate more resources available for the recycling process. Creating an economy that benefits from resource efficiency minimizes the negative impact of waste (Mikhno et al., 2021). If the waste management system is successful, it will contribute to creating a more sustainable world.

1.15 Conclusions

The circular economy in waste management requires collaboration between several stakeholders, including policymakers, manufacturers, retailers, waste management companies, and customers. Everyone needs to play a role in creating a sustainable future and reducing waste. Shared responsibility is crucial, and all stakeholders must work together to achieve the common goal.

As an individual, we can get involved in waste management and the circular economy by adopting a Zero Waste lifestyle, recycling, composting, and reducing our carbon footprint. We can also support businesses that prioritize circular economy principles, such as reusing materials and producing products in a sustainable manner. In doing so, we can take an active role in creating a more sustainable future for our planet. In conclusion, waste management and circular economy are closely related, and adopting circular economy principles can help us address the pressing environmental and social challenges we face today. By designing out waste, keeping materials in use, and regenerating natural systems, we can create a more sustainable and prosperous future. It is up to all of us to take action and collaborate in the transition to a circular economy.

References

Androniceanu, A., Kinnunen, J., & Georgescu. I. (2021). Circular economy as a strategic option to promote sustainable economic growth and effective human development. *Journal of International Studies, 14*(1).

Arena, M., Azzone, G., Grecchi, M., & Piantoni, G. (2021). How can the waste management sector contribute to overcoming barriers to the circular economy? *Sustainable Development, 29*(6), 1062–1071. https://doi.org/10.1002/sd.2202

Ashrafi, M., Magnan, G. M., Adams, M., & Walker, T. R. (2020). Understanding the conceptual evolutionary path and theoretical underpinnings of corporate social responsibility and corporate sustainability. *Sustainability, 12*(3), 760. https://doi.org/10.3390/su12030760

Barros, M. V., Salvador, R., de Francisco, A. C., & Piekarski, C. M. (2020). Mapping of research lines on circular economy practices in agriculture: From waste to energy. *Renewable and Sustainable Energy Reviews, 131*, 109958. https://doi.org/10.1016/j.rser.2020.109958

Castellani, V., Sala, S., & Mirabella, N. (2015). Beyond the throwaway society: A life cycle-based assessment of the environmental benefit of reuse. *Integrated Environmental Assessment and Management, 11*(3), 373–382. https://doi.org/10.1002/ieam.1614

Chisholm, J. M., Zamani, R., Negm, A. M., Said, N., Abdel Daiem, M. M., Dibaj, M., & Akrami, M. (2021). Sustainable waste management of medical waste in African developing countries: A narrative review. *Waste Management and Research, 39*(9), 1149–1163. https://doi.org/10.1177/0734242X211029175

Chuang, S.-P., & Huang, S.-J. (2018). The effect of environmental corporate social responsibility on environmental performance and business competitiveness: The mediation of green information technology capital. *Journal of Business Ethics, 150*(4), 991–1009. https://doi.org/10.1007/s10551-016-3167-x

Desing, H., Brunner, D., Takacs, F., Nahrath, S., Frankenberger, K., & Hischier, R. (2020). A circular economy within the planetary boundaries: Towards a resource-based, systemic approach. *Resources, Conservation and Recycling, 155*, 104673. https://doi.org/10.1016/j.resconrec.2019.104673

Dissanayake, D. G. K. (2022). *'Enabling circular fashion through product life extension.' Sustainable approaches in textiles and fashion: Circular economy and microplastic pollution* (pp. 21–40). Springer.

Elisha, O. D. (2020). Moving beyond take-make-dispose to take-make-use for sustainable economy. *International Journal of Scientific Research and Education, 13*(3), 497–516.

Fatimah, Y. A., Govindan, K., Murniningsih, R., & Setiawan, A. (2020). Industry 4.0 based sustainable circular economy approach for smart waste management system to achieve sustainable development goals: A case study of Indonesia. *Journal of Cleaner Production, 269*, 122263. https://doi.org/10.1016/j.jclepro.2020.122263

Frei, R., Jack, L., & Krzyzaniak, S.-A. (2020). Sustainable reverse supply chains and circular economy in multichannel retail returns. *Business Strategy and the Environment, 29*(5), 1925–1940. https://doi.org/10.1002/bse.2479

Gazzola, P., Pavione, E., Pezzetti, R., & Grechi, D. (2020). Trends in the fashion industry. The perception of sustainability and circular economy: A gender/generation quantitative approach. *Sustainability, 12*(7), 2809. https://doi.org/10.3390/su12072809

Ghasemi, M. K., Yusuff, R. B., & Mohd, B. (2016). Advantages and disadvantages of healthcare waste treatment and disposal alternatives: Malaysian scenario. *Polish Journal of Environmental Studies, 25*, 1.

Ghosh, S. K. (Ed.). (2020). *Circular economy: Global perspective*. Springer.

Gianico, A., Braguglia, C., Gallipoli, A., Montecchio, D., & Mininni, G. (2021). Land application of biosolids in Europe: Possibilities, con-straints and future perspectives. *Water, 13*(1), 103. https://doi.org/10.3390/w13010103

Gonawala, S. S., & Jardosh, H. (2018). Organic waste in composting: A brief review. *International Journal of Current Engineering and Technology, 8*(1), 36–38. https://doi.org/10.14741/ijcet.v8i01.10884

Han, J., Heshmati, A., & Rashidghalam, M. (2020). Circular economy business models with a focus on servitization. *Sustainability, 12*(21), 8799. https://doi.org/10.3390/su12218799

Hossain, R., Islam, M. T., Ghose, A., & Sahajwalla, V. (2022). Full circle: Challenges and prospects for plastic waste management in Australia to achieve circular economy. *Journal of Cleaner Production, 368*, 133127. https://doi.org/10.1016/j.jclepro.2022.133127

Kazancoglu, I., Sagnak, M., Kumar Mangla, S., & Kazancoglu, Y. (2021). Circular economy and the policy: A framework for improving the corporate environmental management in supply chains. *Business Strategy and the Environment, 30*(1), 590–608. https://doi.org/10.1002/bse.2641

Kucukvar, M., Egilmez, G., & Tatari, O. (2014). Evaluating environmental impacts of alternative construction waste management approaches using supply-chain-linked life-cycle analysis. *Waste Management and Research, 32*(6), 500–508. https://doi.org/10.1177/0734242X14536457

Kurniawan, T. A., Dzarfan Othman, M. H., Hwang, G. H., & Gikas, P. (2022). Unlocking digital technologies for waste recycling in Industry 4.0 era: A transformation towards a digitalization-based circular economy in Indonesia. *Journal of Cleaner Production, 357*, 131911. https://doi.org/10.1016/j.jclepro.2022.131911

Maitre-Ekern, E. (2021). Re-thinking producer responsibility for a sustainable circular economy from extended producer responsibility to pre-market producer responsibility. *Journal of Cleaner Production, 286*, 125454. https://doi.org/10.1016/j.jclepro.2020.125454

Mayer, F., Bhandari, R., & Gäth, S. (2019). Critical review on life cycle assessment of conventional and innovative waste-to-energy technologies. *Science of the Total Environment, 672*, 708–721. https://doi.org/10.1016/j.scitotenv.2019.03.449

Mikhno, I., Koval, V., Shvets, G., Garmatiuk, O., & Tamošiūnienė, R. (2021). Green economy in sustainable development and improvement of resource efficiency. *Central European Business Review, 10*(1), 99–113. https://doi.org/10.18267/j.cebr.252

Morseletto, P. (2020). Targets for a circular economy. *Resources, Conservation and Recycling, 153*, 104553. https://doi.org/10.1016/j.resconrec.2019.104553

Musarat, M. A. et al. (2022). *'Circular Economy-Recent Advances in Sustainable Construction Waste Management.' The circular economy-recent advances in sustainable waste management.*

Nanda, S., & Berruti, F. (2021). Municipal solid waste management and landfilling technologies: A review. *Environmental Chemistry Letters, 19*(2), 1433–1456. https://doi.org/10.1007/s10311-020-01100-y

Nguyen, T. K. L., Ngo, H. H., Guo, W., Nguyen, T. L. H., Chang, S. W., Nguyen, D. D., Varjani, S., Lei, Z., & Deng, L. (2021). Environmental impacts and greenhouse gas emissions assessment for energy recovery and material recycle of the wastewater treatment plant. *Science of the Total Environment, 784*, 147135. https://doi.org/10.1016/j.scitotenv.2021.147135

Nižetić, S., Šolić, P., López-de-Ipiña González-de-Artaza, D., & Patrono, L. (2020). Internet of Things (IoT): Opportunities, issues and challenges towards a smart and sustainable future. *Journal of Cleaner Production, 274*, 122877. https://doi.org/10.1016/j.jclepro.2020.122877

Perey, R., Benn, S., Agarwal, R., & Edwards, M. (2018). The place of waste: Changing business value for the circular economy. *Business Strategy and the Environment, 27*(5), 631–642. https://doi.org/10.1002/bse.2068

Perkins, D. N., Brune Drisse, M. N., Nxele, T., & Sly, P. D. (2014). E-waste: A global hazard. *Annals of Global Health, 80*(4), 286–295. https://doi.org/10.1016/j.aogh.2014.10.001

Ragheb, A., El-Shimy, H., & Ragheb, G. (2016). Green architecture: A concept of sustainability. *Procedia—Social and Behavioural Sciences, 216*, 778–787. https://doi.org/10.1016/j.sbspro.2015.12.075

Rajak, S., Vimal, K. E. K., Arumugam, S., Parthiban, J., Sivaraman, S. K., Kandasamy, J., & Duque, A. A. (2022). Multi-objective mixed-integer linear optimization model for sustainable closed-loop supply chain network: A case study on remanufacturing steering column. *Environment, Development and Sustainability, 24*(5), 6481–6507. https://doi.org/10.1007/s10668-021-01713-5

Ramanayaka, S., Keerthanan, S., & Vithanage, M. (2020). Urban mining of E-waste: Treasure hunting for precious nanometals. In *Handbook of electronic waste management*. Butterworth-Heinemann (pp. 19–54).

Ray, S. S., Lee, H. K., Huyen, D. T. T., Chen, S. S., & Kwon, Y. N. (2022). Microplastics waste in environment: A perspective on recycling issues from PPE kits and face masks during the COVID-19 pandemic. *Environmental Technology and Innovation, 26*, 102290. https://doi.org/10.1016/j.eti.2022.102290

Ribić, B., Voća, N., & Ilakovac, B. (2017). Concept of sustainable waste management in the city of Zagreb: Towards the implementation of circular economy approach. *Journal of the Air and Waste Management Association, 67*(2), 241–259. https://doi.org/10.1080/10962247.2016.1229700

Ribul, M., Lanot, A., Tommencioni Pisapia, C., Purnell, P., McQueen-Mason, S. J., & Baurley, S. (2021). Mechanical, chemical, biological: Moving towards closed-loop bio-based recycling in a circular economy of sustainable textiles. *Journal of Cleaner Production, 326*, 129325. https://doi.org/10.1016/j.jclepro.2021.129325

Schröder, P., Lemille, A., & Desmond, P. (2020). Making the circular economy work for human development. *Resources, Conservation and Recycling, 156*, 104686. https://doi.org/10.1016/j.resconrec.2020.104686

Sharma, Y. K. et al. (2019). When challenges impede the process: For circular economy-driven sustainability practices in food supply chain. *Management Decision*

Singh, G. K., Gupta, K., & Chaudhary, S. (2014). Solid waste management: Its sources, collection, transportation and recycling. *International Journal of Environmental Science and Development, 5*(4), 347–351. https://doi.org/10.7763/IJESD.2014.V5.507

Singh, J., Yadav, P., Pal, A., & Mishra, V. (2020). *'Water pollutants: Origin and status.' Sensors in water pollutants monitoring: Role of material* (pp. 5–20).

Smol, M., Marcinek, P., Duda, J., & Szołdrowska, D. (2020). Importance of sustainable mineral resource management in implementing the circular economy (CE) model and the European green deal strategy. *Resources, 9*(5), 55. https://doi.org/10.3390/resources9050055

Sonu, G. M., Rani, G. M., Pathania, D., Abhimanyu, Umapathi, R., Rustagi, S., Huh, Y. S., Gupta, V. K., Kaushik, A., & Chaudhary, V. (2023). Agro-waste to sustainable energy: A green strategy of converting agricultural waste to nano-enabled energy applications. *Science of the Total Environment, 875*, 162667. https://doi.org/10.1016/j.scitotenv.2023.162667

Suárez-Eiroa, B., Fernández, E., Méndez-Martínez, G., & Soto-Oñate, D. (2019). Operational principles of circular economy for sustainable development: Linking theory and practice. *Journal of Cleaner Production, 214*, 952–961. https://doi.org/10.1016/j.jclepro.2018.12.271

Terlau, W., & Hirsch, D. (2015). Sustainable consumption and the attitude-behaviour-gap phenomenon-causes and measurements towards a sustainable development. In *Proceedings of the in Food System Dynamics* (pp. 199–214).

Xavier, L. H., Giese, E. C., Ribeiro-Duthie, A. C., & Lins, F. A. F. (2021). Sustainability and the circular economy: A theoretical approach focused on e-waste urban mining. *Resources Policy, 74*, 101467. https://doi.org/10.1016/j.resourpol.2019.101467

Yeheyis, M., Hewage, K., Alam, M. S., Eskicioglu, C., & Sadiq, R. (2013). An overview of construction and demolition waste management in Canada: A lifecycle analysis approach to sustainability. *Clean Technologies and Environmental Policy, 15*(1), 81–91. https://doi.org/10.1007/s10098-012-0481-6

Zhang, A., Hartley, J., Wang, Y., & Wang, S. (2022). Special issue editorial: Logistics and supply chain management in an era of circular economy. *Transportation Research Part E, 166*, 102911. https://doi.org/10.1016/j.tre.2022.102911

Chapter 2
Valorization of Plastic Wastes in Circular Economy: The Development of an Inter-Organizational Circular System for Valorization of Expanded and Extruded Polystyrene in Brazil

Istefani Carisío de Paula, Ghissia Hauser, Madalena Heinen, Isabella Rodrigues da Cunha e Paula, Victor Phanebecker Calcanhotto, and Fernanda Valentini

2.1 Introduction

The transition from a linear take-make-dispose model of production and consumption to a circular economy model is becoming a commonplace. The design of circular packaging systems may be an alternative to the ever-increasing amount of residues and a way of valorizing waste. The report "The Future of Global Packaging to 2024" forecasted a growth of 2.8% per year and a peak at USD 1.05 trillion of packaging in 2024 (Sastre et al., 2022; The Future of Global Packaging, 2022). Thereafter, the pandemic COVID-19 provoked changes in the population's consumption. The rise in e-commerce and over-protection of food products caused an increase in household packaging waste, most of them made of plastic materials (Kitz et al., 2022).

The question of how to make the transition has been answered by scholars via the development of frameworks. Lieder and Rashid (2016) proposed a framework with a combined view of three aspects: environment, resources, and economic benefits. The authors suggested a concurrent approach that operates through public institutions from the top-down and the industry from the bottom-up. They assume that an inverse motivation exists among the stakeholders of the circular economy (CE), which needs to be aligned and converged. Governmental bodies and policymakers, for

I. C. de Paula (✉) · G. Hauser · I. R. da Cunha e Paula · V. P. Calcanhotto
Industrial Engineering Program, UFRGS/NiProS, Porto Alegre, Brazil
e-mail: istefani@producao.ufrgs.br

M. Heinen · F. Valentini
Trashin Startup Team, Porto Alegre, Brazil

S. A. Bandh and F. A. Malla (eds.), *Waste Management in the Circular Economy*,
https://doi.org/10.1007/978-3-031-42426-7_2

instance, maximize environmental benefits by strict control of industrial businesses. On the contrary, manufacturing companies put their effort into economic benefits and growth, due to competitive pressure. Nevertheless, the scenario has been changing since the environmental, social, and governance (ESG) criteria were set as standards for conscious investors to screen potential investments. King and Locock (2022) developed a framework specifically for plastics in the circular economy, with a focus on interventions broadly across the plastic supply chain, including resources, production, use, recycling, waste, and system analysis. The findings indicate a predominance of papers on system-based approaches and recycling interventions, constituting 53% of all articles examined. The early-stage interventions involving production and use appear as relative gaps, collectively making up only 17% of the articles examined. The framework is not adequate for implementing the transition from linear to circular processes, although it may be used as a support for it.

Moreno et al. (2016) developed a conceptual framework for circular economy design strategies, linking five design strategies and five circular business model archetypes. It provides circular economy design practitioners with a "[…] *holistic view of how to approach circular design, not only from a product perspective, but by taking into account the relevance and importance of the surrounding business models and how to integrate them with the design process. In addition, the framework acknowledges the role of policy and regulation in enabling circular business models, which requires exploring in future research*".

The managerial strategies toward CE, in contrast to the academic efforts, privilege the end-of-pipe initiatives. It is not a surprise that managers in organizations have a sense of urgency to reduce the impact of the portfolio of products that were not designed to avoid environmental or societal impact. Therefore, the contract of consultancy for circularization of products, and reverse logistics of end-of-use or end-of-life materials, is increasing considerably. Such context is fertile for business model strategies like, (i) 'circular supplies', a business model based on industrial symbiosis in which the residual outputs from one process can be used as feedstock for another process, or (ii) 'resource value', a type of business model based on recovering the resource value of materials and resources to be used in new forms of value (Bocken et al. 2016; Moreno et al., 2016).

Despite the dematerialization strategies like the substitution of products by services being preferable to the end-of-pipe ones, all of them demand new methodological approaches. The question of 'how to develop, implement, manage, and improve circular economy systems' remains without a precise answer. A century was necessary to organize linear supply chains, articulating players from organizations, suppliers, consumers, and government, locally and globally. Now, the challenge posed by the shortage of resources, the destruction of the environment, and the social inequalities force us to accelerate the rate of change. Additionally, CE systems, for assessment purposes, may be categorized in different levels of complexity, from a single circular product inside a process (nano-level), the entire organization processes (micro-level), inter-organizational systems (meso-level), and regional, or global circular interchanges (macro-level). Concerning the context of waste management in the circular economy, the inter-organizational level consists of a set of companies from the same segment acting in different roles in the system (WBCSD, 2018). Given the value of recycling within a circular economy (CE) framework, the next question revolves around how to systematically develop, implement, manage, and continuously enhance inter-organizational closed-loop systems in a structured and replicable manner. The objective of this chapter is to demonstrate a structured method designed to develop a pilot inter-organization circular system for the valorization of plastic waste.

2.2 Helice-Trashin Partnership and Research Scope

The Center for Intelligence in Projects and Systems (NIProS[1]) is a research group at the Federal University of Rio Grande do Sul. NIProS is registered in the CNPq[2] research directory platform, and its investigators develop experimental studies to instruct professionals to design innovative solutions for circular economy and sustainability. The research group performs actions via a consultancy organization, named Helice, in cooperation with Brazilian companies/start-ups, municipal and state government representatives, university experts, and societal institutions. The name "Helice" comes from the triple innovation helix. Helice has become an important partner of Trashin (TRASHIN, 2022), a start-up located in Porto Alegre, the capital of Rio Grande do Sul state, South Brazil. The cooperation between Helice and Trashin is a consequence of the mutual benefits this relationship brings. Helice may be seen as a Research & Development (R&D) external branch for Trashin. Its researchers apply industrial engineering tools and methods to design innovative solutions to the complexities of implementing circular systems and reverse logistics. Trashin, in its turn, performs the implementation and operation of the systems designed by Helice, via collaborative work. The bold performance of Trashin is becoming widespread, and large well-known companies have contracted its services lately.

Trashin was contracted by iFood, a Brazilian technology company that is a reference in online delivery. With about 65 million orders per month, iFood works with business intelligence and management solutions to promote and develop an ecosystem of more than 300,000 registered establishments and 200,000 connected delivery people in more than 1700 cities throughout Brazil.

The company holds over 80% market share of the food delivery sector in Brazil. Nevertheless, the Brazilian National Solid Waste Policy (NSWP—Federal law no. 12.305/2010[3]) establishes principles, objectives, instruments, and guidelines for integrated management and solid waste management. Special attention is paid to the shared responsibility of waste producers and amends Act No. 9.605 on criminal

[1] NIProS—Portuguese acronym for Center for Intelligence in Projects and Systems.

[2] CNPq Brazilian research funding agency—Directory Research Group link: http://dgp.cnpq.br/dgp/faces/consulta/consulta_parametrizada.jsf.

[3] Federal Law—NSWP: https://leap.unep.org/countries/br/national-legislation/law-no-12305-national-policy-solid-waste-management.

penalties relating to behavior and activities harmful to the environment. Although the iFood managers are not directly responsible for the packaging their partners use in the meals, they have decided to, proactively, develop a strategy of lowering the amount of packaging of their delivery services by the year 2025.

Trashin was contracted by iFood to develop the reverse logistics and recycling of expanded polystyrene (EPS) packaging in Brazil. Findings of a former study (De Oliveira et al., 2019, p. 562) indicated that "[…] *the EPS recycling demanded sorting and washing operations. The most significant waste generators were public and private institutions, residences, and retailers*". The study also revealed as notable barriers to its recycling the lack of understanding of the nature of expanded polystyrene recycling by Brazilian people, high transport cost due to material low density, and the dispersion of post-consumer expanded polystyrene in several regions of the country, linked with the small number of recycling facilities. These were some of the challenges faced by the Helice-Trashin team at the project's beginning.

2.3 The Method Design

The Helice-Trashin team applied their own method, tested before, in other inter-organizational circular system contexts. The method is anchored in five steps, (i) literature, technical visits, and interviews with experts; (ii) the supply chain characterization and the technical reverse flow mapping; (iii) definition of players for each reverse flow operation and invitation to perform a pilot project; (iv) raise awareness and articulation of those elected for manufacturing the target product, via workshops; and (v) definition of a governance structure and implementation of the pilot circular project. The team developed the project from December 2021 to March 2022 and managed the activities remotely. The Deliver Company was the main funder, and a sponsor made the follow-up of deliverables weekly. We have used as repository and management tools the Monday® Schedule platform, Google Drive®, and the Miro® platform. The workshop members have evaluated the event quality via Google Forms®. A final interview was performed with the Trashin members to evaluate the benefits of the partnership with Helice to design the pilot circular project stage.

2.4 The Design of a Pilot Inter-Organization Circular System for the Valorization of EPS/XPS Packaging Waste

The method's first step included academic and gray literature review, technical visits, and interviews. The team has performed a search in the Web of Science platform using the terms expanded polystyrene (EPS), extruded polystyrene (XPS), packaging, and Isopor® (the brand name). The purpose was to find information on the waste management, production, and use of the material in the packaging supply chain. They explored five premises in the literature review and other sources, as sites of plastic associations and manufacturers (Fig. 2.1).

The first attempt was to define the physical characteristics and amount of each material. The EPS is produced from small flakes that undergo an expansion process with pentane gas (98% composed of gas). The XPS is obtained through an extrusion process, and it is broadly used in food trays at restaurants, packaging, and supermarkets in Brazil. Although PS represented 4.8% of the market (7.7 million tons in 2020) and EPS 0.9% (68 thousand tons in 2020), literature from 2012 indicated that 24% of the material is recycled in Brazil (ABIPLAST & Brazilian Association of Times Higher Education Plastic Industry, 2021; PLASTIVIDA, 2017). We did not find current data to modify this percentage. According to these institutions, the civil construction sector consumes 78% of recycled EPS. The Apparent National Consumption of recycled material was growing at an annual rate of 5.3%, and the production grew at an annual rate of 6.6% (EPSBRASIL, 2022). Both EPS and XPS are fully recyclable, and Fig. 2.2 explains the operations needed.

The EPS and XPS waste value per kilogram increases in the following order, from the cheapest to the most expensive, the pressed material for transportation (US$0,40–1,40); degassed material (US$4,00–5,00), and pelleted material (US$7,00–7,80).

Fig. 2.1 Five premises explored in literature review, interviews, and desk research

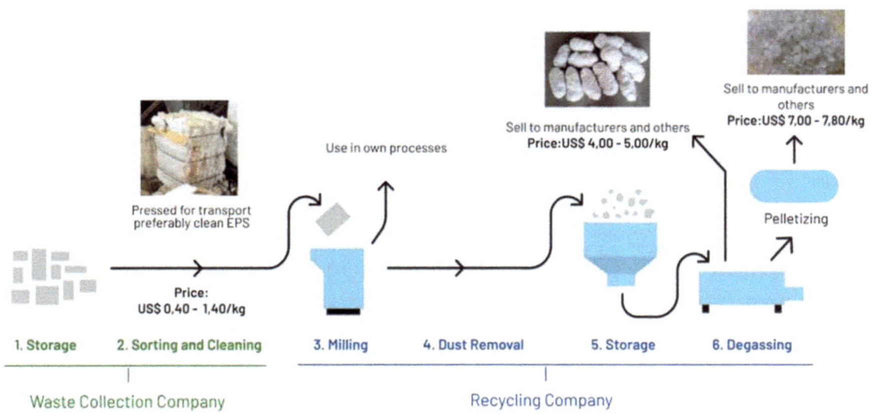

Fig. 2.2 Recycling operations for EPS and XPS

Values in dollars are not corresponding to the local currency, but the proportion was kept for illustration purposes.

Two main recycling organizations named Recycler A and Recycler B were from the packaging segment and civil construction fields, respectively. Recycler A prioritizes buying clean waste material to degas and sell to other recyclers. Recycler B prefers clean material, but also accepts packaging with residual grease from food, for example. They transform the waste into new products for the civil construction industry. This is a piece of important information since a large amount of waste would come from food packaging that should be cleaned before disposal, to improve the material appeal. In this context, the education of the population on adequate discarding of clean packaging arises as an important issue.

Technical visits to waste collection companies revealed a weak technological infrastructure. The predominant technology was hydraulic press for EPS/XPS waste material baling. Despite pressing may reduce the volume for logistics purposes, the degassing methodology is much more efficient (it removes the gas from the EPS/XPS by extrusion and heat). Therefore, Recycler B provided a degassing machine to any collection company interested, on the condition that it processed a minimum of 3–5 tons of EPS and XPS waste per month. Pelletization would increase the value of recycled EPS/XPS, but it demanded more specialized people and more expensive machinery. Interviews revealed that a Brazilian polymer manufacturer could make available a pelletizing equipment for testing purposes.

The next step was the supply chain characterization. The content analysis of interviews with experts was very important for understanding the structure of the EPS/XPS supply chain. De Oliveira et al. (2019) had already designed the supply chain in Brazil, and it was used as a starting point. The Helice team performed 24 interviews with nine manufacturers, five waste collection companies, one Consolidator company, three plastic associations, two education project initiatives, one start-up, and two recycling companies. Figure 2.3 shows an overview of the supply chain mapping, according to the information gathered from interviews.

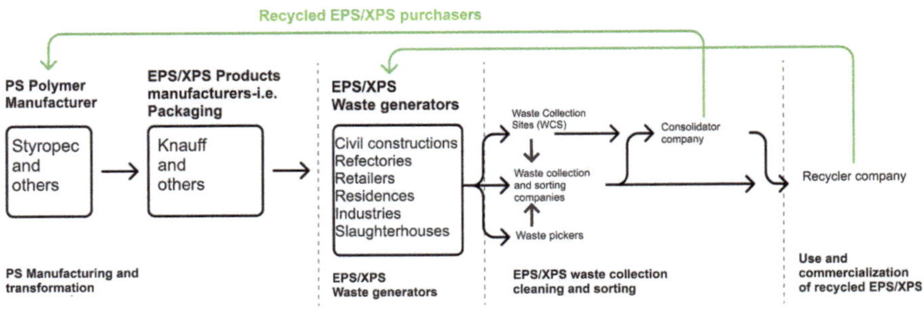

Fig. 2.3 Main players from the EPS/XPS supply chain in Brazil

The team identified the following players: (i) polymer manufacturer, chemical industry, makes polymer PS, EPS (pearls); (ii) product manufacturer transforms polystyrene (PS) via expansion with pentane gas to generate EPS, or by extrusion and butane gas to generate XPS; these players also convert these raw materials in packaging and other objects; (iii) Waste Collector and Sorter company collects and sorts post-consume material from households and large generators. This company separates EPS/XPS from other materials before pressing, degassing, or pelletizing the material; (iv) Consolidator company gathers larger volumes of material processed by Waste Collector and Sorter companies and proceeds the degassing process; (v) recycler uses post-processed EPS/XPS waste and transforms it into a new product. Frequently, the recyclers valorize the EPS/XPS waste by developing products for civil construction like baseboards and ceiling lining boards; (vi) waste generators are home appliance suppliers and similar companies that discard Styrofoam used as equipment protection.

The interviews with these players were useful as the first effort for the next step. Figure 2.3 shows that the reverse flows of EPS/XPS waste depend on either the population discard, the waste collector operations, or the consolidator gathering and degassing of the material to attend low-cost logistics toward the recycler. The main recycling companies A and B are located on the border of Southeast and South Brazil. Therefore, Brazil's continental dimensions may make the logistics costs inviable depending on the distance the waste has to perform to reach the recycler companies. Bearing that in mind, some players were invited to be part of a pilot inter-organization circular system for the valorization of EPS/XPS packaging waste with the aim of testing some hypotheses. The criteria for choosing the organizations were.

- waste collection and sorting companies that were former partners from Trashin in other projects and that were willing to participate and give access to performance data;
- strategic location in logistic terms;
- present some level of management maturity that facilitates pilot implementation;
- companies that already had the necessary environmental licenses;
- physical infrastructure compatible with the needs of the pilot project;
- geographical position in the country to facilitate project scaling.

The hypotheses tested included waste trade costs, logistics costs, the volumes of material collected, cleaned, sorted, and technologies of processing barriers. We selected two dimensions for designing the pilot's strategies. The first was the role played by iFood, which could be an investor in the supply chain or a competitor in the recycling chain. The other dimension was the scale of production that could be local, considering the operation in a single Waste Collection and Sorting company, or regional. In the last case, a Consolidator company would gather pressed material from other waste and sorting companies to make the degassing of larger amounts of EPS/XPS and send them to the Recycling company. Recycler B was selected to be part of the pilot project since they could accept materials with residual contaminants. Figure 2.4 illustrates the scenarios created for each dimension and scale.

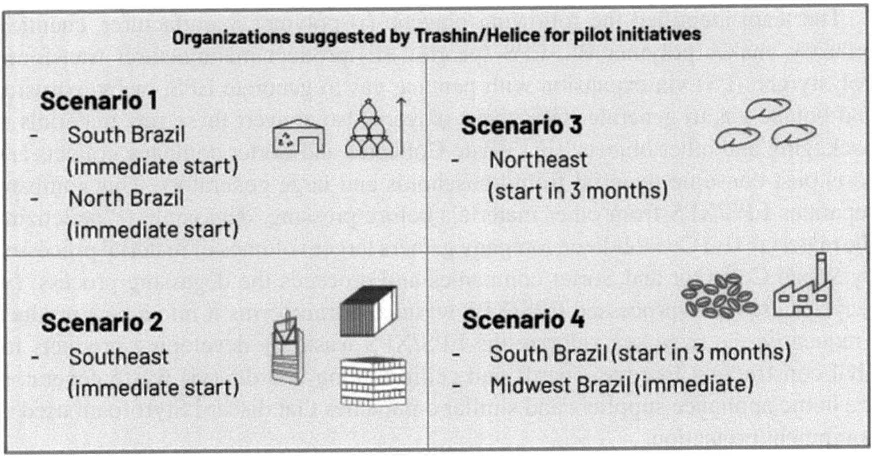

Fig. 2.4 Scenarios and possible players for a pilot inter-organization circular system for the valorization of EPS/XPS packaging waste

As Fig. 2.4 indicates, each scenario demands a given type of technology. Scenario 1 is about a local Waste and Sorting company operating the hydraulic press of EPS/XPS and material baling. Scenario 2 is about a regional consolidator degassing large amounts of EPS/XPS, gathered from other companies. Scenario 3 is about the local degassing of EPS/XPS by a large Collector and Sorting company, and Scenario 4 is about a Consolidator company pelletizing larger amounts of EPS/XPS, gathered from other companies. We suppressed the names of organizations for secrecy purposes, but we have mentioned their geographical positions in Fig. 2.4. All scenarios include the operations of collection > sorting > hydraulic pressing > degassing > recycling. The logistics should be operated between each stage by Trashin, taking the material from one company to another. Trashin operates logistics in all Brazilian regions.

Later the scenarios were simulated in terms of financial values, and scenario 1 was highly sensitive to material volumes input. Communication with EPS/XPS waste generators was mandatory, independent they were the country's population or home appliance companies. Figure 2.5 brings a first risk analysis. Scenarios with technologies not available in Brazil would represent higher risks, caused by the equipment import costs, the need for training people, and the time-consuming learning curves. An example is the densifier equipment that produces EPS bars (Fig. 2.6) not available in Brazil. The higher risks of scenario 2 could be reduced by adopting technologies available such as hydraulic presses. Scenarios with more complex technologies such as degassing and pelletization, despite being riskier also revealed better financial results, since the price paid for the material increased proportionally. The iFood validated the more interesting scenarios for implementation, and the partners were invited to participate in an alignment workshop.

	Considering the Delivery Company role in the system	
Considering scale	**Waste Supply Chain investor** Not active in the market	**Competitor in the recycling chain** Active in the market
Local scale (waste collector and sorter company)	Scenario 1 Lower technical and financial risk	Scenarios 3 e 4 Higher technical risk (technology involved and governance) and financial (investment in equipment) Note: Financial risk can be mitigated by using equipment from recycler (degassing equipment), using equipment from manufacturer (pelletization equipment)
Regional Scale (consolidation company)	Scenario 2 (*) Higher technical (technology and governance) and financial risk - less known situation	

All scenarios demand increase in material volume
(*) Scenario 2 is no longer a high risk if the densifier is replaced by a press

Low Risk — Moderate Ri — High Risk

Fig. 2.5 Preliminary technical and financial risk analysis of scenarios

Fig. 2.6 Bars of EPS produced with densifier equipment. Source: https://tagliabrasil.com.br/compactadores-de-eps

2.5 Workshop for Articulation and Alignment of Players

This is a very important step before implementing and operating the pilot system. We invited 35 members which had accepted formerly to be part of the pilot inter-organizational circular system for EPS\XPS waste. The group included ten members of iFood (sponsor, marketing, consumer communication, sustainability, legislation affairs), six members of Trashin (leadership, operations, consumer communication, project team), six members of Helice (leadership and team, all responsible for the workshop coordination), six members of the Recycler Company B (leadership, operations), three members of Consolidator company (they receive material from 21 other Waste Collector and Sorting companies in Midwest Brazil), two members of Waste Collector and Sorting company in South Brazil, and two members of the Public Ministry (Midwest and South Brazil).

The alignment workshop was developed remotely because of pandemic isolation and due to the geographical location of the participants. We used the ZOOM platform to hold the event and the MIRO platform (www.miro.com) for collaborative activities among the members. The workshop lasted 4 h with the following activities, a brief formal opening by iFood and Trashin-Helice leaderships. Thereafter, there was a short presentation of all companies' business operations, language unification, and EPS\XPS waste chain validation. After a 15-min break, we divided the 35 participants into two mixed groups to make the analysis: "How to increase the volume and quality of the material". The members presented result in the large group after 1 h. The last activity was the analysis: "Risks and barriers" using the qualitative Probability x Impact matrix (PMI, 2017). After the groups' presentations, the workshop was finished with a schedule of further steps of the pilot operation implementation.

The workshop objective was to build a common vision of the circular system among players, with clear limits of operations. The members proposed realistic recommendations to (i) increase the volume of material collected, (ii) to increase the quality of the material collected, and (iii) to increase both volume and quality. The Helice team categorized the recommendations in technical, labor, equipment, innovation, logistics, collection, institutional, material, generator's education, and taxes.

Concerning the 20 recommendations for the volume increase group, the category "Equipment" was the smallest with six items. This suggests that there is an opportunity for improvement of the volume processed by making more pieces of equipment available and/or optimizing their use. The group "Material" (seven items) included suggestions for eliminating EPS/XPS residual contamination. Inside the group of recommendations for increasing volume and quality, the "Collection" (five items) category included opportunities of gathering material from large generators. It should be noted that the players reinforced the education for the population as mandatory along the workshop.

Regarding the 34 risks listed during the workshop, 29 were classified as high risk, with nine receiving the maximum score on the risk scale (0.72). The predominance of high probability and impact risks occurred in the "bureaucratic" and "tax" categories. All participants received an event evaluation form with the following topics: quality of communication prior to the workshop, workshop contents and activities, coordination, and technologies used. The answers were given on a scale from 1 to 5, in which the value 1 means "did not meet expectations" and 5, "exceeded expectations". The return rate was 33%, of which 30% rated four for all items, and 70% rated five for all items.

Regarding the pilot circular system implementation, it is expected to last from 3 to 6 months. Trashin and iFood will communicate to the population and generators information on the discarding of cleaned material. Trashin will be responsible for the operation of the pilot project in three different cities (two cities in the South and one in Midwest Brazil). The start-up will perform the system's governance also. iFood will hold meetings with Trashin to monitor the pilot project, participate in regular meetings with all players, and make technical visits. Trashin will plan and perform the follow-up meetings with the Consolidator and the Waste Collector company, make the analysis of project KPIs, and organize meetings. The players will receive support from Trashin management, collect and share information on operations and KPIs, and participate in regular meetings. From this moment on, Helice will conclude the project reports and wait for future demands.

2.6 Discussion and Practical Implications

Waste management covers all aspects of waste, including waste reduction and the collection, transportation, handling, and disposal process. To a certain extent, the waste management approach is aligned with the circular economy principle of "circulate products and materials", but circular economy goes beyond, recovering the resource value of materials or building new forms of value for the same material. This chapter presented an example of a structured method designed to develop a pilot inter-organization circular system for the valorization of plastic waste.

The five-step method was effective in designing strategies for operating the reverse flow of EPS/XPS and the valorization of the material by a Recycling company (Fig. 2.7) as a starting point for hypothesis evaluations. Figure 2.7 illustrates the portion of the chain that was designed with the method starting with the reverse logistics of EPS/XPS from generators toward the Recycler company that makes new products to put back in the chain (green line). The civil construction is the main consumer of the new product, but the PS manufacturer also consumes degassed and pelletized PS back, as charge for new polymer manufacturing. The good news is that PS is fully recyclable with high levels of recovery. Some countries and cities in the world are eliminating its use as raw material, especially for food. Our findings indicate that PS may be used for food, but preferentially for those that are not prepared with sauce. This kind of food is greasy and may both leave residuals in the packaging or exchange substances with the packaging layer, depending on the food temperature.

The context of this project revealed both the top-down and bottom-up motivation of Lieder and Rashid's (2016) framework. The PNRS act from the Brazilian Government placed pressure on iFood to reduce the impact of its service, even though the company was not directly responsible for the packaging used by its partner restaurants. The iFood's leaders decision was strategic to maintain its leadership in the market and a good image of the brand.

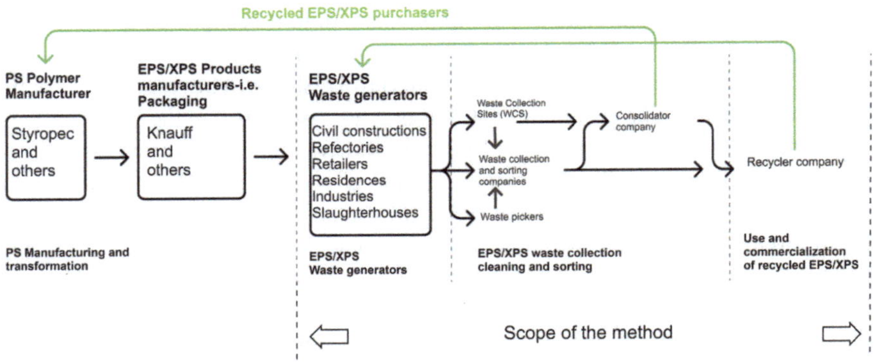

Fig. 2.7 Limits of the scope of the method

Still regarding the scope of this method, it reflects the resource value archetype from Moreno's et al. (2016) framework, a "circular business model based on recovering the resource value of materials and resources to be used in new forms of value". An important premise lies under this kind of circular business model. It is necessary to have a production pulled by a product that is already successful in sales in the market. Otherwise, there is a risk of stocks of unsold products becoming the new residue impacting the environment. This premise was discussed previously in Paula et al. (2021). Both Recyclers A and B fulfilled this prerequisite. The interviews at the beginning of the project revealed that the two organizations were already striving to organize by themselves the reverse logistics of EPS/XPS in Brazil, intending to increase the volume of material as input for their specific operations. Interviews revealed that there was an idle capacity in the recycling processes of both organizations.

We have decided to keep Recycler B in the pilot project firstly because this company added more value to the EPS/XPS waste material than Recycler A, manufacturing civil construction products. Secondly, they accepted the waste material with residual contaminants, and thirdly, it was observed a kind of "competition" between the two recyclers. Considering that the pilot system was preliminary, we decided to start just with one recycler to test the operations hypotheses.

Interviews with Waste collectors and Sorting companies revealed them as the weakest link in the chain. The managers of these companies are frequently vulnerable people that depend on the municipal governments to support the company's operations. The managers are very diligent, and some of these companies have grown and become consolidators. Nevertheless, the smaller Waste Collector and Sorting companies' managers claimed that they did not collect the EPS/XPS because it demanded large stocking spaces and, mostly, the material had low purchasing value.

Concerning the method, the interviews at the beginning of the project were essential in making the whole scenario clearer and bringing out these complexities. Other advantages of the preliminary interviews include.

- identification of elements to design the whole supply chain, the direct and reverse flows more consistently;

- taking the opinion of experts on limitations and operations needed for safely recycling the EPS/XPS;
- elicitation of requirements and limitations of each player, in particular the quality demands of the recyclers and space limitations of waste collectors companies;
- identification of key players in the process;
- identification of sensitive players to participate in the pilot project and workshop.

The interviews were also a strategy for approaching the players, a first step toward building trust. The interviews were useful for designing the reverse flows collaboratively with players, but the workshop has consolidated this effort. The workshop activities were useful to validate the design of the pilot circular system and to reinforce the design of quality into the reverse flows as dependent on all players' cooperation. The team presented the values of circular economy at the workshop beginning and the win–win perspective as the basis for operations and trust. They consider the workshop as a key tool in the five-step methodology.

The workshop remote version was a consequence of the pandemic isolation but proved to be advantageous in saving the costs of logistics and facilities, typical of the face-to-face alternative. Moreover, the use of digital technologies aids in developing players that are less familiar with these tools. The collaborative risk analysis allowed all players to express the elements that bring concern. Some riskier elements were repeated by more than one player and should be the target of mitigation efforts, such as the taxes which are cumulative between the direct and reverse flows. The presence of representatives of the Public Ministry in the workshop was strategic for bringing confidence and pointing solutions to these issues.

The choice of the types of activities proposed for the workshops is central. The clear definition of the workshop objective drives this choice, but we recommend risk analysis as a way of speaking about concern situations and building trust. Finally, the workshop was the starting point for preparing the players for the pilot circular system governance meetings.

2.7 Final Considerations

Despite the waste management approach being considered a less relevant element of the circularization effort by defenders of the circular economy theory, this project has brought some important lessons to this research group. The five-step method presented in this chapter is a consequence of the urgency in making the transition from linear to circular systems and was proposed to fulfill the ever-increasing demand of companies in Brazil. The EPS/XPS waste valorization is one of the cases implemented by the Helice-Trashin partnership. In the perception of the teams, this kind of circular inter-organizational project opens the doors for further collaborations among the players of the systems. It seems easier for managers to share information on data concerning residues and post-consumption materials than sharing information on innovation initiatives. We hypothesize that the operation of end-of-pipe projects with different players and partners will pave the way for developing trust and collaboration to innovative endeavors.

The partnership between the Helice and Trashin teams has proved to bring mutual benefits. People from Helice do not have the vocation to prospect projects in the market. Their objective is to test industrial engineering and research methods to design efficient circular processes. In the meanwhile, they understand the idiosyncrasies of players in each circular system, test hypotheses, and advance the circular economy knowledge area. Trashin people, in its turn, advance the implementation, operation, management, and improvement of circular systems as a consultancy. In the words of Trashin's manager "[…] *Helice gives solidity to the work that Trashin does. The entire theoretical consulting, analytical, and research framework that Helice develops is extremely complementary to everything Trashin does. So when we work together we can build trustable solutions.* […] *Research provides the foundation we need, especially in an environment that is still very uncertain (innovative). These reverse logistics studies, these circular economy studies are still very recent in Brazil. We do this on a slightly larger scale. So Helice comes in doing all this analysis, doing all this research so that the final result is better. All the projects on which we worked together had superior results, not comparing to what Trashin is already doing, but to what the market in Brazil is doing*".

Considering the client's perspective, iFood had access to quality research followed by implementation and cycles of improvement of the system. In the short term, Helice-Trashin proposed the implementation of the pilot circular system for EPS/XPS. In the long term, they recommended analyzing the substitution of plastic with organic alternatives and the reduction of packaging from delivery to attend to circular economy principles. Moreover, Helice-Trashin identified a business opportunity. The pilot circular project started operations in April 2022. In the middle term, iFood as the main funder could operate as an incubator and accelerator of the Waste and Sorting companies in other regions of the country, as a strategy for scaling up the EPS/XPS recycling capacity in Brazil and increasing revenues. Future studies include the analysis of financial viability and the analysis of KPIs as references for improvements of the systems' performance.

Acknowledgements We acknowledge all the players interviewed in this research for the information sharing, iFood for funding the research; the dedication of Camila Borges (iFood's Sustainable Solutions leader), and Trashin for their partnership; Sérgio Finger (Trashin's CEO) and team for their confidence, and open mind.

References

ABIPLAST & Brazilian Association of Times Higher Education Plastic INDUSTRY [Review]. (2021). *The plastic processing and recycling industries in Brazil.* Available in http://www.abiplast.org.br/wp-content/uploads/2022/05/Preview_ABIPLAST_2021_web.pdf. Access in: 7 August 2022.

Bocken, N. M. P., de Pauw, I., Bakker, C., & van der Grinten, B. (2016). Product design and business model strategies for a circular economy. *Journal of Industrial and Production Engineering, 33*(5), 308–320. https://doi.org/10.1080/21681015.2016.1172124

De Oliveira, C. T., Luna, M. M. M., & Campos, L. M. S. (2019). Understanding the Brazilian expanded polystyrene supply chain and its reverse logistics towards circular economy. *Journal of Cleaner Production, 235*, 562–573. https://doi.org/10.1016/j.jclepro.2019.06.319.Available in https://www.sciencedirect.com/science/article/abs/pii/S0959652619322966

EPSBRASIL. (2022). Marketplace. *Market data.* Sectorial Commission ABIQUIM. Available in https://www.epsbrasil.eco.br/mercado.html. Access in: 7 August 2022.

King, S., & Locock, K. E. S. (2022). A circular economy framework for plastics: A semi-systematic review. *Journal of Cleaner Production, 364*, 132503. https://doi.org/10.1016/j.jclepro.2022.132503. Available in https://www.sciencedirect.com/science/article/pii/S0959652622021047

Kitz, R., Walker, T., Charlebois, S., & Music, J. (2022). Food packaging during the COVID-19 pandemic: Consumer perceptions. *International Journal of Consumer Studies, 46*(2), 434–448. https://doi.org/10.1111/ijcs.12691

Lieder, M., & Rashid, A. (2016). Towards circular economy implementation: A comprehensive review in context of manufacturing industry. *Journal of Cleaner Production, 115*, 36–51. https://doi.org/10.1016/j.jclepro.2015.12.042.Available in https://www.sciencedirect.com/science/article/abs/pii/S0959652615018661?via%3Dihub#preview-section-references.

Moreno, M., De los Rios, C., Rowe, Z., & Charnley, F. (2016). Available in. https://www.mdpi.com/2071-1050/8/9/937. Access in. A conceptual framework for circular design. *Sustainability, 8*(9), 937. https://doi.org/10.3390/su8090937

Paula, I. C. et al. (2021). *Decision matrix for development of new demand-driven Interorganizational circular economy systems.* In International Conference on Resource Sustainability. University College Dublin Press (ICRS). Ireland. http://icrs2021dublin.ucd.ie/Media/files/Book-of-Abstracts.pdf. Access in: 9 August 2022

PLASTIVIDA. (2017). *XPS, and EPS for food contact.* Packaging Technology Center—Plastivida, Socio-Environmental Institute of Plastic. http://www.plastivida.org.br/estudos/PS_XPS_EPS_Alimentos.pdf. Access in: 9 August 2022. São Paulo, Brazil.

PMI. (2017). *Project Management Institute. A guide to the project management body of knowledge* [PMBOK guide] (6th ed). PMI.

Sastre, R. M., de Paula, I. C., & Echeveste, M. E. S. (2022). A systematic review of the literature on packaging sustainability: Content, opportunities, and guidelines. *Sustainability, 14*(11), 6727. https://doi.org/10.3390/su14116727. Available in https://www.mdpi.com/2071-1050/14/11/6727

The Future of Global Packaging. (2022). Available in. https://www.smithers.com/services/market-reports/packaging/future-of-global-packaging-to-2024. Access in: 9 August 2022. Smithers, WV.

TRASHIN. (2022) Available in. Smart waste management. *TRASHIN*. https://trashin.com.br/. Access in: 9 August 2022.

WBCSD. (2018). Available in *World business council for sustainable development and EIT climate-KIC. Circular metrics, landscape analysis.* https://www.wbcsd.org/Programs/CircularEconomy/Factor-10/Metrics-Measurement/Resources/Landscape-analysis. Access in: 7 August 2022. European Institute of Innovation and Technology (EIT).

Chapter 3
Valorization of Municipal Solid Wastes in Circular Economy

Kaustubh Chandrakant Khaire, Sanjeet Mehariya, and Bikash Kumar

3.1 Introduction

Anything worthless can be called as waste. Each year, 2 billion tons of municipal solid waste (MSW) are produced from all over the world (Thakur et al., 2020). Food waste classified as post-consumer and pre-consumer waste presents the highest portion of MSW (Pfaltzgraff et al., 2013). Proper management of these wastes is a major concern in the contemporary world, as the open landfills and improper dumping are causing an adverse effect on the environment and health (Pfaltzgraff et al., 2013).

The solid wastes can be classified into different types like industrial wastes (Chandra & Chowdhary, 2015), agricultural wastes (Saini et al., 2015), commercial wastes (Andritsos et al., 2016), and municipal wastes on the basis of their source (Aslani & Taghipour, 2018) and based on physical, chemical, and biological as hazardous waste (Manahan, 2017), non-biodegradable waste (Nielfa et al., 2015), and biodegradable waste (Saveyn et al., 2014) (Fig. 3.1).

K. C. Khaire
School of Energy Science and Engineering, Indian Institute of Technology Guwahati, Guwahati 781039, India

K. C. Khaire · B. Kumar
Department of Biosciences and Bioengineering, Indian Institute of Technology Guwahati, Guwahati 781039, India

S. Mehariya
Department of Chemistry, Umeå University, 90187 Umeå, Sweden

Algal Technology Program, Center for Sustainable Development, College of Arts and Sciences, Qatar University, Doha, Qatar

B. Kumar (✉)
Department of Biosciences and Biomedical Engineering, Indian Institute of Technology Indore, Indore, Madhya Pradesh, India
e-mail: bikash@iiti.ac.in

S. A. Bandh and F. A. Malla (eds.), *Waste Management in the Circular Economy*,
https://doi.org/10.1007/978-3-031-42426-7_3

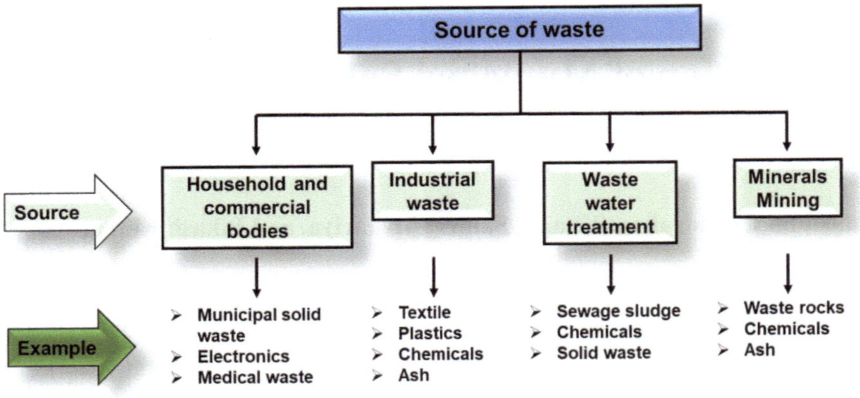

Fig. 3.1 Source and classification of waste

Currently, about 2 billion tons (Bt) of MSW are produced worldwide and it is predicted to enhance up to 9.5 Bt by 2050 (Campuzano & González-Martinez, 2016; Matsakas et al., 2017). The MSW produced in developing countries has a higher plastic content compared to other types of wastes (Solis, 2018). In these countries, the majority of it is openly dumped, while the majority of it is incinerated or landfilled in the developed world. However, that too is having its implications, both on the environment and the human health (Christy et al., 2014; Tyagi et al., 2018) (Fig. 3.2a, b). Region-wise waste generation potential around the globe and percentage contribution by different countries based on their income are summarized in Fig. 3.2.

Therefore, the discovery of new ideas for the proper disposal, processing, and management of MSW becomes imperative to avoid the adverse effects on human health and nature. Among those discoveries, the concept of biorefineries, closely resembling the petroleum-based refinery, that incorporates a transformation process and equipments to change over the natural organic wastes into fuels, energy/power, and economically important chemicals (Cherubini, 2010) is a viable answer to the problem. The ongoing financial model, likewise referred to as the linear economy, is based on the taking, consuming, producing, and discarding. The formation of MSW-based biorefineries is a significant part to start a circular economy, as they are a conceivable answer to the problem and the conservation of value-added products (Malinauskaite et al., 2017). The extensive way of thinking is to utilize the MSW waste as a raw feedstock for making fuels or other value-added products.

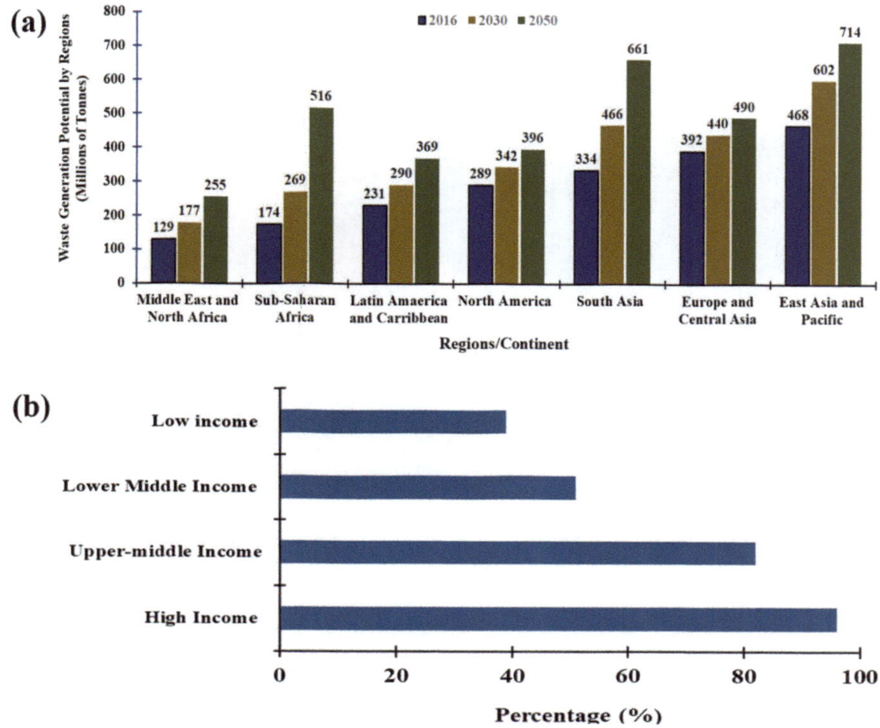

Fig. 3.2 **a** Waste generation potential by region **b** percentage contribution to waste generation by nations based on their income

3.2 Composition of MSWs

MSWs are a heterogeneous mixture of materials like paper, yard decorations, wood, food waste, newspaper, packing material, etc. (Edjabou et al., 2015). The overall constituents and consumption of MSW vary from one country to other (Abu-Qudais & Abu-Qdais, 2000). However, in most of the countries, the main component of MSW is food waste (Fig. 3.3a), e.g., China—57%, Denmark—46%, Jordan—63%, and Malaysia 36–60% (Noor et al., 2013; Chen et al., 2010, 2012; Sahimaa et al., 2015). Other types of wastes present in the MSW include the plastic objects, paper wastes, cardboard, textile wastes, woody wastes, glass, metals containing organic fractions and others (Abdel-Shafy and Mansour, 2018).

Food wastes, wood chips, paper sheets, and paper waste found in MSW are most frequently known as the *Organic Fraction.* Looking toward the MSWs from the lens of circular economy, the plastics contribute the greater part of the calorific worth of MSWs with polystyrene, polypropylene, and polyethylene having a calorific values of 41.4 MJ/kg, 46.4 MJ/kg, and 46.3 MJ/kg, respectively (Gao, 2010). However, because of a greater moisture content, the highest calorific value of MSW is ordinarily somewhere in the range of 5 and 20 MJ/kg (Abu-Qudais & Abu-Qdais, 2000; Tozlu et al.,

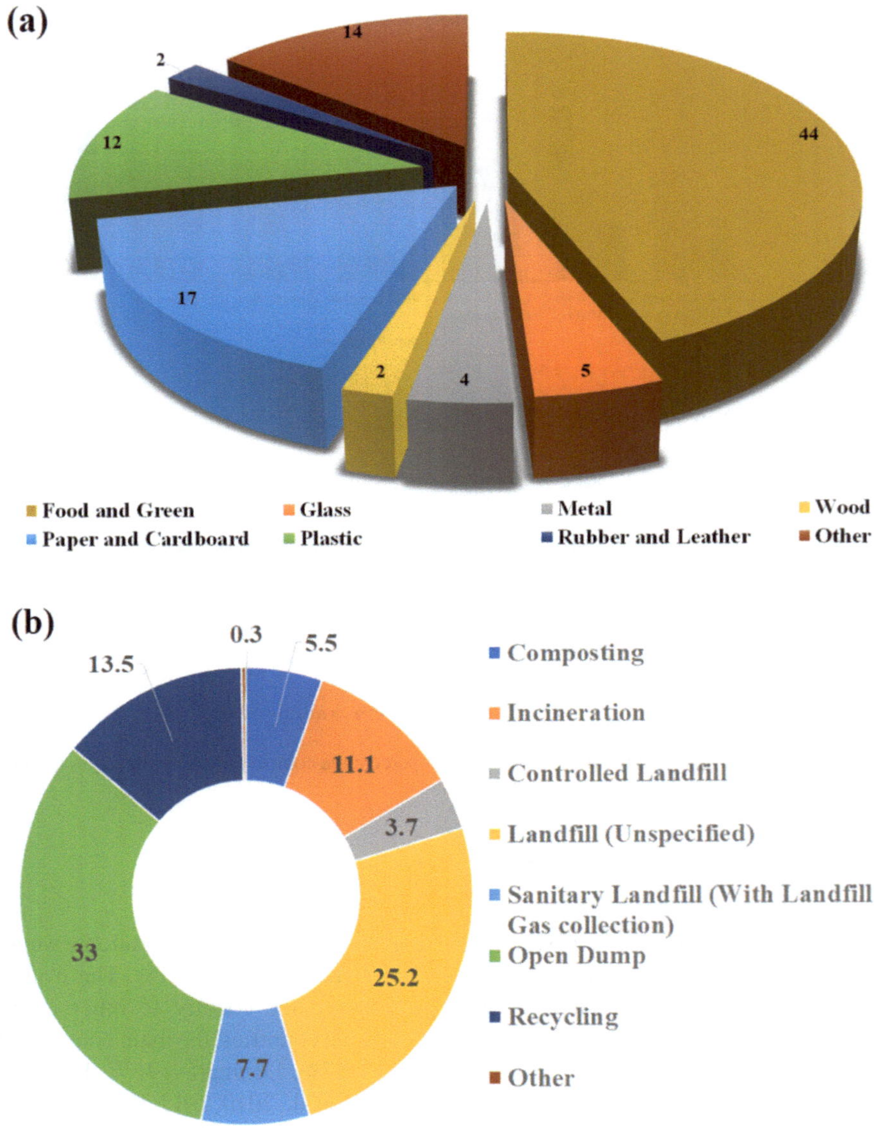

Fig. 3.3 Percentage classification of different available wastes and different strategies to waste management **a** pie chart showing percentage of different waste biomasses available, **b** pie chart showing the percentage of different waste biomass management strategies

2016). The MSW contains two types of polymer groups including biopolymers and synthetic polymers. Carbohydrates (e.g., cellulose and hemicellulose), proteins, and phenolics (e.g., lignin) are the principal biopolymers found in MSW, while polyvinyl chloride, polyethylene, polyethylene terephthalate, polystyrene, and polypropylene are the principal synthetic polymers found in MSW (Al-Salem et al., 2009).

3.3 Sustainability and Business Models

Kofi Annan examined the scholarly impediments of outrageous ways of thinking concerning the monetary turn of events and the climate at the Johannesburg Summit (Annan, 2002). In 2015, the European Union embraced and discussed approaches to reuse and decrease plastic waste under the Circular bioeconomy Package. This plan includes improving plastic waste reuse and a plan for reusability (Solis, 2018). Some European nations are making new arrangements and plans to reduce or utilize the MSW (Solis, 2018). For instance, each maker of plastic bundling is currently committed to guarantee that the waste is gathered, reused, recuperated, or discarded in Sweden (Solis, 2018). Plastic makers are currently liable for the assortment, reuse, and removal of plastic MSW. Today, at least 80% of recyclable plastics are collected in bundles/packages and the collection charges are paid by the waste assortment organizations. Bundling makers have started their collection and reusing businesses, while others are cooperating with the waste assortment organizations (Matsakas et al., 2017). Costa Rica has started a public–private program known as ecolones (https://ecolonescr.com/) to enhance the reuse of MSW. After isolating and cleaning recyclable MSW such as aluminum jars, glass, plastic, tetra pack, and tune are transported to the collection center where the collector gets the "ecolones," virtual cash in returns which can be traded for other products. Kenya and India have banned the use of plastic bags to decrease the use of hard plastic (Singh, 2022; Reuters, 2017). Innovative ideas are additionally being executed in the value-added products and biodiesel industries by using organic MSW. Ideas like total oil management are reforming, in which the whole oil processing, ordering, storing, delivering of crisp cooking oil, and reusing of pre-utilized cooking oil are carried out by a similar organization. The users only have to pay the compensation charge for the utilization of the oil (Tsai, 2019; Mohamed, et al., 2022). The material/textile industry has likewise ideas in reusing the utilized garments. Second-hand shops are, obviously, the first choice that rings a bell. Other sagacious techniques have likewise been utilized, like the shipment of involved garments to unfortunate districts in Africa or South America or to the uneven locales of Asia where there is a requirement for modest comfortable warm garments. Recycling new designs in clothes is likewise turning into a famous trend. Much the same as some vehicle sales centers, different garment stores have a strategy of collecting the used garments in return for an initial investment to buy new ones (Aki et al., 2020; Aus et al., 2021). For instance, H&M has permitted the customers to give packs of dresses in return for a 15% rebate on any item they like (Stål and Jansson, 2017). The development of many circular and non-centralized business ideas where the maker is likewise answerable for the recycling of the item is a logical way to fabricate a worldwide circular economy step by step. These business plans, along with the development of incorporated reusing and recycling, and the making of unified biorefineries, are significantly expected to close the loop in the circular economy.

3.4 Well-Known Innovations Utilized for MSW Management

This part is an examination of the most widely recognized and advanced technologies for overseeing the MSW via burning, landfilling, composting, anaerobic assimilation, and open dumps. The fundamental measures used to choose a given technique for the MSW management are the land region accessibility, properties of the waste, capital expense, designated items, and energy demands. Municipal solid waste management can be achieved by several conventional (Chen et al., 2010) and advanced methods discussed below. The schematics of the same have been summarized in Fig. 3.4.

3.4.1 *Plastic Reuse and Removal*

For non-damaged and reusable wastes, an essential reusing process is used. Mechanical processing (optional reusing) is restricted to single, clean, undamaged plastic, bringing about an item of identical quality. Tertiary reusing depends on the degradation of waste plastics with the help of thermal agents and chemicals. Polyethylene terephthalate is exclusively incorporated with the new polymers with comparable properties to make use of unutilized plastics with the help of Chemolysis methods,

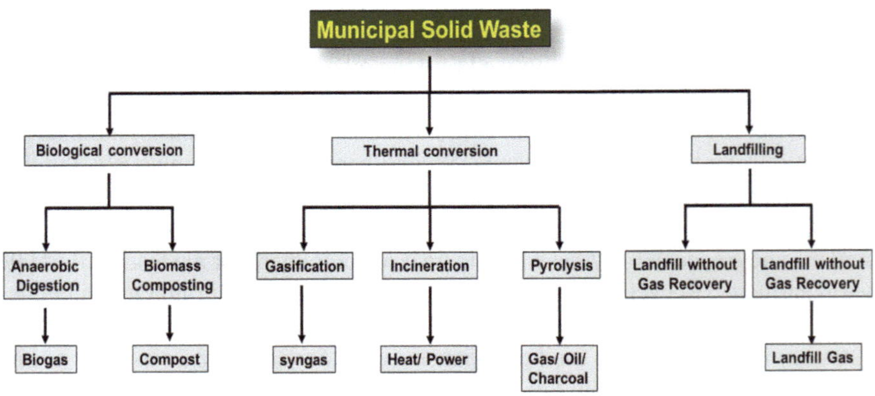

Fig. 3.4 Schematic representation of strategies for the municipal solid waste management

for example, alcoholysis, methanolysis, hydrolysis, and glycolysis. The cremation or incineration of blended waste streams for energy comes under quaternary methodology (Li et al., 2022; Praveenkumar et al., 2024).

3.4.2 *Incineration*

Incineration, a strategy used for MSW management, is characterized as an obliteration of waste within the sight of oxygen at higher temperatures ranging from 850 to 1200 °C in a muffle furnace. The fundamental objective of this technique is to diminish the mass and volume of the MSW generated (Campuzano & González-Martinez, 2016; Ayiania et al., 2019). Countries with space limitations like China, Japan, and Europe regularly use this technique for the disposal of MSWs since it requires less space. However, the requirement of high working and economic expenses is one of its main disadvantages (Annan, 2002; Matsakas et al., 2017). Though this technique conveys about 90% of the absolute mass reduction (Cheng & Hu, 2010; Tozlu et al., 2016; Wilson et al., 2015), but it is responsible for the release of pollutants like CO_x, SO_x, NOx, dioxins, and furans into the air (McKay, 2002). Gypsum (from cleaning the smoke), slags (the uncombusted carbon), and fly ashes are the three primary wastes produced from the incineration of MSW (Li et al., 2012). The Fly ashes contain glass, metals, minerals, unburned organic matter, and paramagnetic metals (Chimenos et al., 1999). The fundamental benefits of incineration of MSW are the production of energy by reducing the mass and volume of the waste (Tozlu et al., 2016). It cannot be utilized for high moisture-containing waste streams with low calorific worth and chlorinated MSW.

3.4.3 *Landfilling*

Landfilling is the most used method for MSW management in underdeveloped countries and requires huge space outside the city limits. The use of organic MSW for landfilling is not recommended due to its anaerobic digestion which results in the emission of methane into the environment (Hettiaratchi et al., 2015; Hoornweg & Bhada-Tata, 2012) which has a 20–23 times higher greenhouse potential compared to carbon dioxide (Matsakas et al., 2017). Moreover, unsavory/unpleasant smells are produced (sulfur organic mixtures) from the landfills (Hoornweg & Bhada-Tata, 2012). However, in modern landfills, the biogas can be collected by using wells and lines and further utilized in boilers or turbines for the generation of power or heat (Campuzano & González-Martinez, 2016). Only 30–40% of the complete gas produced in landfills can be collected by this method which is riskier due to the highly flammable nature of methane. The presence of xenobiotics and heavy metals in landfills' leachate additionally becomes the reason for soil and water pollution (Tyagi et al., 2018). From a financial point of view, landfills could have an advantage over other methods due to the use of available cheap feedstock. The development of the circular economy is essential for the reduction of organic waste use in landfills.

3.4.4 Composting

The aerobic digestion of organic waste, known as composting, is used to oxidize the organic wastes using aerobic microbes. Regularly used composting feedstocks in MSW include agricultural wastes, food wastes, gardening, and agro-industrial wastes. The utilization of MSW for composting has three primary motives.

(i) Production of stable manure products.
(ii) Reduction of methane.
(iii) Reduction of the waste volume.

The final product of the composting process also has a commercial value due to its soil fertility enhancement properties (Tyagi et al., 2018).

3.5 Overview of Recent/Most Promising MSW Management Technologies

The most emerging techniques used for MSW management include the following biochemical and thermochemical processes (Table 3.1).

3.5.1 Pyrolysis

Pyrolysis is an anaerobic process used to depolymerize the MSW to produce a solid or liquid byproduct at temperatures ranging from 350 to 600 °C to reduce the waste by 50–90% by volume (Beyene et al., 2018). Pyrolysis of lignocellulosic material produces bio-oil and biochar as a final product (Khuenkaeo & Tippayawong, 2020; Tozlu et al., 2016). This process is more suitable for C, O, and H-containing polymers, but the presence of other inorganic molecules such as Cl, N, and S causes the emission of pollutants (Gao, 2010). New technologies are necessary for the reduction of harmful emissions during feedstock pyrolysis (He et al., 2010). It is an exceptionally encouraging technology for the handling of lignocellulosic wastes and plastics, but there are very few refineries that are willing to use the biorefinery byproducts (Al-Salem et al., 2009).

Hydrocracking, catalytic cracking, and plasma pyrolysis are evolved from the conventionally used pyrolysis methods (Huang & Tang, 2007). Hydrocracking is carried out in presence of a catalyst and conducted in a hydrogen environment to produce a higher quality product but is an expensive process. Catalytic cracking is like hydrocracking and is carried out in presence of a catalyst for higher bio-oil yield at a lower temperature (Table 3.1).

Catalyst deactivation and MSW contaminations are very sensitive factors for catalytic cracking. Solid acids such as MCM-4, HY, HZSM-5, and $AlCl_3$ are the main catalysts used in the catalytic cracking method (Miandad et al., 2016). The decomposition of toxic compounds present in the gas products was carried out using plasma technologies. For the production of syngas with high heating value and low char content from plastics, this technology is recommended. It is a very expensive and highly electricity-consuming process.

3.5.2 *Gasification*

It is also a type of thermochemical conversion technology similar to pyrolysis, which is carried out in presence of oxygen and a temperature range of 600–1400 °C (Arena, 2012). Steam, CO_2, and O_2 are the main oxidants used for MSW gasification which are also responsible for the caloric value of resulting syngas (Arena, 2012). Syngas produced via gasification in presence of oxygen has higher caloric value as compared to syngas produced in a nitrogen environment. Hydrogen-enriched syngas can be produced in steam gasification via an endothermic process (Arena, 2012). The gasification of MSW produces syngas containing a mixture of H_2, CO_2, and CO which are further used for chemical synthesis of other fuels and value-added chemicals such as acetic acid, formaldehyde, methanol, ethanol, gasoline, olefins, diesel, waxes, ammonia, and other alcohols (Molino et al., 2018; Sikarwar et al., 2017). With a good MSW volume reduction, it also meets the existing GHG emission guidelines (Arena, 2012). The performance of a gasifier is based on the physical property of the waste, waste composition, waste residence time, reaction temperature, and air-to-fuel ratio (Arena, 2012). It can digest the complex MSW mixture containing wood, plastic, and garden waste into valuable products (Arena, 2012).

Migliaccio et al. (2021) demonstrated fluidized bed gasification of sewage sludge at 850 °C in presence of nitrogen/air mixture as a gasification agent maintained at different oxygen/fuel equivalence ratios (ER) ranging from 0.1 to 0.2. Further, computer-aided simulations were performed for devising a better syngas management strategy (Fig. 3.5). However, gasification is a sensitive and expensive process that cannot be justified for the small-scale process.

3.5.3 *Hydrocracking*

This process is carried out in the presence of a catalyst under a pressurized H_2 atmosphere at a temperature of 375–500 °C with a solubilized liquid as a final product

Table 3.1 Recent promising MSW management technologies

Technologies	Waste biomass	Key features	References
Pyrolysis	Mulch film consisting of polylactic acid (PLA) and polybutylene adipate terephthalate (PBAT)	• Efficient disposal of mulch film via CO_2-mediated pyrolysis • The presence of CO_2 resulted in enhanced production of combustible gas and suppressed generation of phenolic and polycyclic compounds during pyrolysis	Kim et al. (2022)
	Sewage sludge	• Pyrolysis conducted at 430 °C pyrolysis reactor • Conversion of sewage sludge conversion into value-added products with 2–threefold higher phosphorus content • Pyrolysis resulted in the removal of toxic components of Sewage sludge	Frišták et al. (2018)
	COVID-19 medical wastes	• The COVID-19 pandemic resulted in huge hospital solid wastes rich in polymers such as nylon, terephthalate, polystyrene, polyethylene, and polypropylene • Pyrolysis can result in the generation of value-added products such as oil, gas, and char	Dharmaraj et al. (2021)
Fluidized bed gasification	Sewage sludge	• Gasification at 850 °C in presence of nitrogen/air mixture as a gasification agent • Different oxygen/fuel equivalence ratio (ER) raging form 0.1 to 0.2 • Characterization of the starting and output products, i.e., solid residues, tar, and syngas • Computer-aided simulations for devising the better syngas management strategy	Migliaccio et al. (2021)
Hydrocracking	Plastic waste	• Catalytic hydrocracking in hydrogen at 225 °C via initial activation over Pt catalyst, followed by sequential cracking and isomerization over WO_3/ZrO_2/HY Zeolite and hydrogenation of intermediates (olefin) over Pt • Efficient yield (85%) of liquid fuels (diesel, jet) and gasoline range from polyolefins • These processes can be tuned to the conversion of common plastic waste to fuels and lubricants	Liu et al. (2021)

(continued)

Table 3.1 (continued)

Technologies	Waste biomass	Key features	References
	Fresh and waste frying oil	• Ni- and Mo-based catalyst supported on sulfated silica (Ni-SS2, NI-M0-SS3)-mediated hydrocracking of fresh and waste frying oil • Ni-SS2 resulted in 71.47% conversion of the liquid product with gasoline fraction (C5-C12) selectivity of 58.73% • Highest activity (51.50 wt%) and gasoline fraction selectivity (43.22 wt%) were demonstrated by NiMo-SS3-mediated hydrocracking	Wijaya et al. (2021)
Hydrothermal liquefaction (HTL)	Simulated food waste (SFW)	• HTL of SFW at different process parameters, i.e., temperatures, pressure, biomass loading, and time of 200–600 °C, 10.2–35.7 MPa, 2–20 wt %, and 1–33 min, respectively • Reaction conditions, i.e., 600 °C, 35.3 MPa, for 30 min resulted in the production of bio-crude with the largest heating value (36.5 MJ/kg) • Bio-crude production was accompanied by 50% and 68% transfer of nitrogen and phosphorus, respectively, into the aqueous phase • Under all reaction conditions, bio-crude consists mainly of saturated fatty acids • Isothermal HTL is better for the generation of heavy compounds as compared to fast HTL	Motavaf and Savage (2021)
Wet oxidation	Sludges from PAKMAYA yeast factory	• Raw activated and digested sludges were subjected to wet oxidation with Cu and Mn group as a catalyst and H_2O_2 as oxidants • Cu-H_2O_2-mediated wet oxidation for 10 min resulted in a 16.5% increment in total organic carbon • Wet oxidation resulted in enhanced settling of sludge solids up to 80% and high treatment efficiency • Ammonia, nitrate, and nitrite concentrations decreased in the supernatant accompanied by an increase in pH from 6.6 to 7.8–8.0 of the sludge	Genç et al. (2002)

(continued)

Table 3.1 (continued)

Technologies	Waste biomass	Key features	References
	Sewage sludge	• Integration of wet oxidation struvite-precipitation for enhanced sewage sludge treatment and phosphorus recovery • Overall yields are affected more by wet oxidation compared to struvite precipitation	Munir et al. (2019)
Anaerobic digestion	Solid food waste biomass	• Solid food waste biomass such as residual meat, noodles, rice, and veggies was subjected to anaerobic digestion using inoculum of animal intestinal waste (ANW) and cattle manure (CM) • Food waste inoculation up to 30% inoculation and mesophilic temperature is the optimum condition for anaerobic digestion • The biogas yield was higher and at a faster rate with ANW as compared to CM in all digester materials	Abbas et al. (2020)

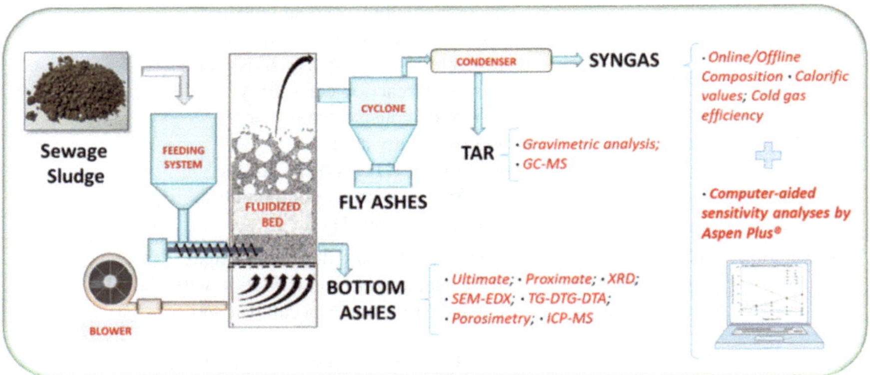

Fig. 3.5 Schematic representation of fluidized bed gasification of sewage sludge (Migliaccio et al., 2021) @ Creative common License

of this process (Solis, 2018). Hydrocracking is mainly used to utilize molten plastics except PVC and process temperature can be reduced with the use of a catalyst. The main hurdle of hydrocracking process deployment is the higher hydrogen cost. Companies like ITC, Bohlen, Freiberg, and Hiedrierwerke are working on the development of the hydrocracking process (Solis, 2018).

3.5.4 *Hydrothermal Liquefaction*

The hydrothermal digestion of food wastes, algae, and sludge is carried out at 300–450 °C temperature in the liquid phase. Similar to the pyrolysis, hydrothermal liquefaction also produces crude solids, gas, bio-oil, and char (Elliott et al., 2015). Unlike pyrolysis, the removal of thermochemical degradation products is carried out by solubilizing in water. Products produced from hydrothermal liquefaction of cellulose and hemicellulose in presence of oxygen are solubilized in the liquid phase (Elliott et al., 2015).

3.5.5 *Wet Oxidation*

Catalytic wet oxidation and wet oxidation are processes carried out in the reactors at a temperature of 130–374 °C in presence of an oxidizing agents like hydrogen peroxide or ozone. Wet oxidation plans to eliminate natural contaminations from fluid effluents with the arrangement of CO_2. More than half a century ago, this innovation was patented by Zimmerman. On an industrial scale wet oxidation of organic MSW, bubbling column reactor is commonly used. The oxidizing agent in the gas-to-liquid form is an essential parameter for the catalytic wet oxidation reaction. Wet oxidation can also be used for the pretreatment of lignocellulosic biomass wastes in presence of oxygen at moderate temperature (Ahring et al., 2015). This can further be utilized for the bioconversion of bioethanol after enzymatic digestion or methane through an anaerobic process.

3.5.6 *Anaerobic Digestion*

The microbial digestion of organic wastes in absence of oxygen is called the anaerobic process. Here, the composition and quality of output biogas depend on the composition of organic wastes and operating conditions (Campuzano & González-Martiinez, 2016). The digested organic waste by this process can be used as a biofertilizer (Chen et al., 2012; Pivato et al., 2016). Anaerobic digestion involves four stages (Mata-Alvarez et al., 2000).

I. Conversion of organic polymers (carbohydrates, proteins) into their monomers.
II. Conversion of monomers into volatile fatty acids through acidogenesis.
III. Conversion of volatile fatty acids into acetate, carbon dioxide, and hydrogen through acetogenesis.
IV. Conversion of acetate, carbon dioxide, and hydrogen to methane through methanogenesis.

Compared to other thermochemical hydrolysis processes, the anaerobic digestion of MSW is a cost-effective and environmentally friendly process that helps in the management of MSW and production of green energy (Hartmann & Ahring, 2006; Mata-Alvarez, 2002). Organic wastes such as lignocellulosic material, food waste, wastewater, sewage sludge and manure can be used as a feedstock for this process (Campuzano & González-Martinez, 2016).

3.6 Motivations and Challenges for the Valorization of Municipal Wastes in a Circular Economy

There are several motivations behind the valorization of municipal waste in turning the dream of a circular economy into reality, but it is also plagued with several challenges which act as inhibitors in the progress of this approach (Björklund & Öhman, 2017; Ceccon & da Silva, 2011; Kalmykova et al., 2018; Khatiwada et al., 2021; Remy, 2018; Rodić & Wilson, 2017).

The motivation behind the valorization of municipal wastes toward a circular economy is as follows:

- Global commitments and memorandums to commit toward limiting the generation of waste.
- Industrial and government are becoming more aware of sustainable methods for industrial developments.
- The negative impact of landfills and incineration plants led to their ban on a huge scale. So, alternative greener and sustainable methods are required.
- Development of institutional arrangement and collaboration of industry, academia, and research is giving a better understanding of opportunities of waste management for the generation of high-value compounds from solid waste.
- Opening of waste management plants for urban waste management and employment opportunities.
- Reduction in waste disposal cost, energy generation, and conservation in newly devised methods.

Major challenges in achieving the objective of valorization of municipal wastes toward a circular economy are as follows:

- Lack of understanding, and budget for initiation of the projects.
- Negligence from government authorities and lack of commitment from political leadership in promoting the waste valorization strategies at the ground level.
- Searching for an appropriate market for the products generated via these approaches.
- The overall cost of the developed process.
- Federal rules and regulations sometimes act as limiting factors in promoting such initiatives, as these approvals and permission lead to unnecessary trouble and cost inflation.
- Public perception and organizational culture also act as a limiting factor in the acceptance of such approaches at grassroots levels.

- Energy intensive nature of existing physical and thermochemical technologies for waste conversion to value-added compounds limits their application at large scale.
- Economic viability and unplanned implementation of the developed processes at the initial stages have resulted in failure of the upscaling and further popularization.

3.7 Conclusions

Based on the above discussion, it is very clear that huge amounts of solid waste are generated around the globe. The availability and waste generation potential of developing and developed countries have been discussed and it clearly showed that developed countries contribute a greater percentage of waste. Municipal waste, textiles, chemicals, electronic, medical waste, and sewage sludge contribute to available waste which was earlier subjected to incineration and dumping to landfills generating huge piles of waste around the globe. Recent studies suggest that MSW can be sustainably converted into value-added compounds and energy. The valorization of MSW can be supported via different advanced technologies such as pyrolysis, hydrocracking, hydrothermal liquefaction, gasification, and anaerobic digestion. These methods are being implemented at a large scale for the generation of energy, but several other limitations limit its fast propagation at the global level. Nevertheless, mankind has realized that waste management is not just the responsibility for a better sustainable future but an opportunity to generate high-value products that can enter the market and generate revenues, thus providing the waste a second opportunity "Waste 2.0" which in turn promises to provide high-value products and a cleaner–greener environment.

References

Abbas, Y., Jamil, F., Rafiq, S., Ghauri, M., Khurram, M. S., Aslam, M., Bokhari, A., Faisal, A., Rashid, U., Yun, S., & Mubeen, M. (2020). Valorization of solid waste biomass by inoculation for the enhanced yield of biogas. *Clean Technologies and Environmental Policy, 22*(2), 513–522. https://doi.org/10.1007/s10098-019-01799-6

Abdel-shafy, H. I., Mansour, M. S. (2018) Solid waste issue: Sources, composition, disposal, recycling, and valorization. *Egyptian journal of petroleum, 27*(4), 1275–1290. https://doi.org/10.1016/j.ejpe.2018.07.003

Abu-Qudais, M., & Abu-Qdais, H. A. (2000). Energy content of municipal solid waste in Jordan and its potential utilization. *Energy Conversion and Management, 41*(9), 983–991. https://doi.org/10.1016/S0196-8904(99)00155-7

Ahring, B. K., Biswas, R., Ahamed, A., Teller, P. J., & Uellendahl, H. (2015). Making lignin accessible for anaerobic digestion by wet-explosion pretreatment. *Bioresource Technology, 175*, 182–188. https://doi.org/10.1016/j.biortech.2014.10.082

Al-Salem, S. M., Lettieri, P., & Baeyens, J. (2009). Recycling and recovery routes of plastic solid waste (PSW): A review. *Waste Management, 29*(10), 2625–2643. https://doi.org/10.1016/j.wasman.2009.06.004

Aki, S. U., Candan, C., Nergis, B., & Önder, N. S. (2020). Understanding denim recycling: A quantitative study with lifecycle assessment methodology. In *Waste in Textile and Leather Sectors*. IntechOpen. https://doi.org/10.5772/intechopen.92793

Andritsos, G., Desideri, A., Gantiez, C., Lemort, V., & Quoilin, S. (2016). *Steady-state and dynamic modelling of a 1 MWel commercial waste heat recovery ORC power plant. Ecos.* In 29th International conference on Efficiency, Cost, Optimization, Simulation and Environmental Impact of Energy systems.

Annan, K. (2002, September 2). *Secretary-general Kofi Annan to the world summit on sustainable development in Johannesburg.*

Arena, U. (2012). Process and technological aspects of municipal solid waste gasification. A review. *Waste Management, 32*(4), 625–639. https://doi.org/10.1016/j.wasman.2011.09.025

Aslani, H., & Taghipour, H. (2018). Seasonal characterization and quantification of municipal solid waste: Energy content and statistical analysis. *Journal of Advances in Environmental Health Research, 6*, 34–43.

Aus, R., Moora, H., Vihma, M., Unt, R., Kiisa, M., & Kapur, S. (2021). Designing for circular fashion: integrating upcycling into conventional garment manufacturing processes. *Fashion and Textiles, 8*, 1–18. https://doi.org/10.1186/s40691-021-00262-9

Ayiania, M., Carbajal-Gamarra, F. M., Garcia-Perez, T., Frear, C., Suliman, W., & Garcia-Perez, M. (2019). Production and characterization of H_2S and $PO_4{}^{3-}$ carbonaceous adsorbents from anaerobic digested fibers. *Biomass and Bioenergy, 120*, 339–349. https://doi.org/10.1016/j.biombioe.2018.11.028

Beyene, H. D., Werkneh, A. A., & Ambaye, T. G. (2018). Current updates on waste to energy (WtE) technologies: A review. *Renewable Energy Focus, 24*, 1–11. https://doi.org/10.1016/j.ref.2017.11.001

Björklund, S., & Öhman, N. (2017). *Biogas opportunities in Curitiba: Analysis of business potential for biogas production from municipal solid waste* [MSc Thesis].

Campuzano, R., & González-Martínez, S. (2016). Characteristics of the organic fraction of municipal solid waste and methane production: A review. *Waste Management, 54*, 3–12. https://doi.org/10.1016/j.wasman.2016.05.016

Ceccon, N. P., & da Silva, C. L. (2011). Institutional arrangement of the intermunicipal consortium of the metropolitan region of Curitiba for management of municipal solid waste. *Urban Public Economics Review*, 46–73.

Chandra, R., & Chowdhary, P. (2015). Properties of bacterial laccases and their application in bioremediation of industrial wastes. *Environmental Science. Processes and Impacts, 17*(2), 326–342. https://doi.org/10.1039/c4em00627e

Chen, L., Zhao, L., Ren, C., & Wang, F. (2012). The progress and prospects of rural biogas production in China. *Energy Policy, 51*, 58–63. https://doi.org/10.1016/j.enpol.2012.05.052

Chen, X., Geng, Y., & Fujita, T. (2010). An overview of municipal solid waste management in China. *Waste Management, 30*(4), 716–724. https://doi.org/10.1016/j.wasman.2009.10.011

Cheng, H., & Hu, Y. (2010). Municipal solid waste (MSW) as a renewable source of energy: Current and future practices in China. *Bioresource Technology, 101*(11), 3816–3824. https://doi.org/10.1016/j.biortech.2010.01.040

Cherubini, F. (2010). The biorefinery concept: Using biomass instead of oil for producing energy and chemicals. *Energy Conversion and Management, 51*(7), 1412–1421. https://doi.org/10.1016/j.enconman.2010.01.015

Chimenos, J. M., Segarra, M., Fernández, M. A., & Espiell, F. (1999). Characterization of the bottom ash in municipal solid waste incinerator. *Journal of Hazardous Materials, 64*(3), 211–222. https://doi.org/10.1016/S0304-3894(98)00246-5

Christy, P. M., Gopinath, L. R., & Divya, D. (2014). A review on anaerobic decomposition and enhancement of biogas production through enzymes and microorganisms. *Renewable and Sustainable Energy Reviews, 34*, 167–173. https://doi.org/10.1016/j.rser.2014.03.010

Dharmaraj, S., Ashokkumar, V., Pandiyan, R., Halimatul Munawaroh, H. S. H., Chew, K. W., Chen, W. H., & Ngamcharussrivichai, C. (2021). Pyrolysis: An effective technique for degradation of COVID-19 medical wastes. *Chemosphere, 275*, 130092. https://doi.org/10.1016/j.chemosphere.2021.130092

Edjabou, M. E., Jensen, M. B., Götze, R., Pivnenko, K., Petersen, C., Scheutz, C., & Astrup, T. F. (2015). Municipal solid waste composition: Sampling methodology, statistical analyses, and case study evaluation. *Waste Management, 36*, 12–23. https://doi.org/10.1016/j.wasman.2014.11.009

Elliott, D. C., Biller, P., Ross, A. B., Schmidt, A. J., & Jones, S. B. (2015). Hydrothermal liquefaction of biomass: Developments from batch to continuous process. *Bioresource Technology, 178*, 147–156. https://doi.org/10.1016/j.biortech.2014.09.132

EPA. (2016). Wastes—Nonhazardous waste—Municipal solid waste. In *Environmental protection agency*.

Frišták, V., & Pipiška, & M., Soja, G. (2018). Pyrolysis treatment of sewage sludge: A promising way to produce phosphorus fertilizer. *Journal of Cleaner Production, 172*.

Gao, F. (2010). *Pyrolysis of waste plastics into fuels* [PhD Thesis], University of Canterbury, Chemical and Process Engineering.

Genç, N., Yonsel, S., Dağaşan, L., & Onar, A. N. (2002). Wet oxidation: A pre-treatment procedure for sludge. *Waste Management, 22*(6), 611–616. https://doi.org/10.1016/s0956-053x(02)00040-5

Hartmann, H., & Ahring, B. K. (2006). Strategies for the anaerobic digestion of the organic fraction of municipal solid waste: An overview. *Water Science and Technology, 53*(8), 7–22. https://doi.org/10.2166/wst.2006.231

He, M., Xiao, B., Liu, S., Hu, Z., Guo, X., Luo, S., & Yang, F. (2010). Syngas production from pyrolysis of municipal solid waste (MSW) with dolomite as downstream catalysts. *Journal of Analytical and Applied Pyrolysis, 87*(2), 181–187. https://doi.org/10.1016/j.jaap.2009.11.005

Hettiaratchi, P., Jayasinghe, P., Tay, J., & Yadav, S. (2015). Recent advances of biomass waste to gas using landfill bioreactor technology—A review. *Current Organic Chemistry, 19*(5), 413–422. https://doi.org/10.2174/1385272819666150119223002

Hoornweg, D., & Bhada-Tata, P. (2012). What a waste: A global review of solid waste management. *Urban Development Series Paper, 15*. World Bank.

Huang, H., & Tang, L. (2007). Treatment of organic waste using thermal plasma pyrolysis technology. *Energy Conversion and Management, 48*(4), 1331–1337. https://doi.org/10.1016/j.enconman.2006.08.013

Kalmykova, Y., Sadagopan, M., & Rosado, L. (2018). Circular economy—From review of theories and practices to development of implementation tools. *Resources, Conservation and Recycling, 135*, 190–201. https://doi.org/10.1016/j.resconrec.2017.10.034

Khatiwada, D., Golzar, F., Mainali, B., & Devendran, A. A. (2021). Circularity in the management of municipal solid waste—A systematic review. *Environmental and Climate Technologies, 25*(1), 491–507. https://doi.org/10.2478/rtuect-2021-0036

Khuenkaeo, N., & Tippayawong, N. (2020). Production and characterization of bio-oil and biochar from ablative pyrolysis of lignocellulosic biomass residues. *Chemical Engineering Communications, 207*(2), 153–160. https://doi.org/10.1080/00986445.2019.1574769

Kim, S., Yang, W., Lee, H. S., Tsang, Y. F., & Lee, J. (2022). Effectiveness of CO2-mediated pyrolysis for the treatment of biodegradable plastics: A case study of polybutylene adipate terephthalate/polylactic acid mulch film. *Journal of Cleaner Production, 372*, 133763. https://doi.org/10.1016/j.jclepro.2022.133763

Li, X.-G., Lv, Y., Ma, B.-G., Chen, Q.-B., Yin, X.-B., & Jian, S.-W. (2012). Utilization of municipal solid waste incineration bottom ash in blended cement. *Journal of Cleaner Production, 32*, 96–100. https://doi.org/10.1016/j.jclepro.2012.03.038

Li, X., Wang, J., Zhang, T., Yang, S., Sun, M., Qian, X., Wang, T., & Zhao, Y. (2023). Sustainable catalytic strategies for the transformation of plastic wastes into valued products. *Chemical Engineering Science*, 118729. https://doi.org/10.1016/j.ces.2023.118729

Liu, S., Kots, P. A., Vance, B. C., Danielson, A., & Vlachos, D. G. (2021). Plastic waste to fuels by hydrocracking at mild conditions. *Science Advances, 7*(17), eabf8283. https://doi.org/10.1126/sciadv.abf8283

Malinauskaite, J., Jouhara, H., Czajczyńska, D., Stanchev, P., Katsou, E., Rostkowski, P., Thorne, R. J., Colón, J., Ponsá, S., Al-Mansour, F., Anguilano, L., Krzyżyńska, R., López, I. C., Vlasopoulos, A., & Spencer, N. (2017). Municipal solid waste management and waste-to-energy in the context of a circular economy and energy recycling in Europe. *Energy, 141*, 2013–2044. https://doi.org/10.1016/j.energy.2017.11.128

Manahan, S. E. (2017). *Industrial ecology: Environmental chemistry and hazardous waste*. Routledge.

Mata-Alvarez, J. (2002). *Biomethanization of the organic fraction of municipal solid wastes*. IWA Publishing.

Mata-Alvarez, J., Macé, S., & Llabrés, P. (2000). Anaerobic digestion of organic solid wastes. An overview of research achievements and perspectives. *Bioresource Technology, 74*(1), 3–16. https://doi.org/10.1016/S0960-8524(00)00023-7

Matsakas, L., Gao, Q., Jansson, S., Rova, U., & Christakopoulos, P. (2017). Green conversion of municipal solid wastes into fuels and chemicals. *Electronic Journal of Biotechnology, 26*, 69–83. https://doi.org/10.1016/j.ejbt.2017.01.004

McKay, G. (2002). Dioxin characterisation, formation and minimisation during municipal solid waste (MSW) incineration: Review. *Chemical Engineering Journal, 86*(3), 343–368. https://doi.org/10.1016/S1385-8947(01)00228-5

Miandad, R., Barakat, M. A., Aburiazaiza, A. S., Rehan, M., & Nizami, A. S. (2016). Catalytic pyrolysis of plastic waste: A review. *Process Safety and Environmental Protection, 102*, 822–838. https://doi.org/10.1016/j.psep.2016.06.022

Migliaccio, R., Brachi, P., Montagnaro, F., Papa, S., Tavano, A., Montesarchio, P., Ruoppolo, G., & Urciuolo, M. (2021). Sewage sludge gasification in a fluidized bed: Experimental investigation and modeling. *Industrial and Engineering Chemistry Research, 60*(13), 5034–5047. https://doi.org/10.1021/acs.iecr.1c00084

Mohamed, M., Sherif, N., Aboelazayem, O., Elazab, H. A., Gadalla, M., & Saha, B. (2022). Waste cooking oil management in Egypt: Production of biodiesel-development of rapid test method. In *Journal of Physics: Conference Series* (Vol. 2305, No. 1, p. 012035). IOP Publishing. https://doi.org/10.1088/1742-6596/2305/1/012035

Molino, A., Larocca, V., Chianese, S., & Musmarra, D. (2018). Biofuels production by biomass gasification: A review. *Energies, 11*(4), 811. https://doi.org/10.3390/en11040811

Motavaf, B., & Savage, P. E. (2021). Effect of process variables on food waste valorization via hydrothermal liquefaction. *ACS ES&T Engineering, 1*(3), 363–374. https://doi.org/10.1021/acsestengg.0c00115

Munir, M. T., Li, B., Mardon, I., Young, B. R., & Baroutian, S. (2019). Integrating wet oxidation and struvite precipitation for sewage sludge treatment and phosphorus recovery. *Journal of Cleaner Production, 232*, 1043–1052. https://doi.org/10.1016/j.jclepro.2019.06.007

Nielfa, A., Cano, R., & Fdz-Polanco, M. (2015). Theoretical methane production generated by the co-digestion of organic fraction municipal solid waste and biological sludge. *Biotechnology Reports, 5*, 14–21. https://doi.org/10.1016/j.btre.2014.10.005

Noor, Z. Z., Yusuf, R. O., Abba, A. H., Abu Hassan, M. A., & Mohd Din, M. F. (2013). An overview for energy recovery from municipal solid wastes (MSW) in Malaysia scenario. *Renewable and Sustainable Energy Reviews, 20*, 378–384. https://doi.org/10.1016/j.rser.2012.11.050

Pfaltzgraff, L. A., De bruyn, M., Cooper, E. C., Budarin, V., & Clark, J. H. (2013). Food waste biomass: A resource for high-value chemicals. *Green Chemistry, 15*(2), 307–314. https://doi.org/10.1039/c2gc36978h

Pivato, A., Vanin, S., Raga, R., Lavagnolo, M. C., Barausse, A., Rieple, A., Laurent, A., & Cossu, R. (2016). Use of digestate from a decentralized on-farm biogas plant as fertilizer in soils: An ecotoxicological study for future indicators in risk and life cycle assessment. *Waste Management, 49*, 378–389. https://doi.org/10.1016/j.wasman.2015.12.009

Praveenkumar, T. R., Sekar, M., Pasupuleti, R. R., Gavurová, B., Kumar, G. A., & Kumar, M. V. (2024). Current technologies for plastic waste treatment for energy recovery, it's effects on poly aromatic hydrocarbons emission and recycling strategies. *Fuel, 357*, 129379. https://doi.org/10.1016/j.fuel.2023.129379

Reuters (2017). *Kenya imposes world's toughest law against plastic bags.* https://swachhindia.ndtv.com/kenya-imposes-worldstoughest-law-against-plastic-bags-11261/

Remy, F. (2018). *Potential for the anaerobic digestion of municipal solid waste (MSW) in the city of Curitiba, Brazil* [MSc Thesis].

Rodić, L., & Wilson, D. C. (2017). Resolving governance issues to achieve priority sustainable development goals related to solid waste management in developing countries. *Sustainability, 9*(3), 404. https://doi.org/10.3390/su9030404

Sahimaa, O., Hupponen, M., Horttanainen, M., & Sorvari, J. (2015). Method for residual household waste composition studies. *Waste Management, 46*, 3–14. https://doi.org/10.1016/j.wasman.2015.08.032

Saini, J. K., Saini, R., & Tewari, L. (2015). Lignocellulosic agriculture wastes as biomass feedstocks for second-generation bioethanol production: Concepts and recent developments. *3 Biotech, 5*(4), 337–353. https://doi.org/10.1007/s13205-014-0246-5

Saveyn, H., Eder, P., et al. (2014). *End-of-waste criteria for biodegradable waste subjected to biological treatment (compost & digestate): Technical proposals.* Publication Office of European Union.

Singh, S. G. (2022). *Single-use plastic ban: Environment lessons India can learn from East Africa.* https://www.downtoearth.org.in/blog/pollution/single-use-plastic-ban-environment-lessons-india-can-learn-from-east-africa-84029

Sikarwar, V. S., Zhao, M., Fennell, P. S., Shah, N., & Anthony, E. J. (2017). Progress in biofuel production from gasification. *Progress in Energy and Combustion Science, 61*, 189–248. https://doi.org/10.1016/j.pecs.2017.04.001

Solis, M. (2018). *Potential of chemical recycling to improve the recycling of plastic waste.*

Stål, H. I., & Jansson, J. (2017). Sustainable consumption and value propositions: Exploring product–service system practices among Swedish fashion firms. *Sustainable Development, 25*(6), 546–558. https://doi.org/10.1002/sd.1677

Thakur, A., Sharma, K., Khaire, K. C., Moholkar, V. S., & Goyal, A. (2020). Enzymes: Key role in the conversion of waste to bioethanol. In *Microbial fermentation and enzyme technology* (pp. 257–268). CRC Press.

Tsai, W. T. (2019). Mandatory recycling of waste cooking oil from residential and commercial sectors in Taiwan. *Resources, 8*(1), 38. https://doi.org/10.3390/resources8010038

Tozlu, A., Özahi, E., & Abuşoğlu, A. (2016). Waste to energy technologies for municipal solid waste management in Gaziantep. *Renewable and Sustainable Energy Reviews, 54*, 809–815. https://doi.org/10.1016/j.rser.2015.10.097

Tyagi, V. K., Fdez-Güelfo, L. A., Zhou, Y., Álvarez-Gallego, C. J., Garcia, L. I. R., & Ng, W. J. (2018). Anaerobic co-digestion of organic fraction of municipal solid waste (OFMSW): Progress and challenges. *Renewable and Sustainable Energy Reviews, 93*, 380–399. https://doi.org/10.1016/j.rser.2018.05.051

Wijaya, K., Nadia, A., Dinana, A., Pratiwi, A. F., Tikoalu, A. D., & Wibowo, A. C. (2021). Catalytic hydrocracking of fresh and waste frying oil over Ni- and Mo-based catalysts supported on sulfated silica for biogasoline production. *Catalysts, 11*(10), 1150. https://doi.org/10.3390/catal11101150

Wilson, D. C., Rodic, L., Modak, P., Soos, R., Carpintero, A., Velis, K., Iyer, M., & Simonett, O. (2015). *Global waste management outlook.* UN Environmental Program.

Chapter 4
Biodegradable Wastes in Bioeconomy

Dolores Hidalgo, Jesús M. Martín-Marroquín, and Francisco Corona

4.1 Introduction

Human activity has been deteriorating environmental conditions and causing negative effects in various areas: the environment, biodiversity, and water and soil resources. For this reason, the search for alternatives to face this situation constitutes an ethical and social responsibility. To meet the objectives of sustainable development and contribute to the improvement of the planet, it is necessary to promote a transition towards a more sustainable economy, for which innovative developments are required in the primary sectors that generate more technologies and efficient methods to increase the agricultural, forestry, and aquaculture productivity without threatening the carrying capacity of the planet or its biodiversity (Lewandowski, 2018). The nine billion inhabitants that are estimated to live on the planet in 2050 will require food, clean water, and renewable energy, even though the area of fertile land for agriculture decreases progressively due to intensive grazing, salinization, desertification, and urban growth, among other factors, and the limitations that climate change has already imposed on some regions must be addressed (Jiménez-Sánchez & Philp, 2015). All these represent great interconnected challenges in a complex way and require new approaches and social, economic, and environmental paradigms (Aguilar et al., 2018).

To achieve maximum benefit, all sectors of the bioeconomy must be coordinated, since they are all interdependent, in order to achieve food security, better nutrition and public health, and cleaner and more efficient industrial processes that contribute significantly to climate change mitigation (Lokko, et al., 2018). Potential benefits of transitioning to a bioeconomy include reduced greenhouse gas emissions, reduced dependence on fossil fuels, wiser resource management, and improved food security. Another positive effect would be the generation of employment opportunities

D. Hidalgo (✉) · J. M. Martín-Marroquín · F. Corona
CARTIF Technology Centre—Area of Circular Economy, Valladolid, Spain
e-mail: dolhid@cartif.es

S. A. Bandh and F. A. Malla (eds.), *Waste Management in the Circular Economy*,
https://doi.org/10.1007/978-3-031-42426-7_4

in rural and urban areas, and the creation of agricultural markets other than those for food and based on bioenergy and the recovery of agricultural and agro-industrial products, by-products and residues, which would represent a source of activities and alternative income suitable for rural areas. The challenge is to increase the scale of activities and, in parallel, respond to the sustainable development goals by introducing radical changes to face social challenges such as climate change, the scarcity of natural resources, and environmental pollution (McCormick & Kautto, 2013). Simultaneously with the changing situation of oil in the world market, the slowdown in agricultural yields, and the effects of climate change, living standards must be improved in various parts of the world, mainly in Asia, South America, and Africa, without necessarily being linked to an increase of meat consumption, especially, since this implies a greater requirement of feed, forage, and water and contributes to deforestation and the emission of greenhouse gases (Jiménez-Sánchez & Philp, 2015). As an example of the variable situation in some consumption patterns, an increase in the demand for animal protein is presented in several countries that have increased their purchasing power and doubled their wealth, in which it is found that gas emissions have increased up to by 80% (Aguilar et al., 2018). These situations have increased awareness of the biosphere's fragility and the urgent need to adopt sustainable consumption and production models based on a cascade, or circular, model of the bioeconomy. It is an economy that uses biomass instead of fossil resources to produce food and other non-food goods, bioproducts for industrial use, pharmaceuticals, and agriculture, among others. The bioeconomy approach aspires to copy the fundamentals of biological intelligence, where a living being constitutes a piece of machinery capable of processing a series of compounds (mainly of organic origin) to transform them into energy, biomass, and other by-products (de Jaramillo, 2018).

Situations such as the variability of the price of oil and gas in the world market (Wang et al., 2022), the slowdown in agricultural yields (Sibani et al., 2022), and the effects of climate change (Albert et al., 2021) have led to an increase in awareness of the fragility of the biosphere and the urgent need to adopt models of sustainable consumption and production, based on a cascade, or circular, model of the bioeconomy.

The sustainable use of renewable biological resources focuses on the three pillars of the bioeconomy: the production of food and feed, the production of bioproducts (preferably with added value), and the production of bioenergy and includes several industries and sectors: agriculture, forestry, fishing, human and animal food, pulp and paper production, as well as part of the chemical, biotechnological, and energy industries. Its strategy highlights the opportunity to achieve economic growth while ensuring the security of biological resources and their efficient and sustainable use (Cristóbal et al., 2016). Using renewable resources, some of the challenges of sustainability including water and soil management, the efficient use of nutrients, and socioeconomic development can be met. This refers to food and nutritional security through sustainable food production, the reduction of waste production through proper use of products and processes, considering the reduction, reuse, and recycling of components of food value chains, as well as the promotion of the efficiency of the use of resources in the production of biomass, and the use by society of all the components of these value chains or networks. This sustainability must be maintained throughout the supply chain to offer food security, a sufficient supply of raw materials and energy, the reduction of the environmental footprint, and the promotion of a healthy and viable rural economy. It is essential to reduce waste generation and promote the efficient recycling of what is generated, with closed cycles of production and reuse of by-products (Lokko et al., 2018).

In this scenario, this chapter provides an overview of current opportunities for biodegradable wastes in the bioeconomy, proposing an integrated bioeconomy model as a way for sustainable development from an environmental, social, and economic point of view.

4.2 Fundamentals of the Bioeconomy

The bioeconomy is based on the transition from dependence on fossil fuels to a situation where agriculture contributes to food security and the production of biomass as a renewable raw material for industry, power generation, and other uses. There is not a single form of bioeconomy but many that adjust to the conditions and possibilities of each situation, as shown in Fig. 4.1. The bioeconomy is defined in different ways worldwide, and the terminology used also differs. Still, bioeconomy policies encompass innovation, sustainability, economic growth, and employment. A shared concept at a global level is that of "the application of research, development and technological innovation for the production and use of resources and innovative biological processes and principles, to supply goods and services in a sustainable way to all sectors of society, commerce and industry" (Barañano et al., 2021). The central elements of the bioeconomy are the biological resources, processes and principles, as well as all the technologies (conventional and modern) associated with their knowledge, development, transformation, or regeneration (Ramcilovic-Suominen, 2022). The aspects common to the different definitions of bioeconomy are its relationship with knowledge and science, technology and innovation, the application of biotechnologies, and the reduction of dependence on fossil fuels, as well as the added value of products, and the concepts of sustainability and eco-efficiency.

The bioeconomy and the circular economy converge in their objectives and must strengthen their complementarities and synergies to be integrated into the sustainable development agendas and to face climate change (Aguilar et al., 2018). The search for a comprehensive global vision on the use of biological resources necessarily leads to greater inclusion of economic aspects in the circular economy and greater visibility of sustainability in the bioeconomy. Just as the circular aspect is inherent in biology, the circular bioeconomy is an integral part of the circular economy and the bioeconomy.

One of the main approaches to the bioeconomy is biorefineries to obtain bioenergy that allows the gradual replacement of fossil fuels by renewable bioproducts or the production of bioproducts. Biotechnological products and processes are, in principle and by definition, cleaner than petrochemical or thermochemical processes.

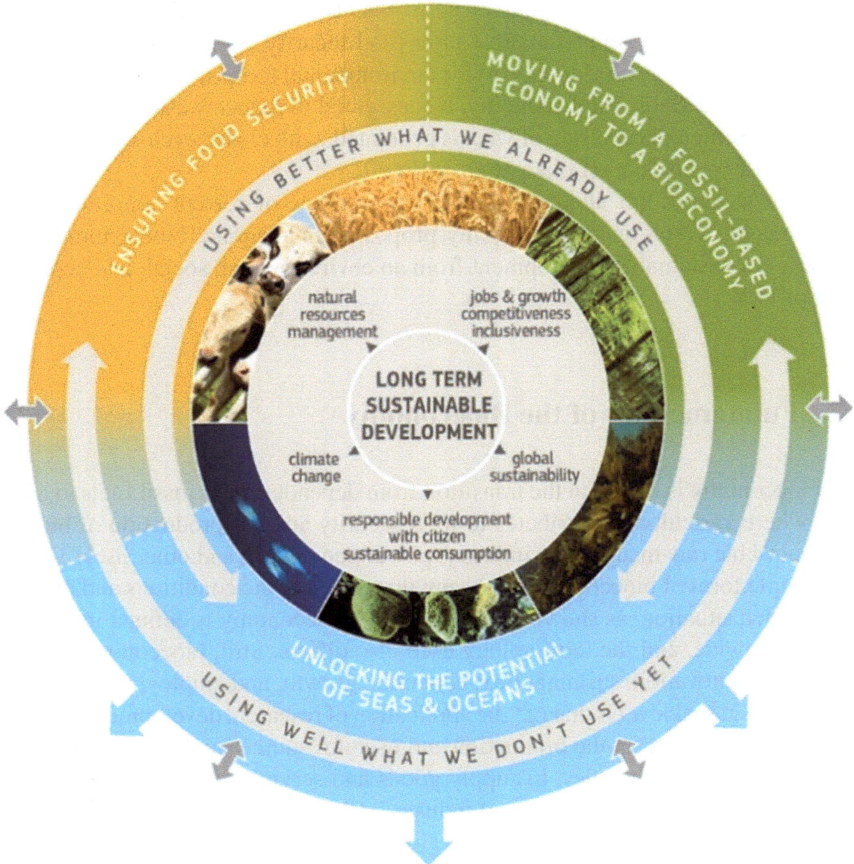

Fig. 4.1 Bioeconomy strategy. *Source* EU (2015)

In this field, there have been several advances in biomass transformation systems. The initial biofuels, called the first-generation biofuels, were obtained from food crops such as wheat, sugar beet, and oilseed plants; the so-called second generation is based on the use of non-food biomass, such as lignocellulosic material, including residues from cereal crops and other crops, and the third generation is derived from algal crops (Shahid et al., 2021). The greatest pressure for the substitution of petroleum products does not stem from scarcity because there is currently no shortage but from political and social pressures related to sustainability, climate change, and energy security. Genomics and synthetic biology have made important advances in generating "microbial factories" that produce various compounds and bioproducts, including plastic monomers, biofuels, and precursors of compounds of industrial importance (Kee et al., 2021). As for synthetic biology, work has intensified to obtain fuels and other aromatic-type compounds from lignocellulose, mainly from non-food crops and agricultural and forestry residues. From biomass and using biotechnological advances, basic compounds are being produced, such as acrylic or succinic acid, or innovative materials, such as biopolymers or ecologically friendly fibres, which allow to respond to the growing social demand for natural products, healthy, and sustainable. This represents a great opportunity for developing countries to process and value their great wealth and abundance of biomass and transform themselves from suppliers of raw materials to suppliers of processed products (Mak et al., 2020).

Marine resources, the so-called blue bioeconomy, are an important source of biomass resources for the food and energy industries. With a wide range of value chains, they are also a source of employment and industrial development opportunities. They also contribute to the population's welfare in the form of fisheries, aquaculture, and food for nutritional security. The contribution of blue bioeconomy with biotechnological products, such as enzymes, biopharmaceuticals, cosmetics, and even bioenergy from marine waste or algae, will be greater as the technologies currently under study advance (Bell et al., 2018). One of the serious crises that threatens the life and health of the oceans is pollution with plastics and waste from non-degradable containers and packaging. Technological developments have allowed the possibility of using biodegradable bioplastics in most industries that use conventional plastic material. These bioplastics have properties similar to conventional ones and reduce dependence on fossil fuels and the carbon footprint (Van Roijen & Miller, 2022).

The idea of the bioeconomy believes in no waste, and in this sense, the possibility of closing cycles through the productive use of waste biomass derived from production and consumption processes should also be mentioned (Scott, 2019). In the contemporary world, there are numerous research groups that explore the potential of this biomass to transform it not only into bioenergy and composting but also into high-value-added products such as polyhydroxyalkanoates (PHAs), biodegradable polymers with various uses, such as bioplastics. The development of new and advanced technologies for processing waste biomass, at some point, allows this material not to be sent to landfills but used for circular processes to produce valuable and innovative products (Bell et al., 2018). Various processing and recovery systems are used for agro-industrial waste that allows the recovery of bioactive compounds and food components (dietary fibres, oligosaccharide pigments, flavonoids, carotenoids, pectins), as well as the obtaining of enzymes, antibiotics, edible fungi, organic acids, and biofuels, and the use of composting as fertilizer and for soil recovery.

4.3 Bioeconomy of Biofuels

The evolution towards greater use of renewable resources has one of its greatest potentials in the energy field. In the energy context, the bioeconomy is presented as a response to the current social and environmental crises derived from the current energy model, to replace fossil fuels with renewable biofuels derived from biomass and biowastes. Bioenergy is called to satisfy an increasing proportion of energy needs, both in the automotive industry and in thermal and electrical demand (Koytsoumpa et al., 2021).

Bioenergy makes a valuable contribution to society in various fields beyond its genuine energy contribution. On the one hand, it plays an important role in preserving the environment, both for its contribution to reducing CO_2 emissions and for its positive impact on ecosystem management, linked, for example, to the reduction of forest fires. On the other hand, its contribution is relevant in terms of social policy; it allows the development of new activity in rural areas based on a market with a continuous demand and without fluctuations, generating stable and well-paid jobs, stopping the depopulation of the rural environment, and contributing to the treatment of biodegradable waste (Ahorsu et al., 2018). The increase in income from local industries and the fixation of the population facilitate the appearance of new infrastructures and services in rural areas. The appearance of a second source of income in the agricultural and forestry industries, through the sale of their residues for energy generation, balances the fluctuations in the markets of the main products of these industries, giving greater security for employers and employees. By stimulating economic investment and technological innovation, it is possible to develop a social and sustainable agriculture model that guarantees the quality of life of small- and medium-sized farmers, based on responsible production methods that promote environmental protection and encourage the conservation of natural resources. Managing land in the best possible way will continue to be a vital and difficult task. Therefore, careful regulation of its use is necessary to ensure that biomass is grown sustainably and does not interfere with the productivity of food agriculture (Perdices, 2018).

It should be clarified that "biomass" encompasses a large group of materials of various origins and very different characteristics. Biomass could be the waste from forestry and agricultural exploitation, from the agroforestry industries, waste of animal or human origin, and crops for energy purposes. In turn, biomass can be transformed into energy through a wide spectrum of thermochemical processes (combustion, pyrolysis, and gasification) or biochemical processes (anaerobic digestion, fermentation). The observatory that monitors the situation of renewable energies in the European Union (Eurobserver, 2022) distinguishes four energy sources within biomass: solid biomass, biogas, organic fraction of solid urban waste, and liquid biofuels. This heterogeneity, in terms of resources and applications, is the main characteristic of biomass, which makes it impossible to address its management from a single perspective since there are as many combinations between types of usable biomass and technologies for its use as energy. Types and technologies of biofuel production are briefly defined below.

4.3.1 Solid Biofuels

Solid biofuels are those fuels obtained from biomass in a solid state. They are generally lignocellulosic and are obtained by physical processes. The origin of these biofuels encompasses different productive sectors, from crops or forestry to waste produced in agri-food or forestry industries. The characteristics of solid biofuels vary according to their composition and moisture, and the energy they can generate per unit of mass or volume depends on these parameters. The best-known and oldest way of using solid biomass is cooking or heating food. In most of these applications, biomass is used as it is collected, without any prior preparation, which entails difficulty in handling and low efficiency in the transformation processes. To facilitate the handling, transport, and feeding of this biomass to the energy conversion system (boiler), it is necessary to reduce the size and shape of this waste, obtaining a solid biofuel suitable for automated fuel supply. These previous transformations can range from simple chipping to more complex processes such as densification (Sacramento-Rivero et al., 2022).

Densification is a process that allows obtaining solid biofuels with higher commercial quality. It consists of the physical–mechanical transformation, with or without additives, of lignocellulosic materials of fine granulometry and low density to obtain solids of regular shape and size and high density. Two types of densified products are distinguished: pellets, whose granule size is less than 30 mm, and briquettes, which are densified products whose size is greater than 30 mm. Generally, briquettes are used in domestic heating (French fireplaces and wood-burning stoves), while pellets are used not only in stoves and boilers for domestic and industrial use but also in the energy production sector. Currently, 46% of renewable energy in the EU comes from solid biomass, exclusively wood. According to the European Renewable Energy Observatory (Eurobserver, 2020), solid biomass consumption represented 102.6 Mtoe in 2020 (94.5 Mtoe leaving aside the UK). Based on national estimates, the European Commission expects that the supply of solid biomass will continue to increase each year with a greater contribution from agricultural biomass (mainly waste and agricultural by-products). Imports from third countries will also increase, mainly in wood chips and pellets (Sikkema et al., 2021).

4.3.2 Gaseous Biofuels

Gaseous biofuels are those fuels obtained from biomass in a gaseous state under normal pressure and temperature conditions. They are used to produce heat and electricity or as fuel for transport. Biogas (consisting of 50–70% methane and carbon dioxide, together with small proportion of other gases such as hydrogen, nitrogen, and hydrogen sulphide) is a fuel gas that is generated by the biodegradation reactions of organic matter in the absence of oxygen and through the action of specific microorganisms (anaerobic bacteria). This process, called anaerobic digestion or biomethanization, can occur in a forced manner in anaerobic digesters or naturally in controlled municipal solid waste (MSW) landfills. The international nomenclature used by Eurostat and the International Energy Agency divides biogas from anaerobic digestion into three subsectors, segmented by origin and waste treatment: (i) biogas from sewage sludge, produced by biomethanization of sewage sludge wastewater treatment plants; (ii) landfill biogas, produced naturally in non-hazardous waste storage facilities, and (iii) other biogas, obtained by anaerobic digestion of livestock waste, from the food industry or by co-digestion. Biogas has a calorific value between 3500 and 4600 kilocalories/Nm^3. This biogas can be used in furnaces, stoves, dryers, and boilers or to produce electrical energy through turbines or gas-to-gas power plants (Rafiee et al., 2021). Most of the current biogas production in the European Union comes from anaerobic digestion (biomethanization) plants of very diverse types and capacities, from small plants installed on farms to large plants in food processing industries. The raw materials are also highly variable: livestock waste, agricultural waste from the agri-food industry, and domestic waste, among others.

In some cases, these biodegradable wastes are supplemented with energy crops (for example, corn) called co-digestion to increase biogas production (Chodkowska-Miszczuk et al., 2021). A recent analysis shows that the contribution of biogas as an energy source at the global level could increase to 95 Mtoe in 2030 and up to 150 Mtoe by 2040 in the stated policies' scenario. Although energy crop residues are included in these predictions, feedstocks grown specifically to produce biogas and biomethane are not included because their sustainability warrants further analysis (IEA, 2020).

Another gaseous biofuel is obtained by gasifying solid biomass (wood, forest residues, household solid waste). The gasification is obtained by subjecting the biomass to very high temperatures (800–1000 °C) in the presence of limited amounts of oxygen. Two different products are obtained depending on whether air or pure oxygen is used as the gasifying agent. In the first case, lean gas (a mixture of carbon monoxide and nitrogen) is obtained, which can be used to obtain electricity and steam. In the second case, synthesis gas (carbon monoxide and hydrogen) is obtained, which can be used as a direct fuel, as a source of hydrogen, or as a chemical raw material to prepare gasoline or diesel through the Fischer–Tropsch process (Yahyazadeh et al., 2021), with ongoing demonstration projects in Finland, Sweden, Norway, and the Netherlands (Dahlgren, 2022).

4.3.3 Liquid Biofuels

Liquid biofuels are defined as those fuels obtained from biomass that are in a liquid state under normal conditions of pressure and temperature. They are used in boilers for the production of heat and electricity or in internal combustion engines. Among them, biodiesel, bioethanol, and pyrolysis oils are the best known.

The term liquid biofuel encompasses all those liquid fuels derived from biomass with characteristics similar to gasoline and diesel, allowing their use in engines without having to make significant modifications. In the initial phase of the development of biofuels, they were classified based on their properties as biodiesel (more recently hydrogenated vegetable oils or HVOs) to replace diesel and bioethanol to replace gasoline. However, concerns arising from biofuels' impact on food security have led to a new classification based on generation categories, which is why they are classified as first-, second-, and third-generation biofuels. Generally, the first generation (1G) is produced from raw materials that can also be consumed as human food, such as sugar, starch, or vegetable oil. Since 1G biofuels are easily extracted using conventional technologies, they are also known as conventional biofuels. Second-generation biofuels (2G), also known as advanced biofuels, are produced from sustainable raw materials, mainly of the lignocellulosic type, which are not used for human consumption. As these lignocellulosic feedstocks are more difficult to transform into biofuels, more complicated conversion technologies are needed. Third generation (3G) is the last subdivision of biofuels and refers to biofuels obtained from algae or food biowaste.

First-generation liquid biofuels based on food crops are produced commercially using mature technologies. They include ethanol from sugary or starchy raw materials and biodiesel from vegetable oils. Figure 4.2 shows the different routes for the production of biofuels from the different feedstocks. The most widely used biofuel worldwide is bioethanol, which is obtained by fermentation of sugary musts that come from vegetables rich in sugar, or from the hydrolysis and fermentation of starch that vegetables store as reserve material. Bioethanol can be used in spark-ignition engines (Otto engines), mixed with gasoline in percentages of 10–15% without engine modifications, or in flexible fuel vehicles if used in greater proportion or as exclusive fuel. In other cases, ethanol can be used as its derivative, ethyl-tert-butyl-ether (ETBE), as an additive to gasoline to improve the octane number (Sandesh & Ujwal, 2021).

The USA is the world's leading bioethanol producer, and its production is linked to the cultivation of corn (Mignone et al., 2022). Corn kernels contain starch that,

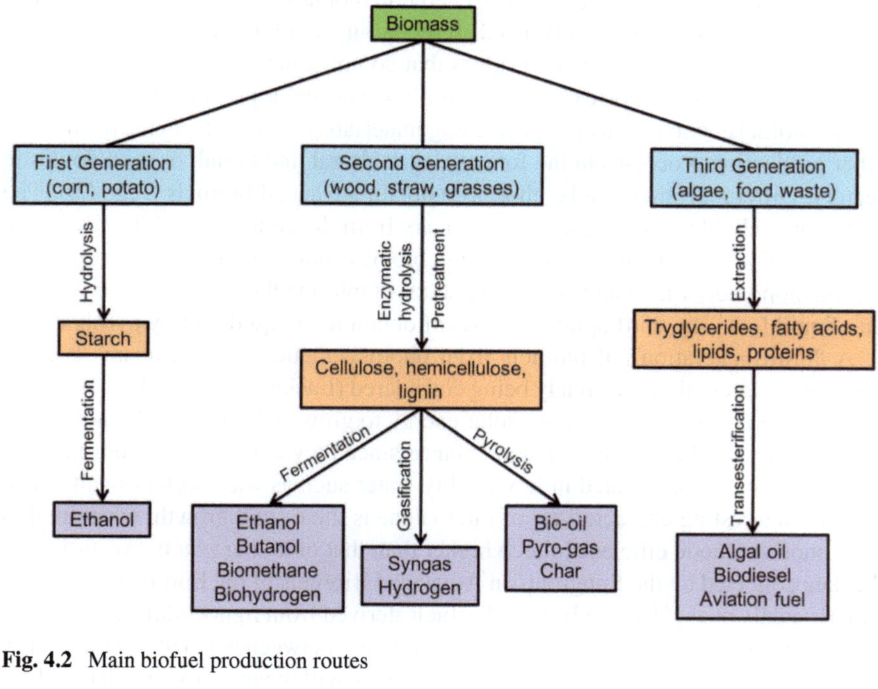

Fig. 4.2 Main biofuel production routes

through hydrolysis, can be transformed into ethanol, involving yeasts. Brazil is the world's second-largest ethanol producer from sugarcane (Karp et al., 2021). This raw material contains sucrose that yeasts can directly transform to generate ethanol. The raw materials currently used in biodiesel production are the oils from the oilseed plants such as palm, soybean, and rapeseed. Using vegetable oil as fuel in a compression ignition engine presents difficulties that derive from the physical–chemical differences between the standard that defines oil and diesel (basically viscosity, cetane number, and freezing point).

For this reason, the characteristics of vegetable oils are modified to make them more similar to diesel through a transesterification process in which triglycerides are transformed into esters. The European Union is the largest producer of biodiesel in the world, with figures in 2020 exceeding 4.2 billion gallons (Kotrba, 2020). Although the technologies associated with bioethanol and biodiesel production are sufficiently developed and have reached a high level of maturity, due to raw material supply problems, these first-generation biofuels cannot respond to the energy demand that the total substitution of petroleum requires. In addition, there is the controversy of competition for food raw materials used for the production of biofuels and the price increase that may result from this competition (Kurowska et al., 2020). The solution that would remove this type of conflict would be to develop new, more productive crops with low production costs that do not compete with the food sector. Some plant species produce oilseeds (Jatropha curcas, Brassica carinata, Camelina sativa, and Cynara cardunculus) or high carbohydrate content (such as Helianthus tuberosus), are well adapted to conditions of low nutritional and water requirements, and could be cultivated specifically for the production of energetic biomass. But the true future development of this sector lies in advanced or second-generation biofuels, that is, those derived from plants or vegetable residues that do not compete with the food sector.

The use of lignocellulosic raw materials offers enormous potential for the production of biofuels, with the advantage of using materials that come from the residues of other production processes in the forestry, agricultural, industrial, or even domestic sectors. Different routes can be chosen to obtain advanced biofuels (Fig. 4.3). The first is biotechnological, extracting the sugars from the cellulose with highly active enzymes. The second consists of gasifying biomass into a mixture of hydrogen and carbon monoxide, which later, through a series of intermediate stages, is transformed into liquid fuel. The third option consists of obtaining a liquid fuel by pyrolysis.

A "third generation" of biofuels from photosynthetic microorganisms such as autotrophic microalgae is already being considered (Lackner, 2022). These microorganisms are cosmopolitan and use solar energy to grow and multiply. An advantage of microalgae is their environmental balance since they consume CO_2 during their growth and can be cultivated in low-quality water such as wastewater or salt water. Another interesting characteristic of microalgae is their rapid growth, which makes their photosynthetic efficiency much higher than that of higher plants. According to the data collected by the Subgroup on Advanced Biofuels of the European Commission (Landälv et al., 2017), advanced biofuels derived from lignocellulosic materials and those obtained from lipids will be able to supply between 6 and 9% of the energy needs of the transport sector in the year 2030. This will mean a growing demand for

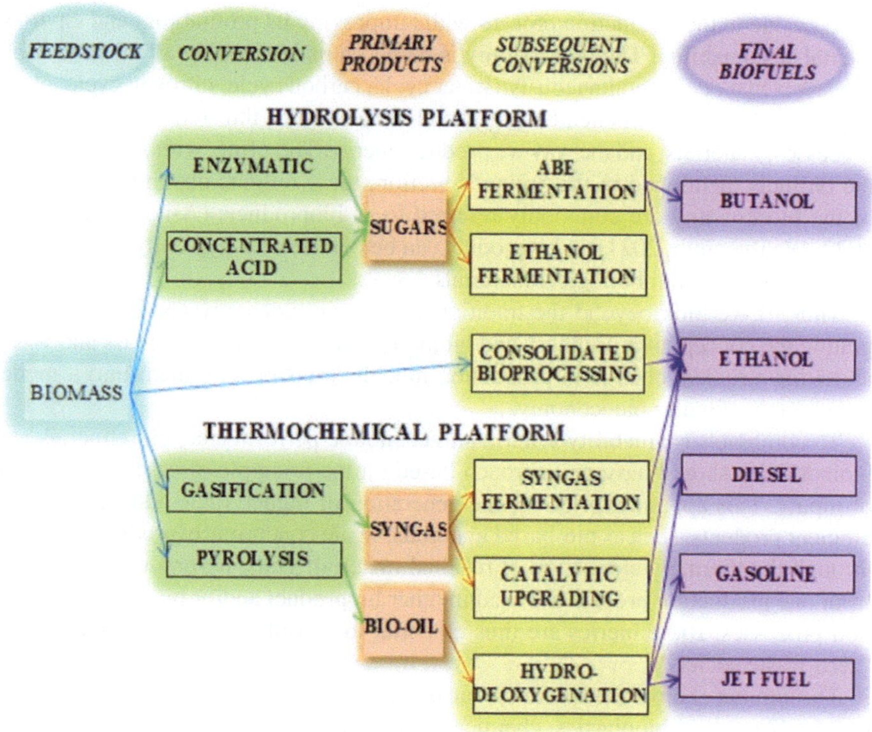

Fig. 4.3 2G Biofuel production technologies

lipids that the current production of vegetable oils will not be able to supply due to their production and price limitations. The production of renewable lipids obtained from microorganisms can significantly impact the future production of biofuels and oleochemicals. Oil microorganisms, such as autotrophic microalgae or heterotrophic yeasts, under certain conditions can produce copious amounts of oil and have higher photosynthetic efficiency than higher plants (Zabermawi et al., 2022). Also a "fourth generation" of biofuels, using genetically modified algae to enhance biofuel production, is already on stage. Lipid and carbohydrate maximization are among the most attractive factors that can enhance the efficiency of the yield of microalgae biomass (Abdullah et al., 2019). (Chinchilla & de la Rúa, 2018).

4.4 Bioeconomy of Bioproducts

In the context of the transition from a linear economy to a circular bioeconomy, as we are experiencing today, the ultimate aim of this cycle is to obtain products or by-products that, taking advantage of the benefits of technologies, provide added value to the production process while improving its profitability and environmental sustainability. The bioeconomy is based on understanding all the renewable processes that occur spontaneously (water cycle, carbon cycle, nitrogen cycle, phosphorus cycle, etc.) from chemistry, at the molecular level, thus redefining industrial processes, materials, and the way we produce energy. Therefore, without considering the biofuels mentioned in the previous section, these outputs obtained in the field of the bioeconomy are commonly referred to as bioproducts (Ashokkumar et al., 2022; Wohlgemuth, 2021). A bioproduct can be defined as a (typically commercial) product derived from biological materials.

Thus, if we look beyond the traditional focus on bioenergy and biofuels, integrating the concept of biorefinery and bioindustry as a paradigm, biomass and waste allow for a radical change in production models, transforming them into a model based on the circular bioeconomy.

To combine sustainability criteria and economic performance, biorefineries and bioindustries should have an approach based on the principle of cascading the use of inputs. This approach priorizes the generation of higher value-added, resource-efficient products, such as bio-based materials, while concurrently producing energy through the thermal valorisation of residual material. Thus, by-products and waste from one production process are used in other bioproduct and/or bioenergy production processes. Biorefineries are thus essential to contribute to the principles of a "zero-waste society".

The number of bio-based products can be vast, including biomaterials, biocomposites, biofibres, and bioplastics, chemical bioproducts such as biofertilizers, biocosmetics, and biopharmaceuticals, food bioadditives, and animal feed bioproducts, among others. Although it is a field in continuous development and growth and the margin for improvement and expansion is enormous, the bioproducts obtained on a more relevant scale and mature level of development can be divided into three main groups: bioplastics, organic acids, and biofertilizers.

4.4.1 Bioplastics

The use of plastics is one of the clearest examples of the drawbacks of the linear economy, where the raw material is petroleum and its derivatives, which are non-renewable. There is a threat of its reserves eventually becoming depleted. In addition, plastic is very resistant to natural degradation, so the waste generated is not assimilated by the environment and accumulates.

Although nowadays, a large part of plastic waste is managed for reuse, recycling, or energy recovery, it is not always managed efficiently. As a result of inefficient management of plastic waste, it is estimated that between 1 and 5% of all plastics in terrestrial and oceanic environments (especially, in the latter) are disposed. According to a recent report by the United Nations Environment Programme (UNEP) (Maes et al., 2021), it is estimated that 11 million tonnes of plastic enter the world's oceans each year. This pollution occurs in developing countries and in industrialized countries such as China and the USA (Jambeck et al., 2015; Lebreton et al., 2017; Schmidt et al., 2017).

To improve this problem, it is necessary to address it from various perspectives, such as improving the plastics management system and optimizing the manufacturing and recycling processes. Regarding the latter, the production of plastics from organic waste (bioplastics) as raw material, which allows for more sustainable and biodegradable plastics in a framework of the development of industrial processes under the prism of the bioeconomy, is considered fundamental to achieve the objectives.

On the one hand, bioplastics are bio-based products that are generated from sustainable sources such as vegetable oils, biomass. On the other hand, bioplastics are those that are naturally degradable, thus eliminating their pernicious and perpetual effect in landfills and oceans. Some bioplastics combine both characteristics, i.e. they are made from sustainable sources, and their natural degradability is high. Bioplastics such as polyhydroxyalkanoate (PHA) or polylactic acid (PLA) meet these characteristics and result from biotechnological developments. Thus, poly(3-hydroxybutyrate) or PHB is one of the most frequently obtained PHA because it has similar characteristics to plastics of non-removable origin, such as polypropylene and polyethylene, and is used as a raw material for automotive parts, biotechnological material, or even packaging (Asiri et al., 2020; Bassi et al., 2021). The yield of the PHA production process depends on multiple operating factors and the characteristics of the waste: concentration of dissolved oxygen or C, the origin of N. However, the main drawback for the industrial scalability of bioplastics production processes is the logistic and pretreatment costs of the waste used as raw material, as well as process yields that are not yet high enough, which gives bioplastics a cost that is not yet competitive with the market (Bhatia et al., 2021).

4.4.2 Organic Acids

Organic acids are other types of bioproducts that have raised interest in recent years. These products are in demand in various sectors of industry. Generally, these bioproducts are obtained from waste with a high organic matter content; therefore, food waste is one of the preferred materials for their production due to its content in a biodegradable matter such as proteins, cellulose, or lipids.

Among these bioproducts, citric acid stands out as a highly valued product worldwide due to its acidifying and preservative properties that ensure the original taste, natural appearance, and consistency of food, pharmaceutical, and cosmetic products. This acid is mainly obtained by fermentation using various microorganisms capable of degrading raw materials such as starch, sucrose, and agro-industrial waste. Therefore, citric acid can be obtained by fermentation using food waste rich in organic matter as substrates (Özüdoğru et al., 2019).

2,3-butanediol can be used in synthesizing flavour enhancers, production of plastics, synthetic rubber, plastics, and fuel additives. The microbiological route to obtain 2,3-butanediol is energetically more favourable than the chemical route. Because of its low toxicity to microorganisms, it results in a higher reaction yield than other bioproducts, allowing a much higher yield than other bioproducts. Therefore, 2,3-butanediol has good short- and medium-term prospects in the biochemical industry (Tinoco et al., 2021).

Succinic acid (butanedioic acid) is a colourless, water-soluble dicarboxylic acid, acidic in taste produced by various methods (Cheng et al., 2012). Applications include production of polybutylene succinate (PBS), production of plasticizers for PVC manufacture, replacement of petrochemical adipic acid derivatives used in the production of polyester polyols for polyurethanes, production of 1,4-butanediol for subsequent production of tetrahydrofuran and polybutylene terephthalate, production of dimethyl succinate. The most common industrial routes include hydrogenation of maleic acid, oxidation of 1,4-butanediol, and carbonylation of ethylene glycol. More recently, however, it has been obtained through fermentation of glucose from renewable feedstocks and the purification of crude biosuccinic acid.

3-hydroxypropionic acid (3-HPA) is a molecule from which many other products of interest can be obtained. For example, it can be converted into acrylic acid (a precursor for producing superabsorbent plastics) or bioplastics. In both cases, these are substitutes for petroleum-based plastics. 3-HPA can be obtained from glycerol (a biodiesel reaction-product) by a two-step enzymatic reaction. In the first stage, glycerol is converted to 3-hydroxypropionaldehyde by an enzyme called glycerol dehydratase, and in the second stage, the oxidation of 3-hydroxypropionaldehyde to 3-HPA takes place by the enzyme aldehyde dehydrogenase (Bhagwat et al., 2021). The advantage of using glycerol as a feedstock for 3-HPA is that it is not toxic to most microorganisms.

Other bioproducts such as 1,3-propanediol (1,3-PDO) or polylactic acid (PLA) are also of interest due to their added value, market demand, and prospects for large-scale implementation of the production process (Fokum et al., 2021; Ghaffar et al., 2014).

Last but not least, volatile fatty acids (VFAs) are elements generally obtained in anaerobic digestion processes of food waste. They can be used as precursors of important bioproducts (acetic, butyric, propionic, and caproic acid), with which biopolymers or even biohydrogen can be produced (Reddy et al., 2018).

4.4.3 Biofertilizers

The paradigm shift from an agriculture based on mineral fertilization to agriculture that promotes organic and biological fertilization is vital in achieving sustainable agriculture in its triple meaning: more productive, less dependent on external inputs, and with a lower environmental impact. However, this achievement depends to a large extent on correct and safe management of organic waste, an adequate process of sanitization and stabilization, and an appropriate application of these products, with the right dosage and at the right time, intending to replace non-renewable mineral fertilizers with a new generation of biofertilizers of organic origin obtained from renewable raw materials.

In the context of the circular economy, it is essential to connect the agricultural sector with other industrial sectors to recover valuable nutrients that can be used as raw materials for producing biofertilizers.

Within the field of agricultural by-products, two current trends can be identified. On the one hand, the production of biocompost and, on the other hand, the production of bio-based biofertilizers. Beyond traditional composting, vermicomposting is a technique that consists of a process of bio-oxidation and stabilization of organic matter, mediated by the combined action of earthworms and microorganisms, from which a stabilized, the homogeneous, and fine-grained final product is obtained, called vermicompost or worm humus, which is highly appreciated on the market. The combined action of earthworms and microorganisms significantly modifies the characteristics and composition of organic waste. The biodegradation and stabilization of organic matter take place under mesophilic and aerobic conditions maintained by the action of earthworms. Therefore, in vermicomposting processes, it is only necessary to maintain adequate moisture of the organic material (by manual irrigation or sprinkling), and other management methods such as aeration, which makes processes such as composting more expensive, are avoided. Vermicompost is used the same way as compost, both as a fertilizer and soil enhancement. It improves the colour, quality, and quantity of fruit, vegetables, flowers, and ornamentals (Chavan et al., 2013).

As far as the production of bio-based biofertilizers is concerned, this is based on the philosophy of nutrient recovery. Several technologies are used to recover nutrients (mainly C, N, and P) from various organic wastes, such as agri-food waste (biomass, livestock waste, food waste) or wastewater (both urban and industrial). Bio-based biofertilizers with good results and yields obtained through nutrient recovery processes are ammonium sulphates or ammonium phosphates obtained in stripping processes, struvite, k-struvite, and other salts obtained through crystallization or algae-based biofertilizers obtained through water purification processes using microalgae (Schoumans et al., 2015).

4.5 Future Trends in Waste-Based Bioeconomy

The prospective analysis of the markets places the waste-based bioeconomy as one of the main global trends in the economy and society for the coming decades (Kircher, 2021). One of the main challenges in implementing the bioeconomy is expected to be the development of innovative processes that adapt current business models to new market realities (Lenze, 2022).

The Organization for Economic Cooperation Development (OECD) developed some years ago a political agenda on the bioeconomy for 2030 (OECD, 2009). According to this agenda, biotechnology offers technological solutions to many problems that society faces today. The application of biotechnology to primary production, health, and industry could result in an emerging "bioeconomy" where biotechnology contributes a significant proportion of the economic potential. Three elements will probably play a role in this bioeconomy: advanced knowledge of genes and complex cell processes, the integration of biotechnology applications in various sectors, and renewable biomass (including biodegradable waste). According to the OECD report, several factors will drive the emerging bioeconomy, creating investment opportunities. Apart from using biotechnology to meet the challenge of environmentally sustainable production, the biggest driver will be the increase in population and per capita income, particularly in developing countries. The growth of countries like China and India indicates that the bioeconomy will be global but that the main markets for biotechnology in primary production and industry could be in developing countries. In addition, the increase in energy demand, if associated with measures to reduce the emission of greenhouse gases, can create large markets for biofuels. The report identifies two new business models that could emerge: collaborative models to share knowledge and reduce R&D costs and integrative models to create and maintain markets. Their adoption and new opportunities for biodegradable waste and biomass could revitalize small biotech companies in primary production and industry.

To estimate the "likely" bioeconomy in 2030, the OECD report adopts a "business as usual" approach to institutional factors, such as regulation, and is based on research into the types of biotechnology products that are produced. It is expected that they will reach the market before 2030. The results suggest that biotechnology could contribute 2.7% of the GDP of the OECD countries for that year, being the largest contribution of biotechnology in industry and primary production, followed by health applications. This contribution could also be greater in developing countries due to the importance of primary production and industry in their economies.

To reap the full benefits of the bioeconomy, the report concludes that purposeful, goal-oriented policies will be needed. This will require leadership, first of all from governments but also from leading companies, to establish the objectives for the application of biotechnology to primary production, industry, and health; to create the structural conditions necessary for success, such as obtaining regional and international agreements; and to develop mechanisms that ensure that this policy can be flexibly adapted to new opportunities.

Also, in recent years, the European Commission has made a strong commitment to the transition and adoption of new forms of production, consumption, and transformation towards a zero-waste scenario (with an emphasis on biowaste) in Europe through the adoption of packages of measures to help to European businesses and consumers in their evolution towards a more solid and circular economy model, where resources are used more sustainably. These measures seek to extract the maximum value and use from all raw materials, products, and waste, promoting energy savings and reducing greenhouse gas emissions by closing the circle of product life cycles through greater recycling and reuse, which brings benefits to both the environment and the economy. These new models that present the possibility of sustainable resource management represent a revolution in how society will obtain vital sources of food, energy, and carbon. They will significantly reduce dependence on oil and other traditional energy sources, adapting to new social needs and environmental conditions. It is, therefore, a true revolution in production systems, implementation and collaboration mechanisms, and economic structures (Chinchilla & de la Rúa, 2018).

The early positioning of countries in the generation of sustainable growth based on knowledge, and leadership in initiatives that promote the bioeconomy, will be a key element for the continued development of the industry and their positioning in a competitive manner. In this context, biotechnology has established itself as the tool that combines the knowledge of different scientific and technological disciplines capable of providing efficient and environmentally compatible solutions to production systems. The progress made in recent years in the field of biotechnology has revealed the potential it offers for the manufacture of products and biomaterials of general interest in a more sustainable, respectful, and efficient way by making it possible to reduce the consumption of resources (raw materials, energy, water, fertilizers) due to greater use of renewable raw materials, as well as waste and by-products from other production processes, thus reducing their generation and their environmental impact and the increase in the recycling of the same. In this sense, it has also promoted the development of clean technologies of biological origin that allow the substitution of chemical production processes for new ones based on enzymatic technologies, which entail lower energy consumption and lower greenhouse gas emissions, by the time that allows greater reuse and use of the raw material, thus increasing the value of the production chain, and, therefore, the performance of the process (Allain et al., 2022).

This opens up a scenario where new agents take on special relevance, as in biorefineries. These infrastructures are positioned as a key instrument to contribute to the establishment of a truly productive sector based on the bioeconomy, in which the conversion of biomass/biowastes into bioproducts is articulated, thus contributing to the transition towards the circular economy and the bioeconomy at the same time that the penalty associated with the generation and/or dumping of waste of biological origin is avoided. The development of integrated biorefineries represents a great advance. It will make it possible to increase the competitiveness of these bioprocesses by establishing new business concepts that respond to the growing demand by many sectors of the population and the private sector through the generation of technological innovations and sustainable solutions, which implicitly save energy and resources, while seeking to protect the environment. However, there are still important obstacles that require an enormous effort and commitment in basic research and state of the art for its development and implementation.

In the current socioeconomic context, the bioeconomy is being driven by many factors, such as population growth, environmental issues, product differentiation, and cost reduction opportunities. The barriers detected for the development of biotechnology, in general, and in this field of bioeconomy from biowaste in particular, are very numerous, ranging from economic barriers (availability of financing or requirement of capital investments), regulatory barriers (demanding regulations and complex regulatory frameworks), barriers at a technical level (cost of raw materials, degree of maturity and availability of technologies, and level of market support for products), structural barriers (little guidance of the public technological offer to the needs of the market), and social barriers (a coherent communication plan is necessary to create awareness about the potential of the use of renewable resources and the benefits that a bioeconomy can provide). Its successful implementation and its economic and social impact will depend on the ability to overcome them.

In short, Europe, in general, faces multiple challenges to improve and maintain a competitive position in the world economy, where it is increasingly necessary to take long-term measures and carry out adequate planning for its development in the context of the bioeconomy through support and encouragement for the business and the generation of more sustainable models in terms of economic, social, and environmental efficiency. Express support is required from public administrations capable of promoting the connection between research and industry, fostering collaboration, and developing stable, efficient, and multilateral communication channels between the different agents of the science-technology-business system with users, decision-makers, politicians, and civil society, and implementing the necessary political-administrative mechanisms that minimize bureaucratic procedures and legislative barriers that may hinder optimal development of biotechnology at the service of the bioeconomy.

4.6 Conclusions

Climate change and the depletion of certain strategic resources have placed the bioeconomy on the public policy agenda of many developed countries. Materials and energy are fundamental inputs of the productive system, and bioproducts and bioenergy are destined to be central elements in this new bioeconomy, as they are important sources of job creation, especially in rural areas, at the local and regional levels. Likewise, bioenergy and bioproducts have a strategic role in the fight against climate change and in reducing dependence on imports of raw materials and fuels. The linking of biomass to the primary sector (agricultural, forestry, and livestock activities), in addition to other industries based in rural areas (agri-food cooperatives, paper, and wood industries, among others), makes the biomass sector (understood as the broad concept sense, which includes forest and agricultural biomass in addition to biodegradable waste streams) into a potential strategic asset for these rural environments, generating new jobs that may help alleviate the high unemployment concentrated in rural areas (closely linked to the primary sector, where in addition the creation of new opportunities is scarce). In short, the bioeconomy is key to the transition from the current economic model based on the use of fossil resources to a new bioeconomy based on renewable organic resources. If we can promote responsible biomass production and effective biodegradable waste management at the source, addressing poverty and climate change, along with establishing an international certification system that verifies their greenhouse gas emissions and assesses their impact on community prosperity and social welfare, the future of the bioeconomy looks promising.

Acknowledgements The authors gratefully acknowledge the support of this work by the European Union's Horizon 2020 Research and Innovation Programme under grant agreement No 101000752 (WalNUT project).

References

Abdullah, B., Syed Muhammad, S. A. F. A. S., Shokravi, Z., Ismail, S., Kassim, K. A., Mahmood, A. N., & Aziz, M. M. A. (2019). Fourth generation biofuel: A review on risks and mitigation strategies. *Renewable and Sustainable Energy Reviews, 107*, 37–50. https://doi.org/10.1016/j.rser.2019.02.018

Aguilar, A., Wohlgemuth, R., & Twardowski, T. (2018). Perspectives on bioeconomy. *New Biotechnology, 40*(A), 181–184. https://doi.org/10.1016/j.nbt.2017.06.012

Ahorsu, R., Medina, F., & Constantí, M. (2018). Significance and challenges of biomass as a suitable feedstock for bioenergy and biochemical production: A review. *Energies, 11*(12), 3366. https://doi.org/10.3390/en11123366

Albert, C., Bustos, P., & Ponticelli, J. (2021). The effects of climate change on labor and capital reallocation. National Bureau of Economic Research (No. 28995).

Allain, S., Ruault, J. F., Moraine, M., & Madelrieux, S. (2022). The'bioeconomics vs bioeconomy'debate: Beyond criticism, advancing research fronts. *Environmental Innovation and Societal Transitions, 42*, 58–73. https://doi.org/10.1016/j.eist.2021.11.004

Ashokkumar, V., Flora, G., Venkatkarthick, R., SenthilKannan, K., Kuppam, C., Mary Stephy, G. M., Kamyab, H., Chen, W., Thomas, J., & Ngamcharussrivichai, C. (2022). Advanced technologies on the sustainable approaches for conversion of organic waste to valuable bioproducts: Emerging circular bioeconomy perspective. *Fuel, 324*, 124313. https://doi.org/10.1016/j.fuel.2022.124313

Asiri, F., Chen, C. H., Hwangbo, M., Shao, Y., & Chu, K. H. (2020). From organic wastes to Bioplastics: Feasibility of nonsterile poly (3-hydroxybutyrate) production by Zobellella denitrificans ZD1. *ACS Omega, 5*(38), 24158–24168. https://doi.org/10.1021/acsomega.9b04002

Barañano, L., Garbisu, N., Alkorta, I., Araujo, A., & Garbisu, C. (2021). Contextualization of the bioeconomy concept through its links with related concepts and the challenges facing humanity. *Sustainability, 13*(14), 7746. https://doi.org/10.3390/su13147746

Bassi, S. A., Boldrin, A., Frenna, G., & Astrup, T. F. (2021). An environmental and economic assessment of bioplastic from urban biowaste. *The Example of Polyhydroxyalkanoate. Bioresource Technology, 327*, 124813. https://doi.org/10.1016/j.biortech.2021.124813

Bell, J., Paula, L., Dodd, T., Németh, S., Nanou, C., Mega, V., & Campos, P. (2018). EU ambition to build the world's leading bioeconomy—Uncertain times demand innovative and sustainable solutions. *New Biotechnology, 40*(A), 25–30. https://doi.org/10.1016/j.nbt.2017.06.010

Bhagwat, S. S., Li, Y., Cortés-Peña, Y. R., Brace, E. C., Martin, T. A., Zhao, H., & Guest, J. S. (2021). Sustainable production of acrylic acid via 3-hydroxypropionic acid from lignocellulosic biomass. *ACS Sustainable Chemistry and Engineering, 9*(49), 16659–16669. https://doi.org/10.1021/acssuschemeng.1c05441

Bhatia, S. K., Otari, S. V., Jeon, J. M., Gurav, R., Choi, Y. K., Bhatia, R. K., Pugazhendhi, A., Kumar, V., Rajesh Banu, J., Yoon, J. J., Choi, K. Y., & Yang, Y. H. (2021). Biowaste-to-bioplastic (polyhydroxyalkanoates): Conversion technologies, strategies, challenges, and perspective. *Bioresource Technology, 326*, 124733. https://doi.org/10.1016/j.biortech.2021.124733

Chavan, V., Joshi, S., & Pejaver, M. (2013). Comparative study of microbial population in vermicompost and biocompost in relation with physicochemical parameters. In *National conference on biodiversity: Status and challenges in conservation* (pp. 245–248).

Cheng, K. K., Zhao, X. B., Zeng, J., & Zhang, J. A. (2012). Biotechnological production of succinic acid: Current state and perspectives. *Biofuels, Bioproducts and Biorefining, 6*(3), 302–318. https://doi.org/10.1002/bbb.1327

Chinchilla, P. C., & de la Rúa, I. A. (2018). Las bioindustrias en el entorno de la bioeconomía: Necesidades, oportunidades y beneficios. *Mediterráneo Económico, 31*, 203–217.

Chodkowska-Miszczuk, J., Martinát, S., & Van Der Horst, D. (2021). Changes in feedstocks of rural anaerobic digestion plants: External drivers towards a circular bioeconomy. *Renewable and Sustainable Energy Reviews, 148*, 111344. https://doi.org/10.1016/j.rser.2021.111344

Cristóbal, J., Matos, C. T., Aurambout, J. P., Manfredi, S., & Kavalov, B. (2016). Environmental sustainability assessment of bioeconomy value chains. *Biomass and Bioenergy, 89*, 159–171. https://doi.org/10.1016/j.biombioe.2016.02.002

Dahlgren, S. (2022). Biogas-based fuels as renewable energy in the transport sector: An overview of the potential of using CBG, LBG and other vehicle fuels produced from biogas. *Biofuels, 13*(5), 587–599. https://doi.org/10.1080/17597269.2020.1821571

de Jaramillo, E. H. (2018). Bioeconomía: El futuro sostenible. *Revista de la Academia Colombiana de Ciencias Exactas, Físicas y Naturales, 42*(164), 188–201. https://doi.org/10.18257/raccefyn.650

European Union. (2015). https://ec.europa.eu/info/sites/default/files/research_and_innovation/research_by_area/images/bioeconomy_graphic_full_2.jpg

Eurobserver. (2020). *Solid biofuels barometer*. https://www.eurobserv-er.org/solid-biomass-barometer-2020/

Eurobserver. (2022). https://www.eurobserv-er.org/

Fokum, E., Zabed, H. M., Yun, J., Zhang, G., & Qi, X. (2021). Recent technological and strategical developments in the biomanufacturing of 1, 3-propanediol from glycerol. *International Journal of Environmental Science and Technology, 18*(8), 2467–2490. https://doi.org/10.1007/s13762-020-03036-w

Ghaffar, T., Irshad, M., Anwar, Z., Aqil, T., Zulifqar, Z., Tariq, A., Kamran, M., Ehsan, N., & Mehmood, S. (2014). Recent trends in lactic acid biotechnology: A brief review on production to purification. *Journal of Radiation Research and Applied Sciences, 7*(2), 222–229. https://doi.org/10.1016/j.jrras.2014.03.002

IEA. (2020). *Outlook for biogas and biomethane: Prospects for organic growth.* IEA, Paris https://www.iea.org/reports/outlook-for-biogas-and-biomethane-prospects-for-organic-growth. Last Access 07/07/2022.

Jambeck, J. R., Geyer, R., Wilcox, C., Siegler, T. R., Perryman, M., Andrady, A., Narayan, R., & Law, K. L. (2015). Marine pollution. Plastic waste inputs from land into the ocean. *Science, 347*(6223), 768–771. https://doi.org/10.1126/science.1260352

Jiménez-Sánchez, G., & Philp, J. (2015). Omics and the bioeconomy: Applications of genomics hold great potential for a future bio-based economy and sustainable development. *EMBO Reports, 16*(1), 17–20. https://doi.org/10.15252/embr.201439409

Karp, S. G., Medina, J. D. C., Letti, L. A. J., Woiciechowski, A. L., de Carvalho, J. C., Schmitt, C. C., Oliveira Penha, R., Kumlehn, G. S., & Soccol, C. R. (2021). Bioeconomy and biofuels: The case of sugarcane ethanol in Brazil. *Biofuels, Bioproducts and Biorefining, 15*(3), 899–912. https://doi.org/10.1002/bbb.2195

Kee, S. H., Chiongson, J. B. V., Saludes, J. P., Vigneswari, S., Ramakrishna, S., & Bhubalan, K. (2021). Bioconversion of agro-industry sourced biowaste into biomaterials via microbial factories—A viable domain of circular economy. *Environmental Pollution, 271*, 116311. https://doi.org/10.1016/j.envpol.2020.116311

Kircher, M. (2021). Bioeconomy—Present status and future needs of industrial value chains. *New Biotechnology, 60*, 96–104. https://doi.org/10.1016/j.nbt.2020.09.005

Kotrba, R. (2020, July). EU biodiesel production remains stable, consumption down 6%. *Biodiesel Magazine*. https://biodieselmagazine.com/articles/2517090/2020-eu-biodiesel-production-remains-stable-consumption-down-6

Koytsoumpa, E. I., Magiri-Skouloudi, D., Karellas, S., & Kakaras, E. (2021). Bioenergy with carbon capture and utilization: A review on the potential deployment towards a European circular bioeconomy. *Renewable and Sustainable Energy Reviews, 152*, 111641. https://doi.org/10.1016/j.rser.2021.111641

Kurowska, K., Marks-Bielska, R., Bielski, S., Kryszk, H., & Jasinskas, A. (2020). Food security in the context of liquid biofuels production. *Energies, 13*(23), 6247. https://doi.org/10.3390/en13236247

Lackner, M. (2022). Third-generation biofuels: Bacteria and algae for better yield and sustainability. In *Handbook of climate change mitigation and adaptation* (pp. 1947–1986). Springer International Publishing.

Landälv, I., Maniatis, K., Heuvel, E. et al. (2017). *Building up the future: Sub group on advanced biofuels [Final report]*. Publications Office. https://doi.org/10.2832/494620

Lebreton, L. C. M., Van Der Zwet, J., Damsteeg, J. W., Slat, B., Andrady, A., & Reisser, J. (2017). River plastic emissions to the world's oceans. *Nature Communications, 8*(1), 15611. https://doi.org/10.1038/ncomms15611

Lenze, S. (2022). Bioeconomy as a normative concept of resilience: Challenges and opportunities. Bioeconomy and Sustainability, 307–321.

Lewandowski, I. (2018). *Bioeconomy: Shaping the transition to a sustainable, biobased economy* (p. 356). Springer nature.

Lokko, Y., Heijde, M., Schebesta, K., Scholtès, P., Van Montagu, M., & Giacca, M. (2018). Biotechnology and the bioeconomy—Towards inclusive and sustainable industrial development. *New Biotechnology, 40*(A), 5–10. https://doi.org/10.1016/j.nbt.2017.06.005

Maes, T., McGlade, J., Fahim, I. S., Green, D. S., Landrigan, P., Andrady, A. L., Turra, A., et al. (2021). *From pollution to solution: A global assessment of marine litter and plastic pollution.*

Mak, T. M. W., Xiong, X., Tsang, D. C. W., Yu, I. K. M., & Poon, C. S. (2020). Sustainable food waste management towards circular bioeconomy: Policy review, limitations and opportunities [Policy review]. *Bioresource Technology, 297*, 122497. https://doi.org/10.1016/j.biortech.2019.122497

McCormick, K., & Kautto, N. (2013). The bioeconomy in Europe: An overview. *Sustainability, 5*(6), 2589–2608. https://doi.org/10.3390/su5062589

Mignone, B. K., Huster, J. E., Torkamani, S., O'Rourke, P., & Wise, M. (2022). Changes in global land use and CO_2 emissions from US bioethanol production: What drives differences in estimates between corn and cellulosic ethanol? *Climate Change Economics, 13*(4), 2250008. https://doi.org/10.1142/S2010007822500087

Organization for Economic Co-operation and Development. (2009). The bioeconomy to 2030: Designing a policy agenda. *OECD Publishing.* https://doi.org/10.1787/9789264056886-en

Özüdoğru, H. M. R., Nieder-Heitmann, M., Haigh, K. F., & Görgens, J. F. (2019). Techno-economic analysis of product biorefineries utilizing sugarcane lignocelluloses: Xylitol, citric acid and glutamic acid scenarios annexed to sugar mills with electricity co-production. *Industrial Crops and Products, 133*, 259–268. https://doi.org/10.1016/j.indcrop.2019.03.015

Perdices, M. B. (2018). El Papel de la bioenergía en la nueva bioeconomía. *Mediterráneo Económico, 31*, 353–369.

Rafiee, A., Khalilpour, K. R., Prest, J., & Skryabin, I. (2021). Biogas as an energy vector. *Biomass and Bioenergy, 144*, 105935. https://doi.org/10.1016/j.biombioe.2020.105935

Ramcilovic-Suominen, S. (2022). Envisioning just transformations in and beyond the EU bioeconomy: Inspirations from decolonial environmental justice and degrowth. *Sustainability Science*, 1–16. https://doi.org/10.1007/s11625-022-01091-5

Reddy, M. V., Hayashi, S., Choi, D., Cho, H., & Chang, Y. C. (2018). Short chain and medium chain fatty acids production using food waste under non-augmented and bio-augmented conditions. *Journal of Cleaner Production, 176*, 645–653. https://doi.org/10.1016/j.jclepro.2017.12.166

Sacramento-Rivero, J. C., Mwampamba, T. H., Navarro-Pineda, F. S., Musule, R., García, C. A., Martínez-Bravo, R. D., Morales-García, A. L., Equihua-Sánchez, M., Fuentes-Gutiérrez, A. F., Gallardo-Álvarez, R. M., Ruiz Camou, C. R., Grande-Acosta, G. K., Manzini, F., Islas-Samperio, J. M., & Camarillo, R. (2022). A methodological framework for assessing the sustainability of solid biofuels systems. *Bioenergy Research, 15*(4), 1797–1819. https://doi.org/10.1007/s12155-021-10365-2

Sandesh, K., & Ujwal, P. (2021). Trends and perspectives of liquid biofuel–Process and industrial viability. *Energy Conversion and Management: X, 10*, 100075. https://doi.org/10.1016/j.ecmx.2020.100075

Schmidt, C., Krauth, T., & Wagner, S. (2017). Export of plastic debris by rivers into the sea. *Environmental Science and Technology, 51*(21), 12246–12253. https://doi.org/10.1021/acs.est.7b02368

Schoumans, O. F., Bouraoui, F., Kabbe, C., Oenema, O., & van Dijk, K. C. (2015). Phosphorus management in Europe in a changing world. *Ambio, 44*(Suppl. 2), S180–S192. https://doi.org/10.1007/s13280-014-0613-9

Scott, M. C. (2019). *Strategies to implement efficient closing cycles* (Doctoral Dissertation). Walden University.

Shahid, M. K., Batool, A., Kashif, A., Nawaz, M. H., Aslam, M., Iqbal, N., & Choi, Y. (2021). Biofuels and biorefineries: Development, application and future perspectives emphasizing the environmental and economic aspects. *Journal of Environmental Management, 297*, 113268. https://doi.org/10.1016/j.jenvman.2021.113268

Sibani, P., Rasmussen, S., & Hansen, P. L. (2022). *Human wealth evolution is an accelerating expansion underpinned by a decelerating optimization process*. arXiv preprint arXiv:2203.17087.

Sikkema, R., Proskurina, S., Banja, M., & Vakkilainen, E. (2021). How can solid biomass contribute to the EU's renewable energy targets in 2020, 2030 and what are the GHG drivers and safeguards in energy-and forestry sectors? *Renewable Energy, 165*, 758–772. https://doi.org/10.1016/j.renene.2020.11.047

Tinôco, D., Borschiver, S., Coutinho, P. L., & Freire, D. M. G. (2021). Technological development of the bio-based 2, 3-butanediol process. *Biofuels, Bioproducts and Biorefining, 15*(2), 357–376. https://doi.org/10.1002/bbb.2173

Van Roijen, E. C., & Miller, S. A. (2022). A review of Bioplastics at end-of-life: Linking experimental biodegradation studies and life cycle impact assessments. *Resources, Conservation and Recycling, 181*, 106236. https://doi.org/10.1016/j.resconrec.2022.106236

Wang, J., Ma, F., Bouri, E., & Zhong, J. (2022). Volatility of clean energy and natural gas, uncertainty indices, and global economic conditions. *Energy Economics, 108*, 105904. https://doi.org/10.1016/j.eneco.2022.105904

Wohlgemuth, R. (2021). Bio-based resources, bioprocesses and bioproducts in value creation architectures for bioeconomy markets and beyond—What really matters. *EFB Bioeconomy Journal, 1*, 100009. https://doi.org/10.1016/j.bioeco.2021.100009

Yahyazadeh, A., Dalai, A. K., Ma, W., & Zhang, L. (2021). Fischer-Tropsch synthesis for light olefins from syngas: A review of catalyst development. *Reactions, 2*(3), 227–257. https://doi.org/10.3390/reactions2030015

Zabermawi, N. M., Alsulaimany, F. A. S., El-Saadony, M. T., & El-Tarabily, K. A. (2022). New eco-friendly trends to produce biofuel and bioenergy from microorganisms: An updated review. *Saudi Journal of Biological Sciences*. https://doi.org/10.1016/j.sjbs.2022.02.024

Chapter 5
Policy Intervention of Waste Management

Salman Shooshtarian, Tayyab Maqsood, and Peter S. P. Wong

5.1 Introduction

Over the past few decades, there has been considerable growth in the construction and demolition (C&D) waste stream because of significant urbanisation driven by a rising population, movement, and migration. In the absence of a standardised definition for C&D waste, this waste stream is typically defined as items taken from demolition and construction sites that are unwanted or discarded. However, the new circular mindset prefers to recognise this stream's materials as resources rather than waste. The National Waste Report (2020) states that, in 2018–2019, Australia generated 27 million tonnes of C&D waste, and increase of 61% since 2006–2007. Comprising 44% of the total waste, this waste source remains the main generator of Australian waste. To respond to this issue, one solution is for Australia to become a circular economy (CE) that would foster a sustainable BE (Elmualim et al., 2018).

To date, numerous definitions of 'CE' have been proposed (Guerra & Leite, 2021; Kirchherr et al., 2017). The Ellen MacArthur Foundation (EMF) explains the CE as '*an industrial economy that is restorative or regenerative by intention*' which is widely acknowledged and adopted by government officials, practitioners, as well as academics (Anastasiades et al., 2020, p. 2). The European Parliament offers a simpler definition, describing CE as a model of production and consumption founded on 'sharing, leasing, reusing, repairing, refurbishing, and recycling existing materials and products for as long as possible', which lengthens the life cycles of products

S. Shooshtarian (✉) · T. Maqsood · P. S. P. Wong
School of Property, Construction and Project Management, RMIT University, Melbourne, Victoria 3000, Australia
e-mail: salman.shooshtarian@rmit.edu.au

T. Maqsood
e-mail: tayyab.maqsood@rmit.edu.au

P. S. P. Wong
e-mail: peter.wong2@rmit.edu.au

S. A. Bandh and F. A. Malla (eds.), *Waste Management in the Circular Economy*,
https://doi.org/10.1007/978-3-031-42426-7_5

and materials (European Parliament, 2022). Within the BE, a CE is an economic system that replaces the standard 'end-of-life' philosophy (Benachio et al., 2020); the emphasis is on minimising, inventively repurposing, recycling, and reclaiming materials from distribution or manufacturing and consumption activities to make use of those materials as long as possible in the cycle, thereby minimising the misuse of natural resources (Benachio et al., 2020; Guerra & Leite, 2021; Kirchherr et al., 2017). In the context of C&D waste management, CE offers an effective approach to mitigate the environmental and economic burdens associated with waste, such as the ongoing maintenance of landfills, the extraction of virgin materials, landfill fees, transportation costs, and illegal dumping (Shooshtarian et al., 2020a).

Moving from a linear economy to CE warrants substantial changes to current production and consumption systems (Peck et al., 2020). One of the major barriers to implementing CE objectives is the lack of a supportive policy framework (Shooshtarian et al., 2020a, 2020b, 2020c, 2020d, 2020e). Hence, one critical improvement area for a society wishing to achieve CE is to develop supportive policies that inspire circularity across the materials supply chain. Three types of policy instruments assist the industry in applying CE principles to BE sector: administrative, economic, and informative. Licences, bans, benchmarks, and voluntary agreements between industry and government are examples of administrative or regulatory instruments, while economic instruments include fees, subsidies, taxes, and other charges. Informative instruments comprise labelling, reporting obligations, certification initiatives and awareness-raising campaigns are examples of (Peck et al., 2020).

5.2 Research Gap and Scope

In Australia, waste regulation occurs at the state and territory levels. As illustrated in (Fig. 5.1) Australia has six states and two territories which have different waste regulatory frameworks.

In this chapter, eight policy instruments that are currently imposed or proposed to be implemented in the Australian context are explored (Fig. 5.2). These include 'levy tax on waste landfilling', 'penalties on illegal waste dumping', 'tax waiver on recycling residual', 'tax on using raw materials' (economic), 'extended producer responsibility' (EPR), 'sustainable procurement' (SP), 'proximity principle' (administrative), and 'recycled product certification' (informative).

5.3 Landfill Tax

The BE's sector's current preferred method for dealing with C&D waste is landfilling. This is because it is an easier and a less expensive option compared with alternative waste management methods, as well as its accessible hours of operation. Udawatta

Fig. 5.1 Australian states and territories

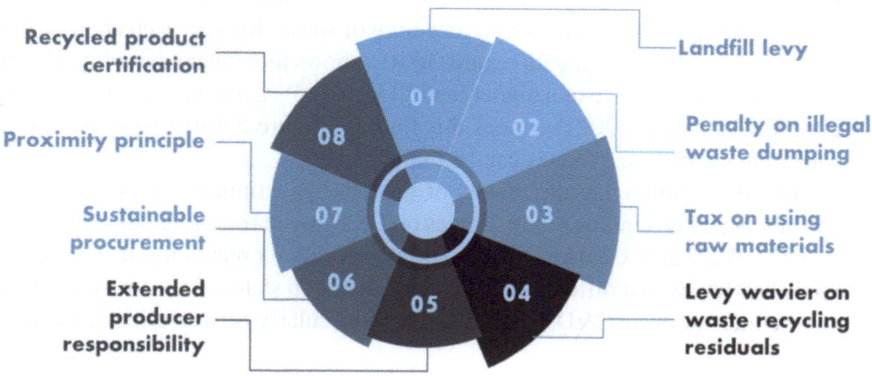

Fig. 5.2 Policy instruments that are explored in this chapter

et al. (2015a, 2015b) contend that poor financial returns and lack of incentives inhibit subcontractors' interest in waste minimisation. Newaz et al. (2020) also point this out, arguing that construction projects' variety makes consideration of waste management difficult during the design phase. Analysis of numerous fit-outs case studies NSW indicates that 78% of waste produced onsite goes to landfill (Fini & Forsythe, 2020). Recycling facility fees applied to waste materials processing compared to those at landfill sites are greater Tam et al. (2009). Ratnasabapathy et al. (2019a, 2019b, 2019c) demonstrate in a recent study that a minor reduction in Australian C&D waste disposal highlights the necessity of devising ways to limit landfilling. The researchers predict that if the trajectory continues, this will result in a 78% diversion rate from landfill by 2025.

Most of the existing studies on this topic emphasise the need for landfill levies. For example, studies suggest that the recovery of waste can be driven by landfill levy fees that are high. Shooshtarian et al., (2020a, 2020b, 2020c, 2020d, 2020e) state that of participants in one survey, 90% recommended landfill levies as generally effective. Landfill levy rates in Australia are decided on by state and territory governments (Davis et al., 2019). This means that there is variation between states, depending on how the rates are formulated (Shooshtarian et al., 2019a, 2019b). A study (Shooshtarian et al., (2020a, 2020b, 2020c, 2020d, 2020e) of states' recent waste strategies found that these documents recommend the revision of current levy practices to dissuade the industry from waste disposal as their second-highest suggestion. The jurisdictions of ACT, SA, QLD, TAS, and WA all suggested this way forward.

Research has been conducted into landfill levies' standards in Australia as well as their suitability (Jayasinghe et al., 2018; Newaz et al., 2020). Stakeholders are concerned about the effective design of this scheme to better mirror practical considerations plus expand its influence (Zhao et al., 2021). Shooshtarian et al., (2020a, 2020b, 2020c, 2020d, 2020e) contend that though the preferred market approach in some scenarios can be landfill levies, while in others, it can discourage best practices, such as by inadvertently leading to a rise in the dumping of waste illegally or its transfer elsewhere, plus less overall recycling of waste. Rameezdeen et al. (2016) research into SA's BE sector, with regards to RL, shows that illegal dumpling can be an unintended consequence of a higher landfill levy. In WA, a government document (as cited in Zhao et al., 2021) echoes this finding. Figure 5.3 displays where in a material's lifecycle these policies can be implemented.

A QLD study highlights the effects of the state government's decision on the establishment and rescinding landfill levies on C&D waste recycling rates (Forghani et al., 2017). Wu et al. (2020) find that the site availability for recycling and landfilling activities and levies cost differential between Australian states are some of the main reasons behind interstate C&D waste transferer particularly between NSW and ACT,

Fig. 5.3 Application of studied policies to the construction material lifecycle

NSW and QLD, SA and VIC. This issue that can be minimised by approaches such as the proximity principle (Jayasinghe et al., 2018), in which waste transfer is restricted over long distances. Newaz et al. (2020) indicate that waste experts in NSW believe that waste levy is ineffective in terms of minimising waste landfilling, due to few practical results from levies and the fact that the state permits waste transfer to QLD, which has comparatively lower landfill levies. In Shooshtarian et al., (2020a, 2020b, 2020c, 2020d, 2020e), participants report that the extant levy schemes could be more effective and require changes to fulfil this goal.

Generally, levies for waste landfilling are informed by various factors such as levy exceptions for particular materials, how waste is composed, levy zones, and the location of landfills. For this reason, they rise regularly (Chileshe et al., 2019; Zhao et al., 2021). Jayasinghe et al. (2018) argue that levies should be decided depending on the waste classification, demographics, financial impact analysis, and levy rebates for waste recovery. Landfill levy revenues are in part spent on bolstering enforcement and compliance, devising good policies, and funding initiatives and methods that foster waste minimisation. However, Australia lacks a national approach to the spending of levies on the aforementioned purposes, with each state government doing so as guided by its specific concerns and preferences (Shooshtarian et al., 2020a, 2020b, 2020c, 2020d, 2020e).

Previous research suggests some strategies to reduce waste disposal. For example, Shooshtarian et al., (2020a, 2020b, 2020c, 2020d, 2020e) indicate that consideration of fees on transport and possible price repercussions for the construction industry when deciding the levy fees; standardisation of levies for landfill; supplementing levies with technology-driven compliance and enforcement systems; improving local infrastructures; directing revenue from landfill levies into resource salvage initiatives via initiatives such as offering monetary enticements or low-interest lending; and resourcing R&D. An additional way forward for having a lower rate of waste disposal is to increase energy recovery from combustible C&D waste resources that would otherwise end up in landfills (Shooshtarian et al., 2019a, 2019b). Tam et al. (2009) suggest that application of a various levels of levy to encourage source separation of C&D waste resources. Additionally, competent waste data and reporting systems that enumerate the precise quantity of landfilled waste would enhance planning and regulation for greater waste minimisation via landfilling or otherwise (Ratnasabapathy et al., 2019a, 2019b, 2019c). Effective waste landfilling is complemented by engineered landfills close to transfer stations and reprocessing centres for optimal transportation (Jayasinghe et al., 2018). Various state and territory guidelines have been produced for best landfilling management, in terms of the location of landfills, plus their planning, building, functioning, upkeep, and shutdown. However, the Australian government suggests the standardisation of best-practice landfill approaches (as cited in Davis et al., 2019). In addition, a blend of sound landfill levies, incorporating the proximity principle and long-distance located landfill sites from construction sites, could encourage recycling instead of landfilling. Finally, the reviewed studies were lacking in landfill operators' opinions as key stakeholders in this issue, accounting for less than 2% in the existing research studies. This emphasises the need for more research to identify waste minimisation prospects and understand resource efficiency in the landfilling phase.

5.4 Penalties for Illegal Waste Dumping

In the Australian, illegal waste dumping and unauthorised stockpiling represent major challenges. While the National Waste Report (2020) suggests that in terms of C&D waste, only 1% of it is from illegal dumping, anecdotal evidence suggests a larger percentage, especially in NT and WA. Researchers find a range of motives and justifications behind illegal dumping and stockpiling. Shooshtarian et al. (2020a, 2020b, 2020c, 2020d, 2020e) suggest that varying jurisdictional rules cause in illegal dumping and stockpiling. Notably, Rameezdeen et al. (2016) and Shooshtarian et al., (2020a, 2020b, 2020c, 2020d, 2020e) find that a higher landfill levy could actually be a causal factor in illegal dumping. Shooshtarian et al., (2020a, 2020b, 2020c, 2020d, 2020e) recognise the lack of levy revenue return to the C&D waste management sector as an obstacle to better prevention of illegal waste practices.

As shown in Shooshtarian et al., (2019a, 2019b), states and territories punish these practices via enforcing substantial penalties. However, penalties for illegal waste dumping vary across Australia's jurisdictions as summarised in Table 5.1.

The state governments use waste strategy documents to present what actions they are taking to reduce illegal dumping (Shooshtarian et al., 2020a, 2020b, 2020c, 2020d, 2020e), including education (awareness-raising, stakeholder engagement); enforcement (evaluation and monitoring, regulations); and encouragement (developing required infrastructures, capacity, and network). Researchers also highlight other approaches to limit illegal dumping and stockpiling, including demolition companies providing better supervision to prevent illegal dumping (Kabirifar et al.,

Table 5.1 Summary of penalties for illegal waste dumping in different Australian states and territories

State/ territory	Penalty
NSW	5 M and/or 7-year imprisonment
NT	Up to 573 K/ 5-year imprisonment (individuals) and 2.9 M (corporations)
TAS	1.59 M (+/5-year imprisonment)
ACT	Up to 200 K (individuals) and 1 M + /7-year imprisonment (corporations)
SA	Up 120 K (individuals), 250 K (corporations), and 500 K + 4-year imprisonment (if causing environmental damage)
WA	Up to 62.5 K (individuals) and 125 K (corporations)
QLD	Up to 239 K (individuals) and 1.19 M (corporations)
VIC	Up to 610 K/7-year imprisonment (individuals) and 1.2 M (corporations)

2021); standardised regulations including identical levy fees and funding for educational programmes (Davis et al., 2019; Laviano et al., 2017); and the adoption of innovations such as image processing, remote sensing, and GIS to oversee activities involving illegal waste (Glanville & Chang, 2015; Ratnasabapathy et al., 2019a, 2019b, 2019c). Another useful strategy to prevent long-term stockpiling, according to Jayasinghe et al. (2018), is a levy obligation that is applied upfront. This is currently being used in NSW, via a fee for waste received at the depot that falls upon disposal.

5.5 Tax on Using Raw Materials

Although most construction materials can technically be recycled, which materials are salvaged and to what extent often largely depends on the material's value. In most circumstances, this is compared to the price of a material when newly extracted or imported. To shift perceptions regarding C&D salvaged and recycled materials, there are some price mechanisms than can be applied so that purchasers are disincentivised regarding the adoption of virgin materials for construction projects. This enticement takes two forms—'removing subsidies for virgin materials' and 'taxing on the use of virgin materials'—with evidence emerging that it has enhanced salvaged and recycled C&D materials' competitiveness in several countries. For example, since 2002, the UK has imposed a regulation (UK's HM Revenue & Customs, 2018) to bolster the position of C&D products made from recycled waste compared to new materials. It comes in the form of a levy (£2 per tonne) applied to use of gravel, sand, and rock for commercial purposes. The purpose is to shift the charge for new materials to incorporate their fundamental impact on the environment. Sweden, Denmark, and France, among other EU countries, have applied a similar tax (Hyder Consulting Pty Ltd., 2011). In the Australian context, waste definitions that exclude clean fill have been adopted by a few states. Coupled with reduced costs, this approach is anticipated to encourage the BE sector to adopt C&D waste rather than raw materials.

5.6 Levy Waiver on Waste Recycling Residuals

Levy waivers lead to greater C&D waste recovery and consequent market development. Creating a scenario where recycling residuals have a levy imposed on their disposal causes them to be less competitive; for example, for each $15/t levy rate rise, the steel recycling industry in VIC sustains an extra $738 k per annum cost (EPA Victoria, 2007). At present, in all jurisdictions, waste recycling facilities' residuals are defined as waste, and therefore, the owners of such businesses incur a landfill fee. Consequently, the relevant industries, construction and waste recovery, have put forward requests to have these residuals no considered longer be perceived by regulations to be waste. Many contend that levies on recycling residuals' disposal limits the position of materials in the global marketplace. Recycling facilities face major spending on disposing residuals, with naturally causes them to seek alternatives that offer a lower charge. Additionally, it leads to the justification of transport of waste to locations interstate that have a more attractive disposal cost.

A levy on the disposal of recycling residuals reduces the competitiveness of materials sold into the international market [**National Waste and Recycling Industry Council**]

When recyclers are liable to pay the levy for the disposal of contaminants that have entered the recycling stream, they see it as a disincentive towards being involved in the recycling industry and instead it encourages shipping unprocessed waste overseas [**The Australian Council of Recycling**]

The disposal of residuals generally represents a significant cost for recycling facilities, which can obviously create commercial incentives to seek lower disposal cost options; it also justifies transport waste to interstate locations with a lower disposal rate [**Re Group**]

Landfill levies penalise the recycling industry for the disposal of residual rubbish that enters the recycling stream [**Visy, Owens-Illinois and SKM Recycling**]

5.7 Extended Producer Responsibility

Extended producer responsibility (EPR), also called the 'polluter pays' or 'take-back' principle, is a key lens through which manufacturers of construction materials can be encouraged to use waste materials in their products. This approach has been proven to be effective in adding to a CE of the BE sector, in a market-based sense, EPR allows for the prevention of waste generation in the first instance. Secondly, it shifts any waste away from landfills towards reuse and recover. Lastly, it fosters the expansion and evolution of C&D waste resources markets. Applied appropriately, EPR can incentivise producers to consider the impacts of their products, the flow-in effect of which is stopping the creation of waste at the source via enhanced product design, technological innovations, and the use of green design and effective waste management initiatives within wide-scale production planning.

Such policies make the material's manufacturer or its supplier accountable—physically and/or monetarily—for the entire lifecycle of their materials, including the waste created (Tam & Lu, 2016). Currently, there does not exist a widespread and harmonised strategy for enacting and benefiting from EPR aims within the C&D industry. Achieving this is not without significant complexity due to the varying stakeholders who participate in creating products, delivery, trade, use, and the management of waste (Shooshtarian et al., 2021a, 2021b). Additionally, the means through which EPR policies are implemented can differ depending on the scenario.

5.7.1 Criteria Used to Develop EPR Policies

Past studies have tried to devise models that consider the complicated factors involved to enhance the outcomes of EPR policies practically (Xu et al., 2021). This section outlines some of them. Dubois et al. (2016) offered five measures of the capability of EPR for C&D waste management development and assessment, as shown in Fig. 5.4 (left). Considering their application to the C&D waste stream in the Netherlands, the researchers found that there is a drive to use EPR for only two criteria (e.g. environmental scope and political priorities). Guggemos and Horvath (2003) put forward a policy framework to optimise achievement of EPR goals for C&D waste management. Based on Thorpe and Kruszewska (1999) model, this framework comprises three forms of policy instrument: regulatory, economic, and information-based (Fig. 5.4—right).

In addition to the models presented in Fig. 5.4, other studies have suggested alternatives that involve a few overlaps and divergences (Forslind, 2005; Langrová, 2002; Lindhqvist, 2000; Nahman, 2010; Widmer et al., 2005). Additionally, interesting data regarding the best optimisation of EPR and relevant initiatives for waste management have emerged from studies exploring the elements that influence the outcomes of EPR. Gupt and Sahay (2015), for example, undertook research that compared 26 case studies to find the elements that are linked with successful EPR application and the key aspects of EPR expansion and execution, in countries that are developing and developed. They found that 'recycling agencies' and 'separate collecting' strongly support positive EPR outcomes, as well as 'financial responsibility of the producers'. The main EPR features were identified as 'regulatory provisions', 'take-back responsibility', and 'financial flow'. A 2016 study on the effectiveness of various environmental policies assessed several policies in Maine, USA. The results revealed that EPR policies are considered to be as highly effective, but that there is uncertainty about their acceptability (Isenhour et al., 2016).

5.7.2 *Main Challenges to Adopting EPR in BE Sector*

The construction industry's widespread embrace of EPR and linked initiatives is held back by a number of obstacles (Guggemos & Horvath, 2003; Shanoff, 1996; Srour et al., 2012). Consequently, not all EPR instruments referred to in Fig. 5.5 are equal to each other in outcomes for the management of C&D waste. The next section outlines the key barriers to the construction industry's successful creation and adoption of an extensive and standardised EPR approach.

There is no mandatory EPR policy legislated in Australia at present for construction material manufacturing; NSW and WA, however, do have voluntary EPR policies (Shooshtarian et al., 2021a, 2021b). Park and Tucker (2017); Ratnasabapathy et al. (2021) point to the lack of a mandatory EPR policy as an obstacle to the reuse of waste materials and creating C&D waste trading systems in Australia, respectively. Shooshtarian et al., (2021a, 2021b) studied the implementation of EPR in Australia and found that there is significant agreement between various stakeholders about the need for EPR. Obstacles to such measures include cost and time implications; complexity of policy establishment and enforcement; responsibility of manufacturers; the range of interested stakeholders; the lifecycle of materials in construction;

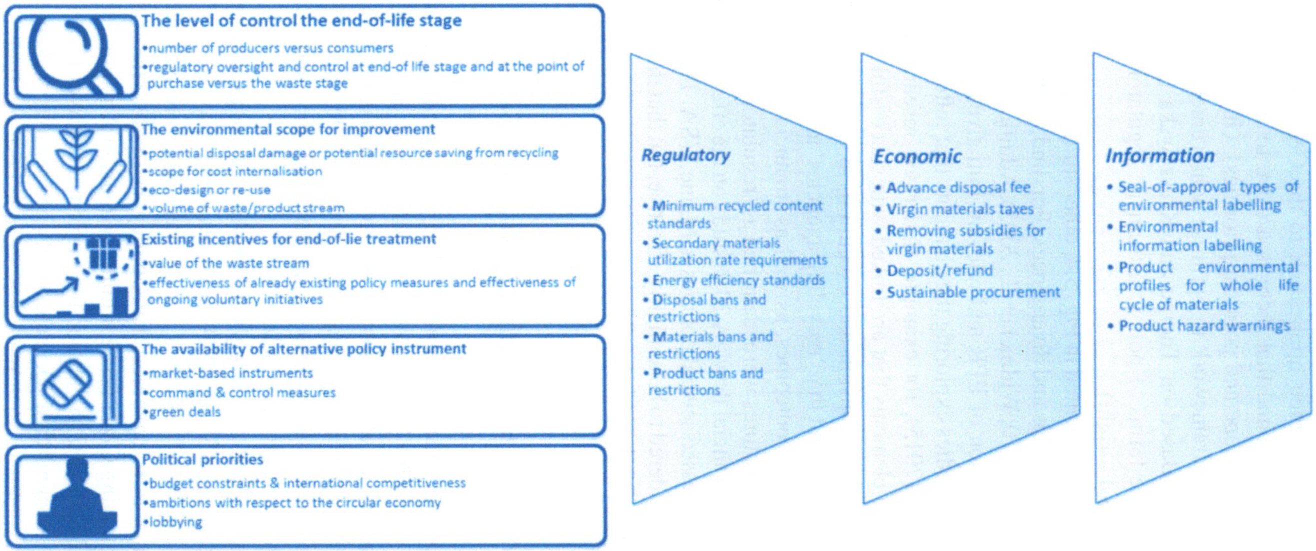

Fig. 5.4 Main criteria proposed for developing an EPR policy (left) and major strategies for facilitating EPR implementation. *Source* materials adopted from Dubois et al. (2016) and Guggemos and Horvath (2003)

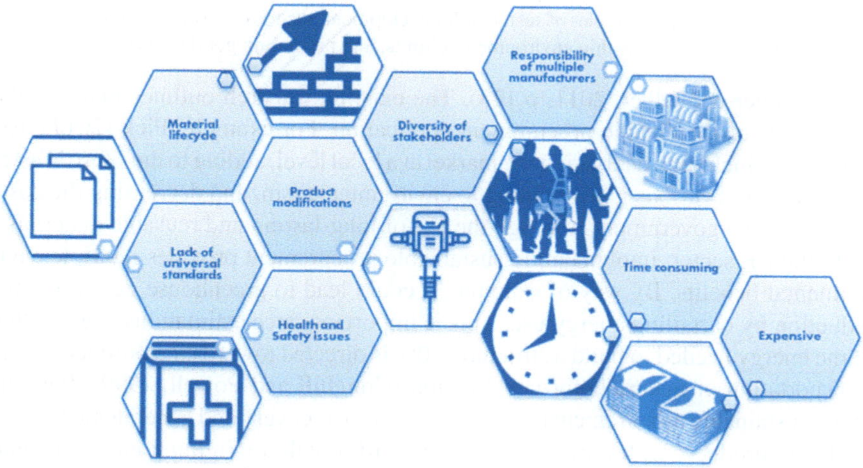

Fig. 5.5 Key obstacles limiting the successful implementation of EPR for the C&D industry

and OH&S considerations. To counter this, suggestions include the establishment of an effective supply chain system; encouraging DfD; product documentation; waste responsibility assignment; and enhanced health and safety risk management.

5.8 Sustainable Procurement

Among existing policies, sustainable procurement can potentially change the status quo of C&D waste management systems for the better. This policy approach will also help to enhance circularity in resources in the BE sector. Various organisations define sustainable procurement for different scenarios and usages. The United Nations (UN), the UK government, and the Australasian Procurement and Construction Council (APCC) accept the following definition (cited in Commonwealth of Australia, 2013, p. 4). Sustainable procurement can also be referred to using terms such as 'green public procurement', 'sustainable public procurement' (Alhola et al., 2019), 'environmental purchasing', 'circular procurement', and 'environmentally preferable procurement'.

> A process whereby organisations meet their needs for goods, services, works and utilities in a way that achieves value for money on a whole life basis in terms of generating benefits not only to the organisation, but also to society and the economy, whilst minimising damage to the environment.

Sustainable procurement adds onto the concepts of CE and the cradle-to-cradle approach evaluated through various measures, including to evaluate product lifecycle. This approach.

is consistent with the principles of sustainable development, such as ensuring a strong, healthy and just society, living within environmental limits, and promoting good governance.

(Brammer & Walker, 2011, p.128). The existing research outlines many of the advantages of sustainable procurement application. For example, Pick (2017) lists advantages including energising the market at a local level, adding to the 'Zero Waste' movement and various goals relating to environmentalism, and decreasing the costs faced by local governments through the use of long-lasting and reusable materials.

In the BE sector, implementing sustainable procurement provides multiple environmental benefits. By way of example, it could lead to greenhouse gas emissions reduction by curtailing energy use. This is important since estimations suggest that of the energy needed to build a structure, 80% is directed towards the production and transportation of materials for construction (Craighill and Powell, 1999). Furthermore, sustainable procurement helps increase waste recovery and generates a demand for C&D products with recycled content (RwRC) with a minimum environmental impact. A research study revealed that among 25 study countries, mandating suppliers to commit to waste reduction goals is one of the five central functions of sustainable procurement policies (Brammer & Walker, 2011). A study in the Netherlands (Zhang et al., 2020) reported that sustainable procurement is the primary approach to stimulating concrete recycling. Sustainable procurement was also the most promising option for establishing the long-term mechanism of using PwRCs in the construction industry (He et al., 2014).

In recent years, only a handful of studies have investigated this topic; however, their scope was defined with some limitations. The limitations are concerned with the study focus or the methodology adopted. For instance, specific aspects such as carbon emission reductions (Lingegård et al., 2021), COVID-19 (Caldera et al., 2022), target particular stakeholders (Ershadi et al., 2021), and market development (Caldera et al., 2020; Shooshtarian et al., 2020a, 2020b, 2020c, 2020d, 2020e) were previously studied. Using methodologies such as case study, inherited constrained generalisability (Sanchez et al., 2014; Shooshtarian et al., 2022a, 2022b, 2022c) provides fractional information about BE sector sustainable procurement. The previous literature analysis (Shooshtarian et al., 2022a, 2022b, 2022c) shows that the relationship between procurement and C&D waste minimisation in Australia is not well studied. Multiple research studies highlight the need to investigate the impact of procurement methods on waste management planning (Park & Tucker, 2017; Udawatta et al., 2015a, 2015b). Furthermore, the previous authors' exposure to real-time stakeholders' concerns expressed in their research projects has highlighted such shortcomings in the Australian BE sector's CE and waste management policy and practice. The stakeholders' concerns were captured in survey analysis (Ershadi et al., 2021; Shooshtarian et al., 2021a, 2021b), a series of interviews (Shooshtarian et al., 2022a, 2022b, 2022c), specialised workshops and industry seminars (SBEnrc P.175, 2021), and other personal communications. Considering all the above, this research was conducted to amend this research gap and offer a more comprehensive insight into this policy approach.

5.8.1 *Sustainable Procurement Principles, Worth, and Benefits in the Australian Context*

Anecdotal evidence shows that the sustainable procurement idea has sparked interest among the main players in the BE sector. This section summarises Australians' views on the principles, worth, and benefits of sustainable procurement concerning C&D waste management.

To enhance the market demand for recovered C&D materials, government agencies need to become more involved in the wider implementation of sustainable procurement practices via the creation of specifications, quality assurance, accreditation, and awareness raising regarding the outcomes (financial, social, environmental, and ethical) of the PwRC and services (Hyder Consulting Pty Ltd., 2011). The findings of a survey study (Shooshtarian et al., 2020a, 2020b, 2020c, 2020d, 2020e) indicated that this policy approach, landfill levies, and investment in technology and infrastructure are three major influential elements that significantly affect market development. In one study, research participants suggest that sustainable procurement is a vital solution for organisational waste management but only if it is a workable alternative or the client agrees to pay more (Davis et al., 2019). In another study, Ershadi et al. (2021) explored the role of the project management office (PMO) in implementing sustainable procurement in the BE sector. The study identified eight primary areas through which the benefits of this approach can be fulfilled. These include task assignment, strategic analysis, goal setting, tendering support, planning support, operation support, maintaining consistency, and post-review. In the wake of the Australian government's 2010 Sustainable Procurement Guideline, a revised guide was released in 2020 (Australian Government, 2020). This government publication offers a framework for government agencies to increase sustainability in future procurement activities. This document offers a list of benefits to purchaser (government), market, and society and the environment that are fulfilled by purchase of PwRC. Figure 5.6 incorporates this list with the Australian ISO 20400 Committee's proposed model.

To identify the primary functions and purposes of sustainable procurement relevant to Australia, the analysis of pertinent literature was conducted. The analytical result shows that sustainable procurement principles are theoretically grounded on eleven items. As displayed in Fig. 5.7, these items denote a range of facets of sustainable procurement and can show public and private organisations how to devise specific sustainable procurement policies.

Table 5.2 summarises the definition of the eleven principles identified in (Local Government NSW, 2021). For each principle, an action is proposed (the last column) that aims to turn the principle into sustainable procurement of PwRC in the built environment sector.

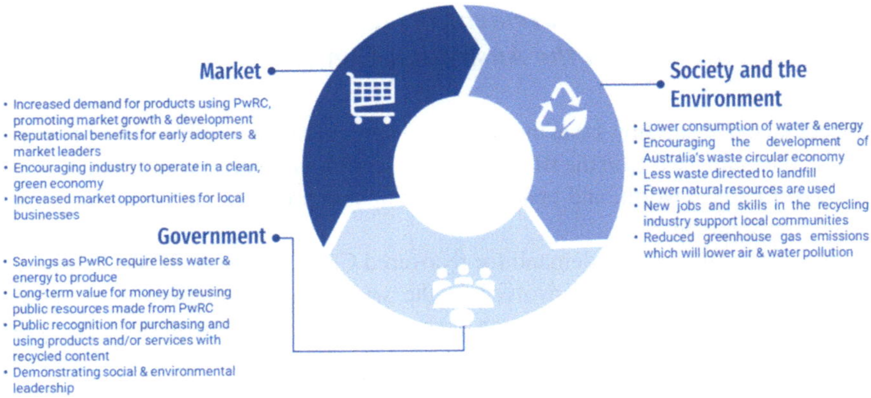

Fig. 5.6 Various benefits of sustainable procurement benefits. *Source* Australian ISO 20400 Committee (2018) and Australian Government (2020)

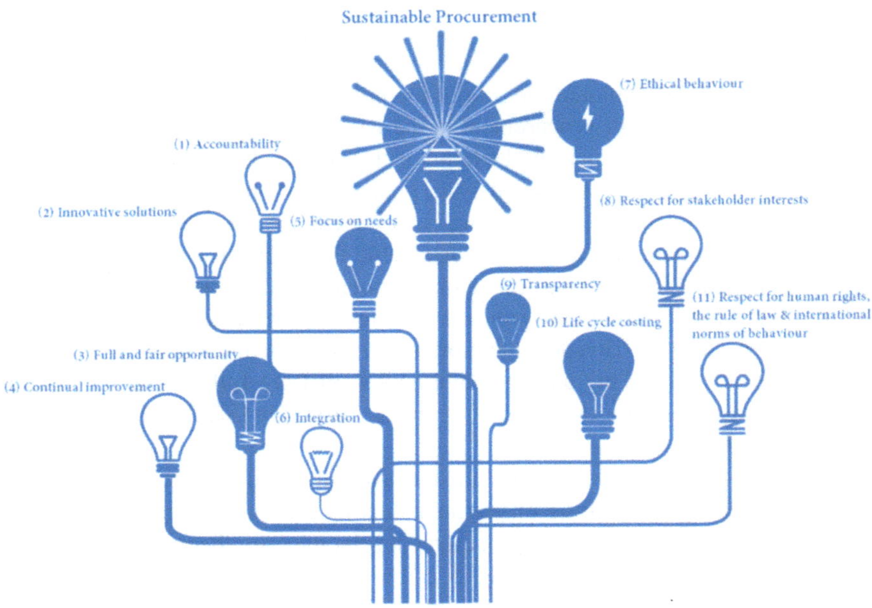

Fig. 5.7 Visual summary of sustainable procurement principles. *Source* Adapted from Local Government NSW (2021)

5.8.2 Challenges of Sustainable Procurement Application

Barriers play a significant part in shaping sustainable procurement practices and determining the success or otherwise of sustainable procurement outcomes. The literature uncovers various elements that hamper the publicity and implementation

Table 5.2 Sustainable procurement principles' functionality and relevance to PwRC procurement

S. no	Principle	Description	Action relevant to PwRC procurement
1	Accountability	Be accountable for its impacts on society, the economy, and the environment, including the impacts of the organisation's supply chain	To increase the cost of utilisation of virgin materials for end-users to be accountable for their procurement policies and practices
2	Innovative solutions	Seek solutions to address its sustainability objectives and encourage innovative procurement practices to promote more sustainable outcomes throughout the entire supply chain	To develop policies that promote innovation in the procurement of PwRC furthering PwRC uptake
3	Full and fair opportunity	Avoid bias and prejudice in all procurement decision-makings	To ensure that PwRC is not compared to virgin materials
4	Continual improvement	Work towards continually improving its sustainability practices and outcomes, and encouraging its supply chains to do the same	To optimise the supply chain in the BE sector to embed in PwRC in procurement planning and practices
5	Focus on needs	Review demand, buy only what is needed and seek more sustainable alternatives first	To prioritise PwRC when specifying materials for low-value applications
6	Integration	Ensure that sustainability is integrated into all existing procurement practices to maximise sustainable outcomes	To motivate the BE sector to build capacity for the application of sustainable procurement
7	Ethical behaviour	Behave ethically and promote ethical behaviour throughout its supply chain	To monitor waste handling and processing activities to minimise the risk of procurement of PwRC
8	Respect for stakeholder interests	Respect, consider, and respond to the interests of stakeholders impacted by its procurement activities	To provide a level playing field for all stakeholders involved in the use of PwRC
9	Transparency	Consider the cost incurred, the value for money achieved, and also the costs and benefits on society and the economy	To ensure transparency in handling, processing, and procurement of PwRC

(continued)

Table 5.2 (continued)

S. no	Principle	Description	Action relevant to PwRC procurement
10	Lifecycle costing	Consider the cost incurred, the value for money achieved, and also the costs and benefits on society, the environment, and the economy resulting from its procurement activities	To consider the social and environmental benefits of procuring PwRC
11	Respect for human rights, the rule of law, and international norms of behaviour	Be aware of any violations throughout its supply chains and actively encourage its suppliers to do the same	To reduce the need for virgin materials through procurement of PwRC

Source Adapted from Local Government NSW (2021)

of sustainable procurement practices. The challenges involved depend greatly on the specific scenario, though there are similarities between various contexts. Table 5.3 outlines studies that explore sustainable procurement barriers and enablers in developed and developing nations and presents 30 barriers reported in the literature. The two key barriers are revealed to be 'the lack of supportive organisational culture' and 'uncertainty about the PwRC quality'.

5.8.3 *Drivers of Sustainable Procurement Application*

Procurement experts contend that sustainable procurement planning must be founded on the realisation of a several mostly context-specific conditions. Previous studies (Table 5.3) have identified a range of enablers that would enable the effective implementation of sustainable procurement, mainly in the public sector in developing (i.e. Australia, New Zealand, Canada, and France) and developed (i.e. South Africa, Saudi Arabia, and Nigeria) nations. From the 20 drivers identified in the literature analysis, the two key enablers were revealed to be 'developing clear & supportive regulations' and 'maintaining transparency & good governance' (Table 5.4).

5.9 Proximity Principle

The basis of this concept is to compel companies that create waste to deliver it to a location for processing that is within a particular distance from where it originated. The proximity principle (PP) can aid in stopping waste transport from one location to another to avoid and minimise levy liabilities. This notion was supported by submissions to The Environment and Communications References Committee

Table 5.3 Main challenges of global application of sustainable procurement policy when procuring PwRC

		1	2	3	4	5	6	7	8	9	10	11	12	13	14	15
Reference	Study context	Lack of long-term planning	Higher costs of PwRC	Lack of methods to measure sustainablity	Lack of knowledge about sustainable products	Obligation to procure from local suppliers to lower	Unavailability of certian PwRC	Lack of governement support	Misconception about PwRC production cost & time	Lack of motivation	Lack of supportive organisational culture	Lack of inclusion of PwRC in construction specification	Lack of mandatory imposition	Uncertainity about PwRC quality	Lack of operation	Lack of an organisational policy
DaCosta and DaMotta(2019)	Brazil	■	■		■	■		■			■					
Delmonica et al.(2018)	Brazil									■	■				■	
Pick (2017)	Canada											■		■		
Bougrain (2017)	France									■	■			■	■	
McMurrat et al. (2014)	Malaysia	■			■			■								
Islam et al. (2017)	Saudi Arabia										■					
Adell and Schaefe (2019)	South Korea					■	■							■		
Hasselbalch et al. (2015)	UN		■										■			■
United Nations Environmental Programme (2016)	Worldwide									■						

		16	17	18	19	20	21	22	23	24	25	26	27	28	29	30
Reference	Study context	Struggle to procurement behavioural change	Lack of market & market barriers	Traditional procurement procedure	Oversupply of PwRC	Lack of sustainable suppliers	Time pressure to complete projects, minimising the	Difficulty in assessign the non - environmental	The complexity of cost benefit analysis for PwRC	Lack of training and education for suppliers	Involvement of various stakeholder	Storage of relevant resources	Suppliers focus on price than quality	Desire to diversiy contractors	Economic uncertainty	Lack of aptitude for sustainable procurement by
DaCosta and DaMotta(2019)	Brazil									■						
Delmonica et al.(2018)	Brazil		■												■	
Pick (2017)	Canada										■		■	■		
Bougrain (2017)	France				■		■									
McMurrat et al. (2014)	Malaysia											■				
Islam et al. (2017)	Saudi Arabia															■
Adell and Schaefe (2019)	South Korea															
Hasselbalch et al. (2015)	UN		■	■												
United Nations Environmental Programme (2016)	Worldwide	■						■	■							

Table 5.4 Main drivers of global application of sustainable procurement when procuring PwRC

Reference	Study context	1 Developing supportive and clear	2 Project owners commitment to	3 Project team capability	4 Strategic partnership	5 Customer expectaion/pressure	6 Providing incentives and rewards	7 Strong technical expertise	8 Quality assurance to maintain the	9 Corporate reputation	10 Proper palnning and anticipating the	11 Improving sustainability	12 Setting a measurable target	13 Maintaining rransparency and	14 Mandatory application of	15 Conducting a cost-benefit analysis to	16 Potential to export locally manufactured	17 Screening out suppliers with a lack	18 Knowledge sharing among project	19 Effective coordination to	20 Procurement post review
Shooshtarian et al. (2022a.2022b,2022c)	Australia								■		■		■	■		■		■		■	■
Ershadi et al. (2021)	Australia																				
Pick (2017)	Canada		■											■							
Zhu et al. (2013)	China	■				■	■														
Bougrain (2017)	France												■								
Bohari et al. (2017)	Malaysia	■	■		■		■					■			■				■		
Sajjad et al. (2020)	New Zealand					■				■											
Ogunsanya et al. (2019)	Nigeria	■										■		■							
Gounden (2016)	South Africa	■			■			■									■				

(2018) (Laviano et al., 2017). A contentious example of PP implementation is occurring in NSW, in which companies that create waste are allowed to get rid of it in a location 150 km from its origin. The Waste Contractors and Recyclers Association of NSW argues that the adoption of a nation-wide PP would lead to better outcomes. In turn, the Law Council of Australia (LCA) argues that any nation-wide PP must consider the Constitution's Sect. 92 stipulation that commerce and trade between Australian states must be completely free. Meanwhile, the Waste Management Association of Australia submits that long-distance waste transportation must cease, stating that.

> …we do not agree with long-distance transportation; we actually agree there has to be a proximity principle in place to stop the excessive and unnecessary movement of waste across distances, particularly if there is the infrastructure in place. You can't actually invest and develop infrastructure if you haven't got certainty about what's coming through the front gate. In Europe you do have a proximity principle, so we need to solve how we do that. (p. 59)

Another point of contention tied to the PP is that authorities must be aware of the ramifications of the application policies like this. Some argue that the environmental advantages that arise from the transport of waste could be weakened through the PP's application. In terms of the creation of a C&D waste market on the domestic level, the exchange of waste materials (both recovered and unrecovered) across various regions is needed to support and maintain sectors and companies within the economy. Another motivator for the encouragement of adoption of the PP in Australia is China's National Sword Policy, which caused the waste and resource recovery sector in Australia to pledge that it will create an effective, long-term national market for the trade of Australian waste products. The key obstacle preventing the establishment of Energy from Waste (EfW) amenities and waste energy recovery services is a lack of appropriate waste feedstock. As such, until the benefits of waste management in proximity can be fully realised, Australia will need to allow the transportation of recovered/unrecovered waste materials over a reasonable distance.

5.10 Recycled Product Certification

Recycled product certification (RPC) offers proven data on a PwRC quality, safety, performance, and environmental standards. Recycled materials that have been awarded an RPC after stringent material evaluation and quality control can be perceived more positively and thus attain a higher uptake. Companies within a supply chain can gain advantages by using RPC to demonstrate their products attributes and features as required by potential purchasers (Shooshtarian & Maqsood, 2021). For instance, for builders, RPC is efficient management of information reducing the workload in a tender process (Fig. 5.8). Utilising RPC will allow builders to manage any risk that may arise from the usage of recycled materials. In this circumstance, clients and contractors will bear the risk equally. Such programmes also enable sustainable procurement policies in private and public sectors domestically and develop the

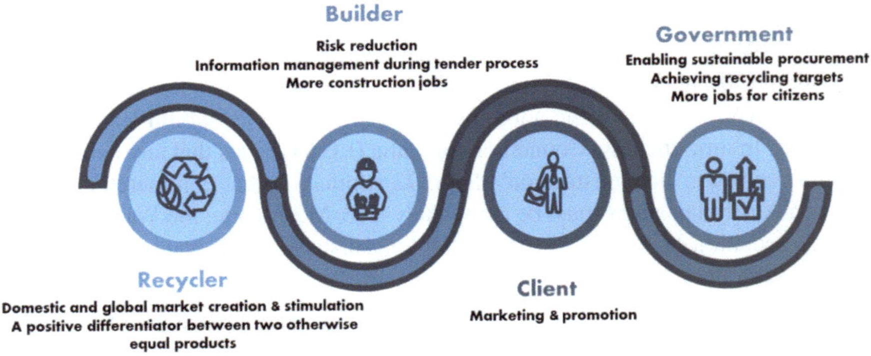

Fig. 5.8 Advantages of RPC for recycled products supply chain participants

overseas market for states that recognise certain certificates. For clients, it offers an option for those who are willing to work towards sustainable development (WSP Environmental and TRL Ltd., 2005). For suppliers and clients, RPC schemes will serve as a marketing tool to demonstrate the sustainability of their products and projects, respectively.

It is highlighted that classification of PwRC and aggregates into easily comprehensible categories, together with correct certification, facilitates future client purchases (Silva et al., 2017), as they will be able to acquire a product suited to its intended use (e.g. structural concrete may use a recycled aggregate of class A, whereas recycled aggregate of class D may be used in road construction for subgrade). Currently, there are two categories of marketable PwRC: non-certified and certified. Non-certified PwRC constitutes most of recycling plants' output. However, certification is crucial due to consumers' increasingly stringent needs for PwRC of a specified and assured quality. It is important to understand that product standards and eco-labels are distinct. Eco-labels are supposed to tell consumers about the environmental impact of their purchases. Most eco-labels do not give consumers real information about the product, but rather suggest that it is among the top 10–20% of its product category from an environmental standpoint (WSP Environmental and TRL Ltd., 2005).

5.10.1 Main Barriers to the Application of Product Certification

Despite the long history of product certifications (WSP Environmental and TRL Ltd., 2005), their full implementation in the BE sector is rather recent. Consequently, it is notable that there are few publications detailing the problems associated with their actual use. The review of pertinent literature found that the implementation of product certification procedures for recycled C&D waste materials is hindered by several obstacles. Figure 5.9 depicts that these include the process being costly, the costs

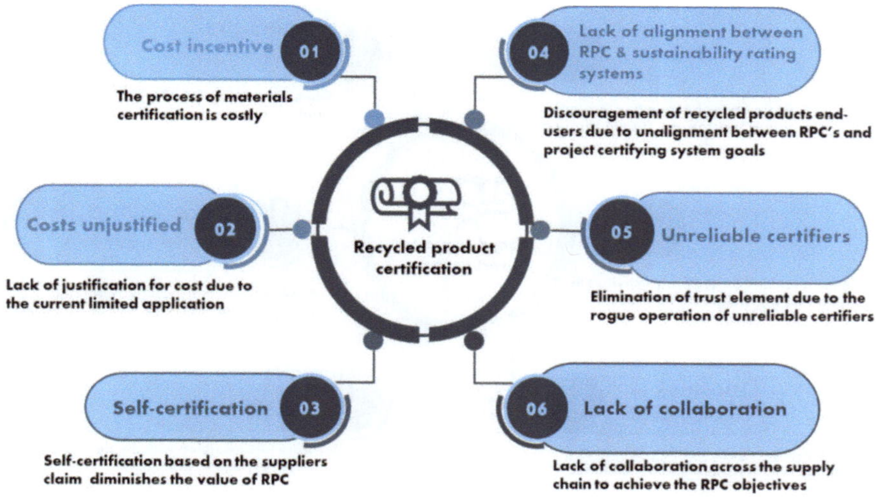

Fig. 5.9 Main barriers to the application of product certification for PwRC

being currently unjustified, self-certification, the unaligned RPC and sustainability rating systems, and the operation of unreliable certifiers.

A report by Equilibrium (2019) indicated that, in Australia, support testing of products containing recycled content is inadequate, meaning that such products are not independently validated. Uptake of product certification is low, likely due to the high costs involved. Further, the scarce opportunities to use these products are inadequately advertised, making it hard for stakeholders to justify the expense of certifying such products. Gaffar (2019) indicated that mandatory product certification expenses in the EU add to the price of the material and may negate any cost savings from reusing it. Self-certification is also utilised to establish the quality of recycled materials. Recyclers and providers of these materials often perform self-certification. Although it is less expensive, the certification does not provide the assurance that end-users may safely utilise recycled materials.

5.10.2 Main Enablers of Application of Product Certification

Six categories of RPC enablers are identified following a comprehensive literature review. These classifications are illustrated in Fig. 5.10. Effective quality monitoring and certification of PwRC by suppliers are required to instil and maintain stakeholder trust in the materials. However, this must be accompanied by more vigorous government action in the form of legislation and standardisation (Silva et al., 2017). The government also has a part to play in changing the general public's perceptions towards PwRC via promotional activities through the media. Ongoing training

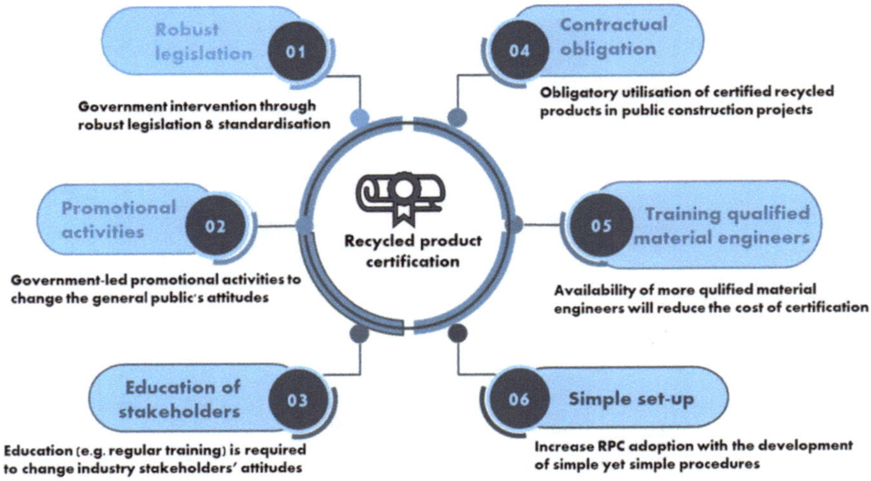

Fig. 5.10 Main enablers of the application of product certification for PwRC

offered by the government could contribute to shifting the attitudes of industry stakeholders (Bao & Lu, 2020). Furthermore, contractual obligation or otherwise client requirement to use certified PwRC is another enabler for the implementation of RPC (WSP Environmental and TRL Ltd., 2005). Currently, in the EU, it is illegal to use recycled content in construction materials that are not certified (Gaffar, 2019), which has resulted in more frequent uptake of these programmes in construction projects across the European territory. According to a report by WSP Environmental and TRL Ltd. (2005), the advantages of mandatory RPC include the eradication of poor actors, a better perception towards the data's reliability among all stakeholders, fewer opportunities for manipulation, and providing the credibility needed to cement the RPC's value.

The fifth category of enablers refers to the availability of qualified material engineers at reasonable prices. According to anecdotal evidence, these professionals are scarce and expensive to employ. By educating more individuals to become trained materials inspectors, not only will jobs be created, but audit costs will also be reduced. The final category addresses the simplicity of certification methods for PwRC. The waste recovery industry's interest in RPC will be piqued by certification processes that are simple yet effective and require a fair amount of administrative work (WSP Environmental and TRL Ltd., 2005).

5.11 Summary

A successful circular-built environment heavily relies on the existence of supportive policy frameworks. A policy framework that targets waste landfilling will ensure that these resources' value is retained in the economy in an ongoing way. This chapter analysed eight policies that share this goal and can be used by policymakers and the industry practitioners. Some of these policies currently exist in the Australian context.

The successful implementation of these policies requires education provided for all stakeholders concerned with C&D waste management. Furthermore, these stakeholders should enjoy a level playing field in order to willingly contribute to the successful implementation of these policies. Hence, the government must take the lead not just in creating and applying these policies, but also to offer educational opportunities and a level playing field.

This chapter contributes to the theory and practice of CE in the BE sector. The effectiveness of these proposed policies can be further investigated in different contexts. Furthermore, where the policies are being imposed in the Australian context, the suggestions and the context-specific context lessons learnt in this chapter can be used in risk management planning for other contexts, i.e. other locations, industries, or waste streams.

References

Adell, A., & Schaefe, B. (2019). *Green public procurement in the republic of korea: A decade of progress and lessons learned.* KEITI.

Alhola, K., Ryding, S.-O., Salmenperä, H., & Busch, N. J. (2019). Exploiting the potential of public procurement: Opportunities for eircular economy. *Journal of Industrial Ecology, 23*(1).https://doi.org/10.1111/jiec.12770

Anastasiades, K., Blom, J., Buyle, M., & Audenaert, A. (2020). Translating the circular economy to bridge construction: Lessons learnt from a critical literature review. *Renewable and Sustainable Energy Reviews, 117*, 109522. http://www.elsevier.com/locate/rser, https://doi.org/10.1016/j.rser.2019.109522

Australian Government. (2020, November). *Sustainable procurement guide: A practical guide for commonwealth entities* Department of Agriculture (Ed.). CC BY 4.0.

Australian ISO 20400 Committee. (2018). *Building the business case for sustainable procurement in Australia—Guidance.*

Bao, Z., & Lu, W. (2020). Developing efficient circularity for construction and demolition waste management in fast emerging economies: Lessons learned from Shenzhen, China. *Science of the Total Environment, 724*, 138264. https://doi.org/10.1016/j.scitotenv.2020.138264

Benachio, G. L. F., Freitas, M. C. D., & Tavares, S. F. (2020). Circular economy in the construction industry: A systematic literature review. *Journal of Cleaner Production, 260,* 121046. https://doi.org/10.1016/j.jclepro.2020.121046

Bohari, A. A. M., Skitmore, M., Xia, B., & Teo, M. (2017). Green oriented procurement for building projects: Preliminary findings from Malaysia. *Journal of Cleaner Production, 148,* 690–700. https://doi.org/10.1016/j.jclepro.2017.01.141

Bougrain, F. (2017). *Impact of green public procurement on the market of recycled concrete*. Delft University of Technology.
Brammer, S., & Walker, H. (2011). Sustainable procurement in the public sector: An international comparative study. *International Journal of Operations & Production Management, 31*(4), 452–476. https://doi.org/10.1108/01443571111119551
Caldera, S., Mohamed, S., & Feng, Y. (2022). Evaluating the Covid-19 impacts on sustainable procurement: Experiences from the Australian built environment sector. *Sustainability, 14*(7), 4163. https://doi.org/10.3390/su14074163
Caldera, S., Ryley, T., & Zatyko, N. (2020). Enablers and barriers for creating a marketplace for construction and demolition waste: A systematic literature review. *Sustainability, 12*(23), 9931. https://doi.org/10.3390/su12239931
Chileshe, N., Jayasinghe, R. S., & Rameezdeen, R. (2019). Information flow-centric approach for reverse logistics supply chains. *Automation in Construction, 106*, 102858. https://doi.org/10.1016/j.autcon.2019.102858
Commonwealth of Australia. (2013). *Sustainable procurement guide*.
Craighill, A., & Powell, J. C. (1999). *A lifecycle assessment and evaluation of construction and demolition waste*.
Da Costa, B. B., & Da Motta, A. L. T. (2019). Key factors hindering sustainable procurement in the brazilian public sector: A delphi study. *International Journal of Sustainable Development and Planning, 14*(2), 152–171.
Davis, P., Simon, L., Sher, W., Tang, P., & Newaz, M. T. (2019). *Key solutions for construction and demolition (c&d) waste management in NSW, Australia* (pp. 612–628). CQ University.
Delmonico, D., Jabbour, C. J. C., Pereira, S. C. F., de Sousa Jabbour, A. B. L., Renwick, D. W. S., & Thomé, A. M. T. (2018). Unveiling barriers to sustainable public procurement in emerging economies: Evidence from a leading sustainable supply chain initiative in Latin America. *Resources, Conservation and Recycling, 134*, 70–79. https://doi.org/10.1016/j.resconrec.2018.02.033
Dubois, M., de Graaf, D., & Thieren, J. (2016). *Exploration of the role of extended producer responsibility for the circular economy in the Netherlands*. London, UK.
Elmualim, A., Mostafa, S., Chileshe, N., & Rameezdeen, R. (2018). *Construction and the circular economy: Smart and industrialised prefabrication. In: Unmaking waste in production and consumption: Towards the circular economy*. Emerald Publishing Limited, Bingley, pp. 323–336 https://www.emerald.com/insight/content/doi/10.1108/978-1-78714-619-820181025/full/html
EPA Victoria. (2007). *Impact of landfill levy on the steel recycling sector in victoria*.
Equilibrium. (2019). *Review of standards and specifications for recycled content products* (p. 20). The Australian Department of Agriculture, Water and Environment.
Ershadi, M., Jefferies, M., Davis, P., & Mojtahedi, M. (2021). Achieving sustainable procurement in construction projects: The pivotal role of a project management office. *Construction Economics and Building, 21*(1), 45–64. https://doi.org/10.5130/AJCEB.v21i1.7170
European Parliament. (2022). *Circular economy: Definition, importance and benefits*. News: European Parliament.
Fini, A. A. F., & Forsythe, P. (2020). Barriers to reusing and recycling office fit-out: An exploratory analysis of demolition processes and product features. *Construction Economics and Building, 20*(4), 42–62.
Forghani, R., W. Kanjanabootra, S. S., & Totoev, Y. (2017). *Consequence of waste levy revocation: Case study Queensland, Australia*. Sydney University of Technology.
Forslind, K. (2005). Implementing extended producer responsibility: The case of sweden's car scrapping scheme. *Journal of Cleaner Production, 13*(6), 619–629.
Gaffar, S. (2019). How we can recycle more buildings. *The Conversation*.
Glanville, K., & Chang, H.-C. (2015). Mapping illegal domestic waste disposal potential to support waste management efforts in Queensland, Australia. *International Journal of Geographical Information Science, 29*(6), 1042–1058.

Gounden, K. (2016). *Factors influencing sustainable procurement within the private and public sector in South Africa.* University of Pretoria.

Guerra, B. C., & Leite, F. (2021). Circular economy in the construction industry: An overview of united states stakeholders' awareness, major challenges, and enablers. *Resources, Conservation and Recycling, 170*, 105617. https://doi.org/10.1016/j.resconrec.2021.105617

Guggemos, A. A., & Horvath, A. (2003). Strategies of extended producer responsibility for buildings. *Journal of Infrastructure Systems, 9*(2), 65–74.

Gupt, Y., & Sahay, S. (2015). Review of extended producer responsibility: A case study approach. *Waste Management & Research, 33*(7), 595–611.

Hasselbalch, J., Costa, N., & Blecken, A. (2015). Investigating the barriers to sustainable procurement in the united nations. In M. Klumpp, S. de Leeuw, A. Regattieri, & R. de Souza (Eds.), *Humanitarian logistics and sustainability* (pp. 67–86). Springer.

He, Q., Shi, S., & Hu, M. (2014). *Exploring a long-term mechanism of construction and demolition waste recycling: A case of Chongqing* (pp. 1109–1120). Springer.

Hyder Consulting Pty Ltd. (2011). *Construction and demolition waste status report: Management of construction and demolition waste in Australia.* Department of Energy and Environment, Department of Sustainability, Environment, Water, Population and Communities.

Isenhour, C., Blackmer, T., Wagner, T., Silka, L., & Peckenham, J. (2016). Moving up the waste hierarchy in maine: Learning from "best practice" state-level policy for waste reduction and recovery. *Maine Policy Review, 25*(1), 15. https://doi.org/10.53558/VZUB1903

Islam, M. M., Murad, M. W., McMurray, A. J., & Abalala, T. S. (2017). Aspects of sustainable procurement practices by public and private organisations in Saudi Arabia: An empirical study. *International Journal of Sustainable Development & World Ecology, 24*(4), 289–303. https://doi.org/10.1080/13504509.2016.1209794

Jayasinghe, R. S., Rameezdeen, & Chileshe, N. (2018). *Waste management practices in Australia: Comparison of strategies.* Proceedings of 22nd International Conference on Advancement of Construction Management and Real Estate. Melbourne, Australia.

Kabirifar, K., Mojtahedi, M., & Wang, C. C. (2021). A systematic review of construction and demolition waste management in Australia: Current practices and challenges. *Recycling, 6*(2), 34. https://doi.org/10.3390/recycling6020034

Kirchherr, J., Reike, D., & Hekkert, M. (2017). Conceptualizing the circular economy: An analysis of 114 definitions. *Resources, Conservation and Recycling, 127*, 221–232. https://doi.org/10.1016/j.resconrec.2017.09.005.

Langrová, V. (2002). *Comparative analysis of EPR programmes for small consumer batteries: Case study of the Netherlands, Switzerland and Sweden.*

Laviano, H., Jordan, B., Monica, T., & Noor, D. (2017). *Waste and recycling industry in Australia: A submission to the senate inquiry.*

Lindhqvist, T. (2000). *Extended producer responsibility in cleaner production: Policy principle to promote environmental improvements of product systems.* Lund University.

Lingegård, S., Olsson, J. A., Kadefors, A., & Uppenberg, S. (2021). Sustainable public procurement in large infrastructure projects—policy implementation for carbon emission reductions. *Sustainability, 13*(20), 11182. Available from https://www.mdpi.com/2071-1050/13/20/11182

Local Government NSW. (2021). *Sustainable procurement guide: For local government in NSW.*

McMurray, A. J., Islam, M. M., Siwar, C., & Fien, J. (2014). Sustainable procurement in Malaysian organizations: Practices, barriers and opportunities. *Journal of Purchasing and Supply Management, 20*(3), 195–207. https://doi.org/10.1016/j.pursup.2014.02.005

Nahman, A. (2010). Extended producer responsibility for packaging waste in South Africa: Current approaches and lessons learned. *Resources, Conservation and Recycling, 54*(3), 155–162. https://doi.org/10.1016/j.resconrec.2009.07.006

National Waste Report. (2020). *National waste report.* Department of Climate Change, Energy the Environment and Water.

Newaz, M. T., Davis, P., Sher, W., & Simon, L. (2020). Factors affecting construction waste management streams in Australia. *International Journal of Construction and Management, 22*(13), 2625–2633.https://doi.org/10.1080/15623599.2020.1815122

Ogunsanya, O. A., Aigbavboa, C. O., Thwala, D. W., & Edwards, D. J. (2019). Barriers to sustainable procurement in the nigerian construction industry: An exploratory factor analysis. *International Journal of Construction and Management*, 1–12. Available from https://doi.org/10.1080/15623599.2019.1658697

Park, J., & Tucker, R. (2017). Overcoming barriers to the reuse of construction waste material in Australia: A review of the literature. *International Journal of Construction Management, 17*(3), 228–237. https://doi.org/10.1080/15623599.2016.1192248

Peck, P., Peck, J., Richter, L., & Delaney, K. (20200. *Circular economy—Sustainable materials management: A compendium by the international institute for industrial environmental economics (IIIEE) at lund university.* The International Institute for Industrial Environmental Economics (IIIEE).

Pick, F. (2017). *Waste as a resource: Tools for construction and demolition waste management.* Metro Vancouver in conjunction with the National Zero Waste Council.

Rameezdeen, R., Chileshe, N., Hosseini, M., & Lehmann, S. (2016). A qualitative examination of major barriers in implementation of reverse logistics within the south Australian construction sector. *International Journal of Construction Management, 16*(3), 185–196. https://doi.org/10.1108/SASBE-05-2013-0028

Ratnasabapathy, S., Alashwal, A., & Perera, S. (2021). Exploring the barriers for implementing waste trading practices in the construction industry in Australia. *Built Environment Project and Asset Management.* 11(4), 559-576. https://doi.org/10.1108/BEPAM-04-2020-0077

Ratnasabapathy S., Perera, S., & Alashwal, A. (2019a). *A review of construction waste data and reporting systems used in Australia* (pp. 396–405). CQ University.

Ratnasabapathy, S., Perera, S., & Alashwal, A. (2019b). A review of smart technology usage in construction and demolition waste management. In Y. G. Sandanayake, S. Gunatilake, & Waidyasekara (Eds.), *Proceedings of the 8th World Construction Symposium* (pp. 45–55).

Ratnasabapathy, S., Perera, S., Alashwal, A. M., & Lord, O. (2019c). Assessment of waste generation and diversion rates in residential construction projects in Australia. In: *CIB world building congress: Constructing smart cities.* The Hong Kong Polytechnic University.

Sajjad, A., Eweje, G., & Tappin, D. (2020). Managerial perspectives on drivers for and barriers to sustainable supply chain management implementation: Evidence from News Zealand. *Business Strategy and the Environment, 29*(2), 592–604.

Sanchez, A. X., Lehtiranta, L., Hampson, K. D., & Kenley, R. (2014). Evaluation framework for green procurement in road construction. *Smart and Sustainable Built Environment, 3*(2): 153–163. https://doi.org/10.1108/SASBE-05-2013-0028

SBEnrc P.175. (2021). *Project 1.75 creation and stimulation of end markets for construction and demolition waste.* Perth, Australia.

Shanoff, B. (1996). Proposed recycling rules create obstacles. *World Wastes, 39*(6), 14–17.

Shooshtarian, S., & Maqsood, T. (2021). How does product certification help construction and demolition waste resource recovery?. In *The fifth estate.*

Shooshtarian, S., Caldera, S., Maqsood, T., & Ryley, T. (2020a). Using recycled construction and demolition waste products: A review of stakeholders' perceptions, decisions, and motivations. *Recycling, 5*(4), 31. https://doi.org/10.3390/recycling5040031

Shooshtarian, S., Caldera, S., Maqsood, T., Ryley, T., & Khalfan, M. (2021a). An investigation into challenges and opportunities in the Australian construction and demolition waste management system. *Engineering, Construction and Architectural Management, 29*(10), 4313–4330. https://doi.org/10.1108/ECAM-05-2021-0439

Shooshtarian, S., Caldera, S., Maqsood, T., Ryley, T., Wong, P. P. & Zaman, A.(2022a). Analysis of factors influencing the creation and stimulation of the Australian market for recycled construction and demolition waste products. *Sustainable Production and Consumption, 34*, 163–176. https://doi.org/10.1016/j.spc.2022.09.005

Shooshtarian, S., Khalfan, M., Maqsood, T., Wong, P. S., & Yang, R. J. (2020b). Market development for construction and demolition waste stream in Australia. *Journal of Construction Engineering, Management & Innovation, 3*(3), 220–231. https://doi.org/10.31462/jcemi.2020.03220231

Shooshtarian, S., Le, T., Feng, Y., & Bettini, L. (2022b). *Analysis of sustainable procurement in supplying recycled content: A case study in western Australia.* RMIT University.

Shooshtarian, S., Maqsood, T., Barrett, C., Wong, S. P., Yang, J. R., & Khalfan, M. (2020c). Opportunities to reduce brick waste disposal. In A. Ghaffarianhoseini (Ed.), *The architectural science association.*

Shooshtarian, S., Maqsood, T., Caldera, S., & Ryley, T. (2022c). Transformation towards a circular economy in the Australian construction and demolition waste management system. *Sustainable Production and Consumption, 30*, 89–106. https://doi.org/10.1016/j.spc.2021.11.032

Shooshtarian, S., Maqsood, T., Khalfan, M., Wong, S. P., & Yang, J. R. (2019a). *Managing construction and demolition (C&D) waste in Australia.* CIB World Building Congress 2019. The Hong Kong Polytechnic University.

Shooshtarian, S., Maqsood, T., Khalfan, M., Yang, J. R., & Wong, S. P. P. (2020d). Landfill levy imposition on construction and demolition waste: Australian stakeholders' perceptions. *Sustainability, 12*(11), 1–15. https://doi.org/10.3390/su12114496

Shooshtarian, S., Maqsood, T., Wong, P., Khalfan, M., & Yang, R. (2019b). Review of energy recovery from construction and demolition waste in Australia. *Journal of Construction Engineering, Management & Innovation, 2*(3), 112–130. https://doi.org/10.31462/jcemi.2019.031 12130

Shooshtarian, S., Maqsood, T., Wong, P. S., Khalfan, M., & Yang, R. J. (2021b). Extended producer responsibility in the Australian construction industry. *Sustainability, 13*(2), 620. https://doi.org/10.3390/su13020620

Shooshtarian, S., Maqsood, T., Wong, S. P., Yang, J. R., & Khalfan, M. (2020e). Review of waste strategy documents in Australia: Analysis of strategies for construction and demolition waste. *International Journal of Environmental Technology and Management, 23*(1), 1–21. https://doi.org/10.1504/IJETM.2020.110147

Silva, R., De Brito, J., & Dhir, R. K. (2017). Availability and processing of recycled aggregates within the construction and demolition supply chain: A review. *Journal of Cleaner Production, 143*, 598–614. https://doi.org/10.1016/j.jclepro.2016.12.070

Srour, I., Chong, W. K., & Zhang, F. (2012). Sustainable recycling approach: An understanding of designers' and contractors' recycling responsibilities throughout the life cycle of buildings in two us cities. *Sustainable Development, 20*(5), 350–360. https://doi.org/10.1002/sd.493

Tam, V.W.-Y., & Lu, W. (2016). Construction waste management profiles, practices, and performance: A cross-jurisdictional analysis in four countries. *Sustainability, 8*(2), 190. https://doi.org/10.3390/su8020190

Tam, V. W., Kotrayothar, D., & Loo, Y.-C. (2009). On the prevailing construction waste recycling practices: A south east Queensland study. *Waste Management & Research, 27*(2), 167–174. https://doi.org/10.1177/0734242X08091864

Thorpe, B., & Kruszewska, I. (1999). *Strategies to promote clean production: Extended producer responsibility. Clean production action.* Institute for Sustainable Futures, UTS Appendix A.

Udawatta, N., Zuo, J., Chiveralls, K., & Zillante, G. (2015a). Attitudinal and behavioural approaches to improving waste management on construction projects in Australia: Benefits and limitations. *International Journal of Construction Management, 15*(2), 137–147. https://doi.org/10.1080/15623599.2015.1033815

Udawatta, N., Zuo, J., Chiveralls, K., & Zillante, G. (2015b). Improving waste management in construction projects: An Australian study. *Resources, Conservation and Recycling, 101*, 73–83. https://doi.org/10.1016/j.resconrec.2015.05.003

UK's HM Revenue & Customs. (2018). *Aggregate levy guidance.* U. G. Her Majesty's Revenue and Customs (Ed.).

United Nations Environmental Programme. (2016). *Sustainable public procurement.*

Widmer, R., Oswald-Krapf, H., Sinha-Khetriwal, D., Schnellmann, M., & Böni, H. (2005). Global perspectives on e-waste. *Environmental Impact Assessment Review, 25*(5), 436–458. https://doi.org/10.1016/j.eiar.2005.04.001

WSP Environmental and TRL Ltd. (2005). *Evaluating options for declaring recycled content in construction products and projects.*

Wu, H., Zuo, J., Yuan, H., Zillante, G., & Wang, J. (2020). Cross-regional mobility of construction and demolition waste in Australia: An exploratory study. *Resources, Conservation and Recycling, 156*, 104710. https://doi.org/10.1016/j.resconrec.2020.104710

Xu, J., Ye, M., Lu, W., Bao, Z., & Webster, C. (2021). A four-quadrant conceptual framework for analyzing extended producer responsibility in offshore prefabrication construction. *Journal of Cleaner Production, 282*, 124540. https://doi.org/10.1016/j.jclepro.2020.124540

Zhang, C., Hu, M., Yang, X., Miranda-Xicotencatl, B., Sprecher, B., Di Maio, F., Zhong, X., & Tukker, A. (2020). Upgrading construction and demolition waste management from downcycling to recycling in the Netherlands. *Journal of Cleaner Production, 266*, 121718. https://doi.org/10.1016/j.jclepro.2020.121718

Zhao, X., Webber, R., Kalutara, P., Browne, W., & Pienaar, J. (2021). Construction and demolition waste management in Australia: A mini-review. *Waste Management and Research, 40*(1), 34–46.https://doi.org/10.1177/0734242X211029446

Zhu, Q., Geng, Y., & Sarkis, J. (2013). Motivating green public procurement in china: An individual level perspective. *Journal of Environmental Management, 126*, 85–95. https://doi.org/10.1016/j.jenvman.2013.04.009

Chapter 6
Electronic Waste Appreciation—Strategies Targeting the Circular Economy

Mirian Paula dos Santos, Ivan Aldaya, Lucio Cardozo-Filho, Rafael A. Penchel, Denise Crocce Romano Espinosa, and José Augusto de Oliveira

6.1 What is Waste of Electric and Electronic Equipment?

Electronic devices' lifetime is becoming shorter due to the accelerated technological evolution, which has led to a subsequent increase in the amount of the waste of electric and electronic equipments (WEEE) that has been generated by both residential and industrial consumers. These wastes represent a wide class of devices ranging from toys to complex industrial equipments. Some authors, like Forti et al. (2020) and Habib et al. (2022), proposed a division into the following six categories: cooling and freezing equipment, screens and monitors, lamps, large household appliance, small household appliance, and small information technology (IT) equipments, encompassing more than 50 subclasses (Table 6.1). The Global E-waste monitor (Forti et al., 2020) ascertains that only 17.4% of the WEEE generated in 2019 was properly collected and recycled, whereas the remaining 44.3 million metric tons (82.6%) was disposed of in unsuitable ways. Furthermore, the authors predict the generation of 74.7 million metric tons of e-waste for 2030. Asia, the USA, and the European Union are the largest producers of WEEE, with Europe generating 16.2 kg per capita, followed by the USA (13.3 kg per capita) and Asia (5.6 kg per capita).

M. P. dos Santos (✉) · I. Aldaya · L. Cardozo-Filho · R. A. Penchel · J. A. de Oliveira
Center for Advanced and Sustainable Technologies—CAST, São Paulo State University (UNESP), Campus of São João da Boa Vista, Sao Joao da Boa Vista, Sao Paulo, Brazil
e-mail: mirian.santos@unesp.br

L. Cardozo-Filho
Department of Chemical Engineering, State University of Maringá, Maringá, Paraná, Brazil

D. C. R. Espinosa
Department of Chemical Engineering, Polytechnic School of the University of Sao Paulo (USP), Sao Paulo, Brazil

S. A. Bandh and F. A. Malla (eds.), *Waste Management in the Circular Economy*,
https://doi.org/10.1007/978-3-031-42426-7_6

Table 6.1 Characteristics and examples of each WEEE category

CATEGORY	EQUIPMENT
01	TEMPERATURE EXCHANGE EQUIPMENT - TEE *Refrigerators, freezers, and air conditioners.*
02	SCREENS, MONITORS AND EQUIPMENT THAT CONTAIN SCREENS WITH A SURFACE GREATER THAN 100 CM² – S&M *Televisions, monitors, laptops, notebooks, and tablets.*
03	LAMPS *Fluorescent lamps, flash lamps, and LED lamps.*
04	LARGE HOUSEHOLD APPLIANCE CONTAINING ONE EXTERNAL DIMENSION EXCEEDING 50 CM - LE *Washing machines, clothes dryers, dishwashing machines, electric stoves, large printing machines, copying equipment, and photovoltaic panels.*
05	SMALL HOUSEHOLD APPLIANCE PRESENTING EXTERNAL DIMENSION SHORTER THAN 50 CM - SE *Electronic and electrical tools, toys, and medical devices. Such as video game consoles, electric stoves, and electric screwdrivers.*
06	SMALL IT AND TELECOMMUNICATION EQUIPMENT PRESENTING EXTERNAL DIMENSION SHORTER THAN 50 CM - IT *Mobile phones, global positioning system (GPS) devices, pocket calculators, routers, personal computers, printers, and telephones.*

6.2 Main Composition of Waste of Electric and Electronic Equipment

Cell phones, computers, and other IT equipment account for around 30% of the constituents of e-waste, refrigerators and televisions represent another 30%, and monitors alone compose another 10% (Ankit et al., 2021). According to Liu et al. (2022) in China, 60.8% weight of WEEE was composed of common household appliances, including personal desktop computers, air conditioners, washing machines, refrigerators, and televisions in 2010. This percentage experienced a significant rise, reaching 77.3% in 2018. In 2019, the highest fraction of WEEE was composed of small devices, followed by large equipment and temperature exchange equipment, representing 17.4, 13.1, and 10.8 Mt, respectively (Forti et al., 2020).

Rapid obsolescence is observed for personal and domestic equipments such as televisions, smartphones, notebooks, and toys/games with a prospect of doubling by 2045 (Parajuly et al., 2020). In this context it is worth noting that from here to there we face some changes in the consumption trends, Industry, Internet of Things (IoT), and Artificial Intelligence (AI) technologies are becoming more accessible offering uncountable smart devices. Figure 6.1 shows how the WEEE generation is envisaged to increase in the next years caused by the popularization of smart household

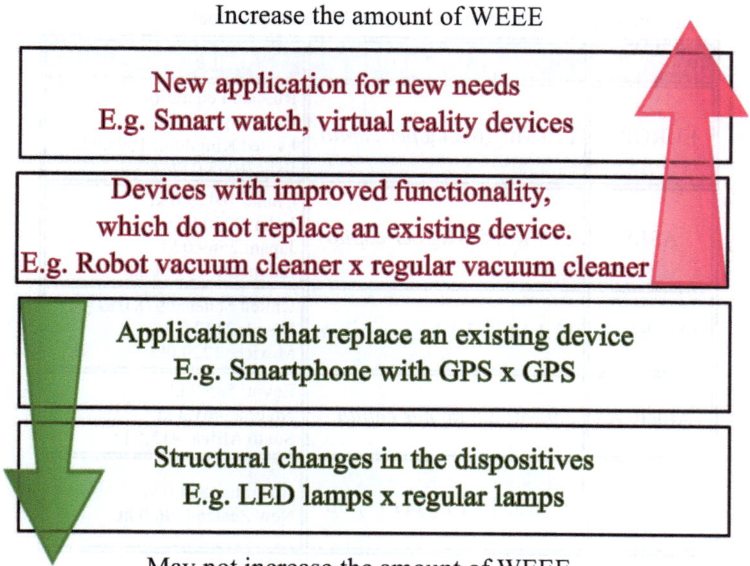

Fig. 6.1 Trends in the production of the WEEE. *Source* Adapted from Bacher et al. (2017)

appliances, such as televisions, refrigerators, smartphones, the development of novel dispositive for virtual and augmented reality based on sensors and wearables, and the new application of radio frequency identification (RFID) systems. For example, it was seen that the sale of drones increased from 2.5 million units in 2016 to 7 million units in 2020 in the USA. On the other hand, other improvements may contribute to decrease the e-waste propagation (Bacher et al., 2017).

Electromobility is an entire sector that is not included in the electric and electronic equipment (EEE), but cars, bikes, trucks, and buses are responsible for the employment of a vast number of lithium-ion batteries. Even vehicles that do not have electrical traction are being increasingly supplied with electric and electronic components. Safety and navigation systems, entertainment stations, and on-board computers are a few examples of devices that use screens, microcontrollers, and sensors that are also present in the EEE. These features push the automotive industry closer to the WEEE chain supply because these devices should follow the same disposal or collection route of the e-waste, being addressed with the same End-of-Life (EoL) solutions (Bacher et al., 2017).

Table 6.2 shows the country-wise and continent-wise generations of WEEE.

Despite their variety, the electronic equipment is mostly composed of different materials like plastics (in the structural part), glass (screens and displays), metals (ferrous and non-ferrous), and other materials with ferrous metals tending to constitute the largest proportion of a typical WEEE in size and weight (Shittu et al., 2021). According to Williams (2016), WEEE contains about 40–60% metal, with iron/steel,

Table 6.2 WEEE generation by continent in 2019, Mt = million metric tons

Continent	WEEE generation	Main producers
EUROPE	120 Mt; 16.2 kg *per capita*	Russian Federation: 1631.0 kt Germany: 1607.0 kt United Kingdom: 1598.0 kt Italy: 1063.0 kt
ASIA	24.9 Mt; 5.6 kg *per capita*	China: 10129.0 kt India: 3230.0 kt Japan: 2569.0 kt Indonesia: 1618.0 kt
AMERICA	13.1 Mt; 13.3 kg *per capita*	United States: 6918.0 kt Brazil: 2143.0 kt Mexico: 1220.0kt
AFRICA	2.9 Mt; 2.5 kg *per capita*	Egypt: 585.8 kt Nigeria: 461.3 kt South Africa: 415.5 kt
OCEANIA	0.7 Mt; 16.1 kg *per capita*	Australia: 554.0 kt New Zealand: 96.0 kt

Source Forti et al. (2020)

copper, and aluminum being the main metallic components. The constituents of a typical sample of WEEE are shown in Fig. 6.2.

There are different compositions of WEEE with other metals reported in the literature, as for example de Souza et al. (2022), identified Al, Fe, Sn, and Mg in significant amounts. The National Conference of State Legislatures (NCSL, 2018) of the USA pointed out that the production of electronic devices requires a significant number of resources—metals, plastics, and glass. For instance, the production of one

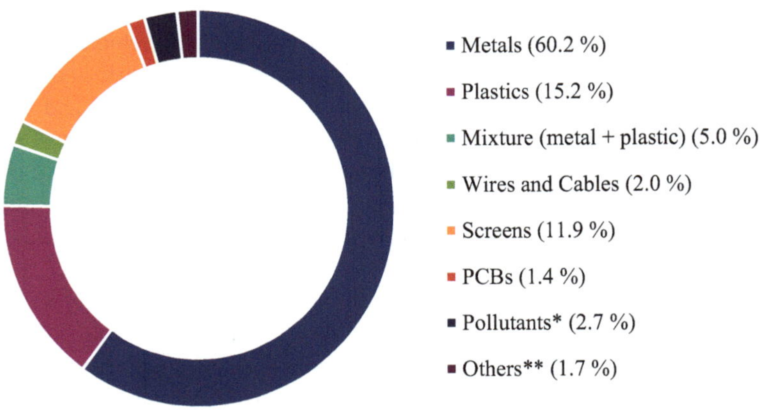

Fig. 6.2 Overall composition of the WEEE. *Source* Adapted from Ramprasad et al. (2022). *Pollutants encompass phosphates, nitrates, and other organics. **Others include 55 different metals that are Cu, Zn, Ni, Pb, Hg, As, Au, Ag, and rare earth elements

desktop computer takes at least 240 kg of fossil fuels, 22 kg of chemicals, and 1.5 tons of water. Another example, the manufacture of a liquid–crystal display (LCD) television requires 34% of its weight of iron, 30% plastics, 22% glass, 6% electronics, 5% aluminum, 2% copper, and 1% other (Rajesh et al., 2022).

6.2.1 Plastics

Plastics are responsible for 20–30% of the total e-waste stream. They take up differentiated proportions of WEEEs, accounting for a range ranging from 3.5% to 45.0% weight (Liu et al., 2022). Generally, plastics are employed in the structural part of the devices, but some types of plastics are also used with defined functions based on their mechanical and thermal properties. Several studies reported the composition and application of plastic in the e-waste (Bacher et al., 2017; Cardamone et al., 2021; Lahtela et al., 2022; Liu et al., 2022; Peeters et al., 2015) as summarized in Fig. 6.3. As there are many components, it is difficult to define an exact proportion of different polymers in the WEEE. Even within a specific category, the composition of the devices can be similar or quite diversified.

The main concerns about plastic recovering of WEEEs is, the brominated flame retardant impregnation, added to devices to meet the safety specifications and, the metallic contamination due to characteristic improvement. These aspects hinder the re-insertion of the plastics in the productive chain because according to some European Union (EU) directives, such as Restriction of Hazardous Substances (RoHS), plastics are containing more than 0.1 wt% of flame retardants or metallic content, and more than 0.01 wt% of cadmium. These plastics are hence suitable for manufacturing any products, including EEEs, and the compliant recyclers typically adopt the controlled incineration to avoid the release of dioxins and furans. Besides the flame retardants and metals, several additives, such as halogen rich pigments, stabilizers, and plasticizers, are often dangerous for the environment and for the human health (Forti et al., 2020; Lahtela et al., 2022; Stenvall et al., 2013).

Another serious issue that has been recently reported is the presence of microplastics, which are small pieces of plastics ranging from 100 nm to 5 mm. According to their provenience, they have been classified into primary ones, produced in this size range for a specific purpose, and secondary ones, arising from the degradation of plastics by environmental circumstances (Rodrigues et al., 2022). An increasing concern owing to the presence of microplastics in marine waters and other natural ecosystems is their potential long-term adverse impacts on marine life and, eventually, human health (Liu et al., 2022). Due to their minute size, the microplastics have particular features at the surface that confers the capability to adsorb heavy metals and other organic pollutants from the soil and water, thus spreading the contaminants not only through the ecosystems, but also to transport the contaminants to living tissues. According to de Goswami et al. (2020), microplastics can affect the system soil–plant, soil–microorganisms and are able to accumulate in the soil detrital food chain. In a similar way, the aquatic biota is also affected by microplastics, since

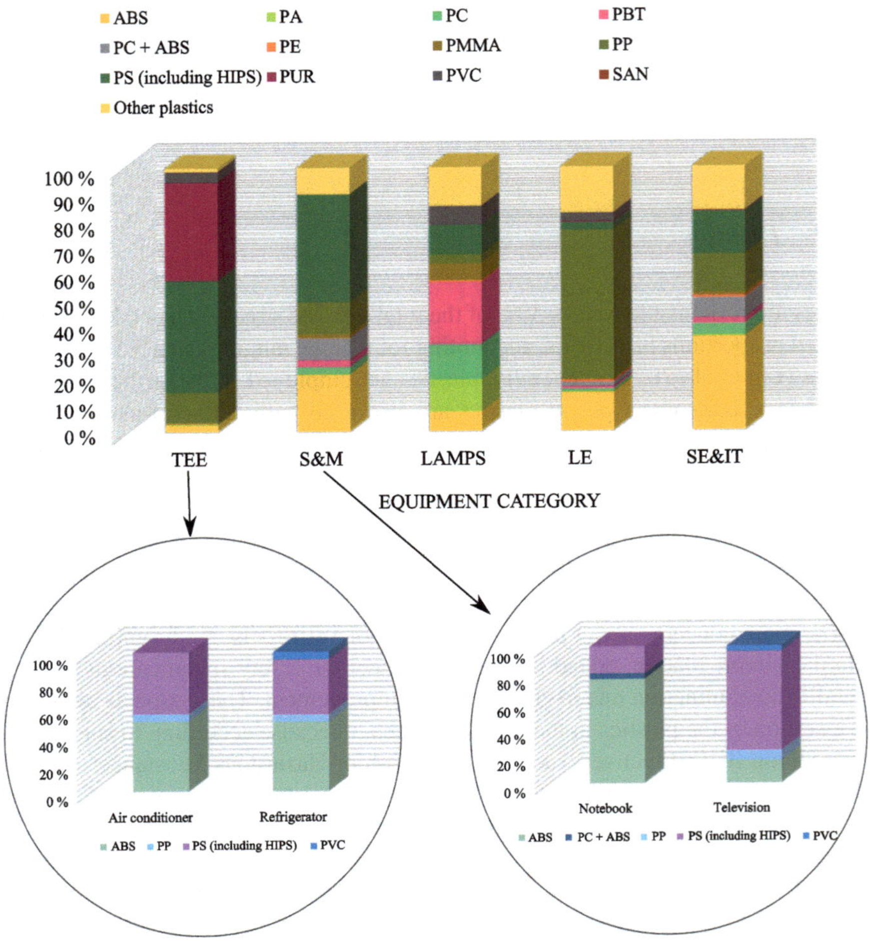

Fig. 6.3 Plastic composition related to each WEEE category. ABS, Acrylonitrile–Butadiene–Styrene; PA, Polyamide; PC, Polycarbonate; PBT, Polybutylene; Terephthalate; PC + ABS, Blends of Polycarbonate and Acrylonitrile–Butadiene–Styrene; PE, Polyethylene; PMMA, Poly(Methyl Methacrylate); PP, Polypropylene; PS, Polystyrene; HIPS, High Impact Polystyrene; PUR, Polyurethane; PVC, Poly(vinyl Chloride); SAN, Styrene–Acrylonitrile Resin. *Source* Adapted from Cardamone et al. (2021) and Liu et al. (2022)

they can lead to the accumulation of harmful plastic additives, such as plasticizers, flame retardants, pigments, and stabilizers (Chai et al., 2020; Rodrigues et al., 2022). Considering that microplastics have a potential for bioaccumulative effect along trophic levels (Goswami et al., 2020; Pelamatti et al., 2022), the severity of their adverse impacts becomes even more critical.

6.2.2 Glasses

The glass content in the WEEE arises mainly from screens, monitors, and photovoltaic panels. Among other glass sources, the obsolete Cathode-Ray Tubes (CRTs) and, more recently, the Flat Panel Displays, as well as the photovoltaic panels, present high aggregated value because they are impregnated with precious or critical metals. The proper collection and management of this machinery, therefore, are attracting an increasing interest. In this context, these materials are generally recycled through mechanical and thermal processes, but new alternatives are being reported in the literature (Arduin et al., 2020; Mostaghel & Samuelsson, 2010; Padoan et al., 2019).

6.2.3 Valuable Metals

The electrical and electronic devices contain at least 69 chemical elements, including precious metals (e.g., gold, silver, copper, platinum, palladium, ruthenium, rhodium, iridium, and osmium), Critical Raw Materials (CRM) (e.g., cobalt, indium, germanium, bismuth, and antimony), and non-critical metals such as aluminum and iron (Forti et al., 2020). CRM are those for which the supply chain is under risk, and economical aspects are greater than the other raw materials. These materials tend to leak from the official collection channels and are rarely recovered properly from the WEEE (Williams, 2016). In this way, to recover rare earth elements and other CRMs from WEEE, it is essential to suppress the anticipated shortage of natural resources (Aminoff & Sundqvist-Andberg, 2021). In addition, high-tech e-waste also often contains metals such as hafnium, indium, lead, nickel, rare earth metals, tantalum, tin, and zinc in combination with valuable reusable plastics, including Polycarbonates, Polyethylene, polyesters, Polypropylene, and phenol formaldehyde (Williams, 2016).

6.2.4 Hazardous Materials

The WEEE is a complex mixture of compounds ranging from precious and valuable elements, to contents of extreme toxicity. Hazardous materials and components of e-waste include asbestos, batteries, CRTs, gas discharge lamps, components containing mercury, chlorinated, and fluorinated organics, plastics containing halogenated fire retardants, polychlorinated biphenyl present in capacitors, printed circuit boards, toner cartridges, and toners (Williams, 2016). Ankit et al. (2021) compiled the information of the main hazardous substances contained in the WEEE, identifying their effects (Fig. 6.4). On the other hand, Ramprasad et al. (2022) focused on rare earth metals and their impact on the environment and health (Fig. 6.5).

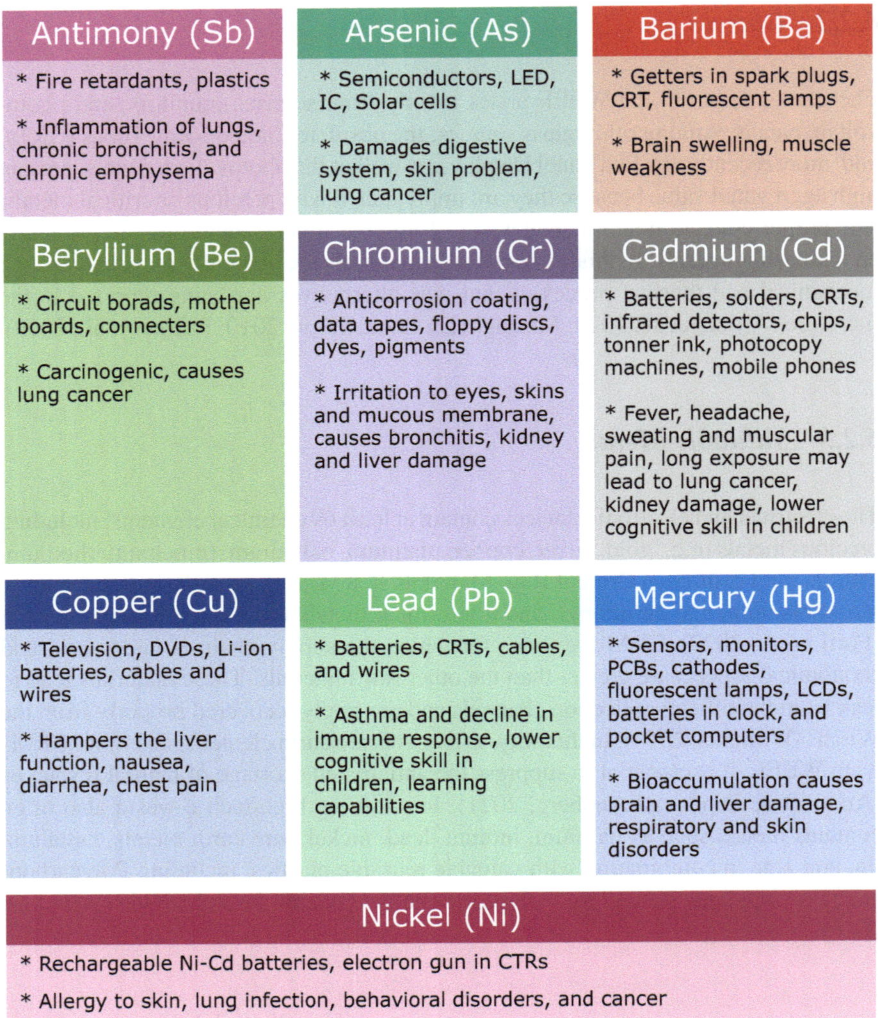

Fig. 6.4 Hazardous heavy metals present in the WEEE. *Source* Adapted from Ankit et al. (2021)

6.2.5 Harmful Organic Compounds

In all the six categories of WEEE, harmful organic compounds are present in a major or minor amount. Among these harmful organic compounds, chemicals such as chlorofluorocarbons (CFCs), hydrochlorofluorocarbons (HCFCs), and brominated flame retardants (BRF) are the most common noxious organic compounds found in the WEEE.

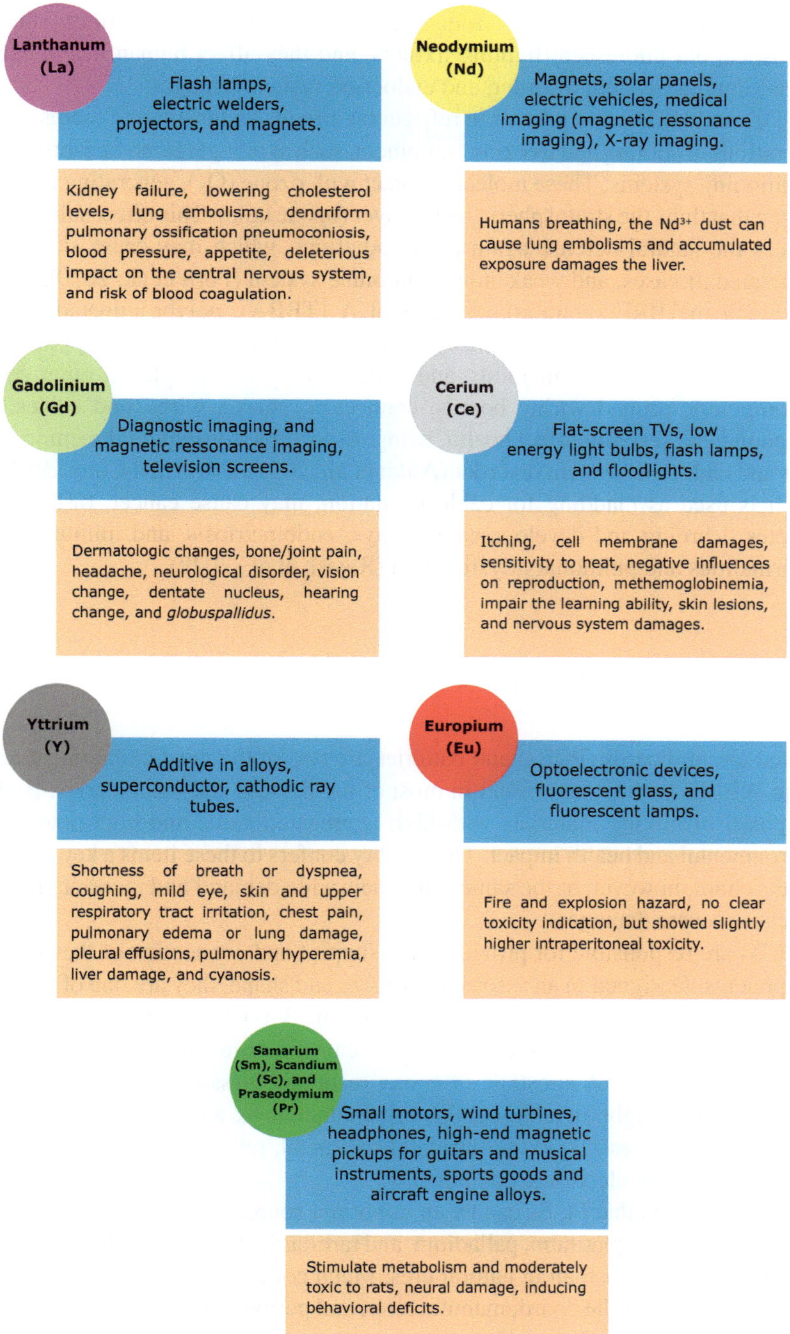

Fig. 6.5 Main sources of rare earth elements in WEEE and their health damage. *Source* Adapted from Ramprasad et al. (2022)

Halogenated compounds are found in PCBs condenser, transformer, adhesive, in plastics, old fluorescent lighting fixtures, and they affect human health causing cancer, immune system disorder, and endocrine system damages.

CFCs and HCFCs are present in refrigerant circuits and insulating foams of older generations of cooling and freezing equipment, such as refrigerators, freezers, and air-conditioning systems. These molecules react with ozone (O_3), generating molecular oxygen that thins the stratospheric ozone layer (ozone hole). This effect subsequently reduces the natural defense against UV radiations, which may cause skin cancer, eye-related diseases, and weakening of immune system (Forti et al., 2020).

Among the BRFs, tetrabromobisphenol A (TBBA), polybrominated biphenyl (PBB), and polybrominated diphenyl ethers (PBDEs) are used in appliances to reduce the product's flammability and appearance, for example, in outer casings of computers, printed wiring boards, connectors, relays, wires, and cables. This compound can severely affect health, being responsible for impaired memory functions and endocrine system disorder (Ankit et al., 2021). Polyvinyl Chloride (PVC), which is used as cladding for cable insulation, may cause cancer, birth defects, diabetes, learning and developmental delays, endometriosis, and immune system abnormalities and pulmonary dysfunction (Rajesh et al., 2022).

6.2.6 PCBs and Batteries

Printed circuit boards (PCBs) and batteries are two of the most common structures in the WEEE, as they are present in most of the devices and machineries. Both are composed of mixing materials with high aggregated value and high potential for environmental and health impact. This duality confers to these items a key role in the WEEE chain; however, at the same time, their wide demand and incorrect handling can cause severe damages.

PCBs are responsible for providing the support and connection of the electronic components. Designed in an assortment of size and shape, they are one of the fastest growing streams in waste generation, representing 3–6 wt% of the total e-waste (Lu et al., 2022). They are made of plastics, silicates, non-ferrous, and ferrous metals (Işıldar et al., 2018). A substrate of epoxy resin and glass fiber represents 23% of the total board weight. In addition, this substrate includes a copper foil covered with a colored solder mask layer that insulates the metal, revealing only the conductive tracks between the electronic components.

It was reported that PCBs can be an important source of valuable metals such as gold, copper, silver, platinum, palladium, and rare earth metals, sometimes presenting concentrations higher than in natural ores. Metal concentrations of PCB depend on their source, type of the board, manufacturer, and period of production (Işıldar et al., 2018; Lu et al., 2022). Figure 6.6 represents the main metals found in the PCB and their function based on the data described by Işıldar et al. (2018).

Despite these important functions, the hazardous potential of PCBs must be examined. As shown in Fig. 6.4, toxic metals such as lead and chromium, besides the

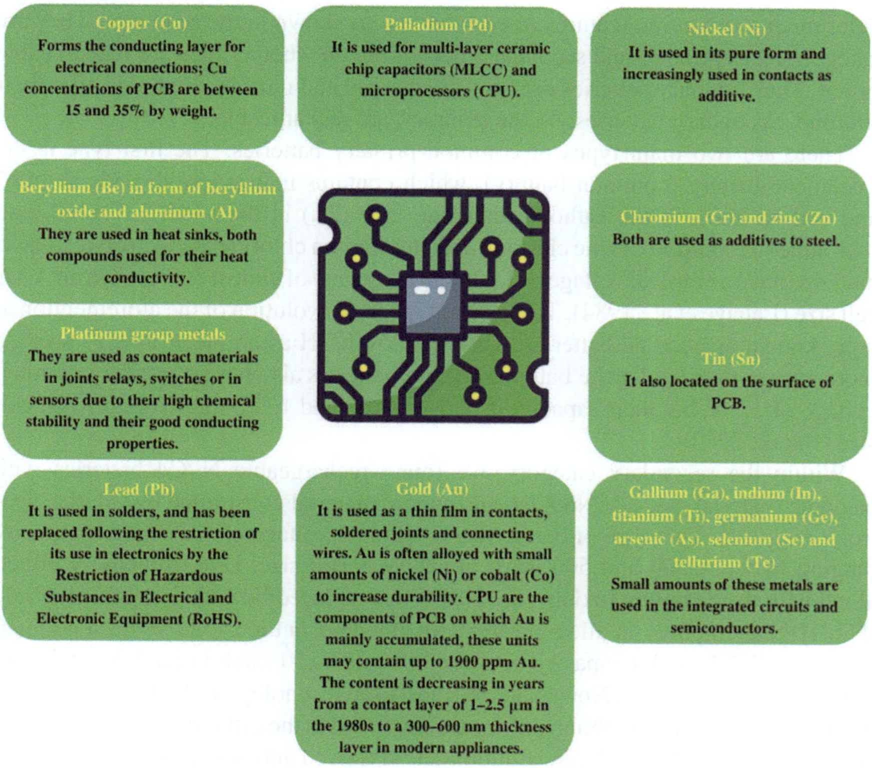

Fig. 6.6 Main metals detected in the PCB and the health threat

brominated organic compounds that act as flame retardants (TBBA being the best known), are among the components of the PCBs. Therefore, improper handling or uncontrolled burning can cause severe disorders to the health and the environment. Furthermore, these hazardous substances hinder the manual disassembly, which is an important step in the removal of the electronic components of the board, and consequently, it becomes necessary to develop machines capable of melting the solder and removing the components in an automated way.

In spite of batteries not being included directly in the WEEE processing chain, they are regularly found in the e-waste, arising from laptops, phones, power tools, hybrid electrical vehicles, and other battery powered appliances. In the same manner of PCBs, batteries are composed of valuable and critical metals but harmful components. The architecture of these devices encompasses different shapes and energy storage capacities.

In a general way, batteries deliver energy when the electrons released by chemical reactions migrate from anode (negative electrode) to cathode (positive electrode) through the electrolyte. Additionally, it is necessary to promote the separation between cathodic and anodic materials. The association of different materials determines the specific features of each battery as shown in the sequence. They are categorized as primary and secondary, depending on whether they are rechargeable or not. Thus, primary batteries cannot be recharged and must be discarded after their lifetime. Secondary batteries, on the contrary, are rechargeable.

There are two main types of common primary batteries. The first type is the Leclanché battery (common battery), which contains manganese dioxide (MnO_2) and carbon black in the cathodic material, zinc (Zn) in the anodic material, and uses aqueous solution of zinc chloride and ammonium chloride as electrolytes These batteries have a nominal voltage of 1.5 V and a capacity of almost 300 mAh in an AAA cell size (Lahaye et al., 1984). The second type is an evolution of the aforementioned type, known as alkaline batteries. They employ KOH as an electrolyte to provide more energy than Leclanché batteries. These batteries also have a standard nominal voltage of 1.5 V, but their capacity is between 800 and 1,200 mAh for an AAA cell size (Barak, 1980).

Within the secondary category, we found rechargeable Ni/Cd batteries with cathodic material made of NiOOH, and anodic material constituted by Cd, the electrolyte is KOH. This type of battery has a nominal voltage of 1.2 V and a capacity ranging between 300 and 500 mAh in an AAA cell size. Due to memory effect, allied with the cadmium toxicity, the Ni/Cd was replaced by nickel metallic hydride (NiMH) that employs an alloy of rare earth elements in the anode. It has a nominal voltage of 1.2 V and a capacity between 600 and 1250 mAh in an AAA cell size (Binnemans et al., 2013). Nowadays, the most used technology is the Li-ion batteries, where different kinds of lithium oxides are employed as the cathodic material, carbon/graphite is used in the anode, and lithium salts dissolved into organic solvents are used as electrolyte. To improve the performance and minimize the toxicity, the organic solvents were replaced for a semisolid polymer with high conductivity, leading to the batteries denominated LiPo, lithium-ion polymer (Amanor-Boadu & Guiseppi-Elie, 2020).

According to Bacher et al. (2017), the amount of collected batteries is lower than the quantity deposited on the market annually. Rechargeable batteries tend to be less recovered than non-rechargeable batteries, which may be attributed to the longer lifespan of the former batteries. Practically, 80% of the rechargeable batteries market is shared by Li-ion and NiMH.

Considering the valuable components, nickel metal hydride batteries contain 33% nickel, 30% iron, 10% rare earth elements (La, Ce, Pr, and Nd), 3% cobalt, 1% manganese, 1% zinc; and Li-ion batteries present 15–25% aluminum, 5–15% copper foil, 1–5% lithium salt form, 25–45% lithium cobalt oxide (Tunsu & Retegan, 2016).

As shown earlier, if the batteries do not receive proper handling, they can release harmful organic compounds, as well as heavy and toxic metals in the environment.

6.3 E-Waste Disposal Methodologies

As previously exposed, WEEE is composed of an intricate combination of valuable and toxic materials. In this context, the proper management of these residues becomes crucial to allow the maximum exploration of the valuable resources without damaging the environment and human health. Ankit et al. (2021) reported that the contamination of water, air, and soil caused by the irregular manipulation of e-wastes can lead to the entry of noxious substances in the human food chain. The rich pollutant content, associated with long-term exposition, may cause several impacts on the soil microbe and health of higher organisms (Fig. 6.7).

According to the Global e-waste monitor 2020 (Forti et al., 2020), currently the options in the management of the WEEE can be split into four different scenarios.

1st—Formal collection: In this scenario, the harvesting happens in accordance with the legislation, e-waste is collected by designated organizations, producers, and/or the government, and through the reverse logistics, the final disposal occurs in a specialized treatment facility. Therefore, proper EoL strategies are applied to recover the valuable materials in an environmentally controlled way and manage the hazardous substances. Residuals then go to incineration or controlled landfills.

2nd—Collection with regular trash, in a waste bin: Here, the WEEE is disposed of in normal waste bins alongside the household garbage, treated as a regular domestic mixed-waste, and disposed into landfills or incinerated without pre-treatment to remove hazardous and/or valuable materials. Both options are inappropriate due to the potentially negative impact to the environment, and to the losses of resources.

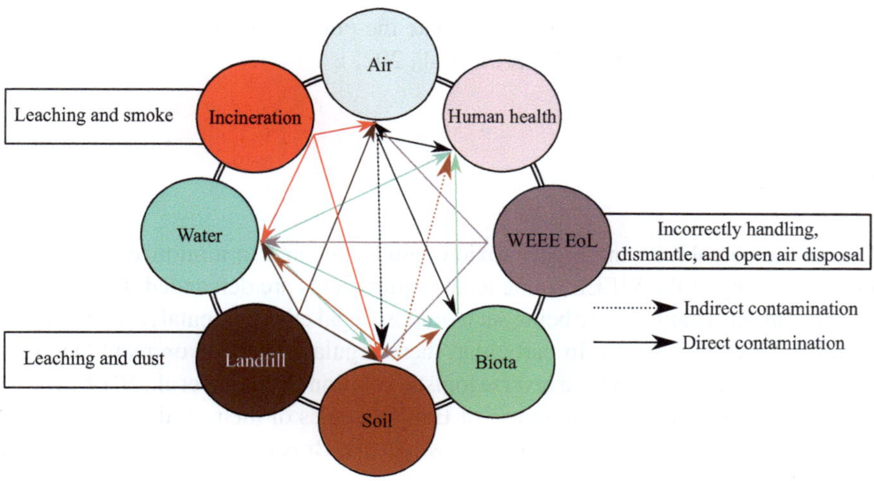

Fig. 6.7 Net of contamination caused by incorrect management of WEEE. *Source* Adapted from Ankit et al. (2021)

3rd—Collection outside of formal systems in countries with a developed WEEE management infrastructure: In this case, individual dealers or companies are responsible for collecting and trading the residues, which can be destined to metal and plastic recycling without depollution of the hazardous substances that might also be exported to other countries.

4th—Collection outside of formal systems in countries without a developed e-waste management infrastructure: This is a frequent scenario found in developing countries. Self-employed people oversee informally collecting and recycling the WEEE. The collection happens from door-to-door by buying or collecting the e-waste from households, businesses, and public institutions. Before the final disposal, the equipment is manually dismantled to recover the interesting materials. Then, the informal recyclers burn and melt the remaining to extract some valuable metallic residue or to obtain energy. This "backyard recycling" causes severe damage to the environment and human health.

Socio-economic inequalities, allied to a massive exploitation of natural resources to attend an increasing consumer demand, are feeding a bad side of reuse and recycling: the uncontrolled dumping in unprotected countries. From a globalized view, reuse, reduce, and recycle are receiving less attention than the waste disposal. Around 65% of the e-waste generated in the European Union (EU) leaves the continent as part of undocumented mixed exports (Frazzoli et al., 2022), being an example of unequal distribution of human and environmental costs. Millions of tons of WEEE generated globally have been exported to developing countries (Nigeria, Ghana, Brazil, Mexico, China, India, Vietnam, the Philippines, or Pakistan). In these countries, crude recycling, open burning, and dumping into landfills are frequently used, with major impact on people, animals, and environment. The disparity in realities highlights the urgency of actions to regulate the policies adopted in relation to the WEEE in different countries (Frazzoli et al., 2022).

6.4 Legislation

Due to the valuability and highly toxic components, and to minimize the impacts in the handling of the WEEE, some legal requirements are demanded. International regulations and standards are being set to impose legal environmental responsibilities on all the involved agents. In particular, these regulations focus on manufacturers, which are required to provide a reverse logistic solution (Ghadimi et al., 2019), as well as a planning for the appropriate use of EoL strategies of their products (Directive, 2011/65/EU; Directive, 2012/19/EU; Brasil, 2010; Brasil, 2022; MMA, 2019; Wu, et al., 2020).

Since the 1980s, in addition to the more rigorous governmental legislations, other actors have begun to press for the companies' good environmental performance, such as Non-Governmental Organizations (NGOs), communities, and customers. This has occurred, among other factors, due to the greater diffusion of environmental issues in society and its consequences on people's lives (Oliveira et al., 2021).

Across the world, important movements toward the regulation and legislation about the management and generation of WEEE are arising.

6.4.1 Europe

The European Directive on WEEE (Directive, 2012/19/EU) was created to contribute to sustainable production and consumption by setting the prevention of WEEE generation as a priority. The Directive, then, contributes to the efficient use of resources, and the retrieval of secondary raw materials through reuse, recycling, and other forms of recovery that improves the environmental performance of everyone involved in the life cycle of EEE. To achieve these objectives, the Directive requires the separate collection and proper treatment of WEEE and sets targets for their collection, as well as for their recovery and recycling. In addition, it helps European countries fight illegal waste exports more effectively by obstructing the exporters to disguise illegal shipments of WEEE and reduces the administrative burden by calling for the harmonization of national EEE registers and the reporting format.

Simultaneously, the Restriction of Hazardous Substances Directive (RoHS) bans the use of certain hazardous substances (such as lead, mercury, cadmium, hexavalent chromium, and some polybrominated flame retardants) in WEEE. Nevertheless, RoHS allows possible exemptions (Directive, 2011/65/EU).

6.4.2 North America—USA and Canada

The USA has no federal legislation that governs the recycling of e-waste or that prohibits it from being exported to developing countries. In spite of this, different schemes and initiatives exist in the USA for WEEE management like the National Strategy for Electronics Stewardship (NSES). This program enables the promotion of environmentally safe EoL management of WEEE, reduction of WEEE exports to developing countries, and encourages concepts such as ecodesign in electronics manufacturing (US-EPA, 2022).

Canada either has no federal WEEE legislation and the management is handled by the private sector under a Stewardship Program: Electronic Product Stewardship Canada (EPSC) that has resulted in the proliferation of WEEE management organizations in the provinces (Shittu et al., 2021).

6.4.3 Asia

Asia is the largest WEEE generator at global scale. The economic prosperity experienced in recent decades by many Asian countries catalyzed the amounts of generated

WEEE. As a result, some countries, including China, Hong Kong, Japan, India, and South Korea, adopted legislation based on the principle of extended producer responsibility (EPR), which provides the legal framework for official collection and treatment of WEEE (Shittu et al., 2021). Other countries that are implementing specific regulations and actions are: Vietnam and the United Arab Emirates.

6.4.4 Africa

In the African continent, two Conventions have been adopted in an attempt to moderate the noxious substances mobility across the continent. The Basel Convention took effect in 1992 to control transboundary movement of hazardous waste, and the Bamako Convention came into force in 1998, aiming to complement the Basel Convention. In addition, Ghana, Kenya, Nigeria, South Africa have developed their own specific legislation about WEEE management, yet Kenya and South Africa are signatory of the mentioned Conventions.

6.4.5 Latin America and Oceania

Among the Latin American countries, Argentina, Brazil, Bolivia, Chile, Colombia, and Mexico have developed their own laws and regulatory directions. Some of them are signatory of the mentioned Conventions or yet present partnership with other more developed countries to encompass the extended producer responsibility in the management of WEEE.

In Oceania, only the Australian Government created specific legislation to address the WEEE management.

6.5 Circular Economy and WEEE

Although sustainable development is needed badly in the face of climate change and social impacts, the practice of sustainability in developing countries is still more important due to international pressure, legislation, and society (Oliveira et al., 2021).

Concern about the impacts of electronic waste disposal is evident from the survey[1] of publications related to the subject. From Fig. 6.8, which is limited to the period between 2000 and mid-2022*, we can see the exponential increase in the number

[1] The survey was carried out in a database of English-written research articles, review articles, book chapter, book review, editorials, mini reviews, short communications, and conference abstracts. The analyzed period ranges from 2000 to June 2022. The terms used in the searches "WEEE", "e-waste", and "electronic waste" were each independently combined with the following terms: "Management", "Circular Economy", "End-of-Life", and "Recycling".

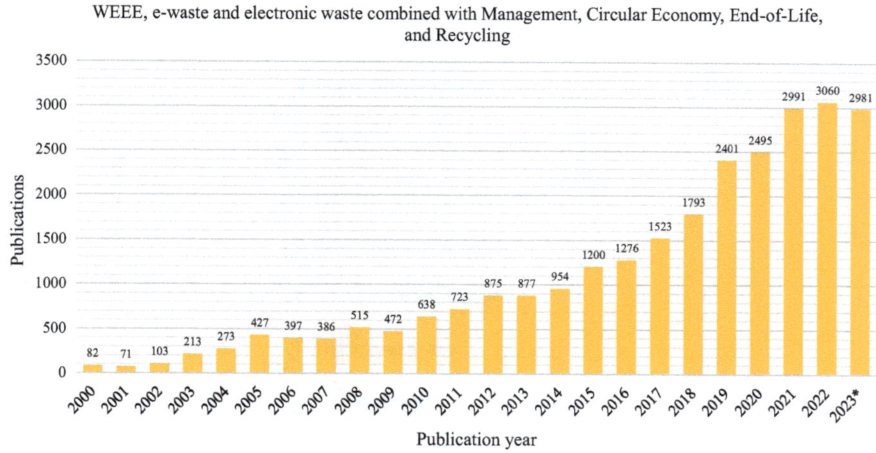

Fig. 6.8 Evolution of the number of published articles related to e-waste handling since 2000

of publications that combine the terms WEEE, e-waste, and electronic waste, with the terms Management, Circular Economy, End-of-Life, and Recycling, exceeding 2000 publications in the last three years.

In Fig. 6.9, the numbers of documents employing each term are shown separately. It is possible to see that the Circular Economy emerged around 2010 and has an overwhelming growth, almost doubling the number of publications each year. The embedded log scale graph allows us to perceive how expressive the Circular Economy contribution in the literature is.

6.6 Logistic Reverse and EoL Strategies

According to Ellen MacArthur Foundation (2022), the Circular Economy is based on three principles: design out waste and pollution; keep products and materials in use; and regenerate natural systems. As previously discussed, legislation requires the manufacturers to be responsible for providing reverse logistics solutions, as well as planning for the appropriate use of EoL strategies of their products. In this sense, recycling emerges as one of the most important EoL tools for the promotion of the Circular Economy. However, there are other EoL strategies to be applied before recycling in the management of the electronic products. Figure 6.10 shows that each one can be employed in an increasing level of relevance for circularity.

During the design stage, there are three actions that can be applied to the electronic product:

- *Refuse* is the first initiative. The discard of obsolete equipment whose function is covered by another device, such as GPS and digital cameras that were embedded in the cell phones;

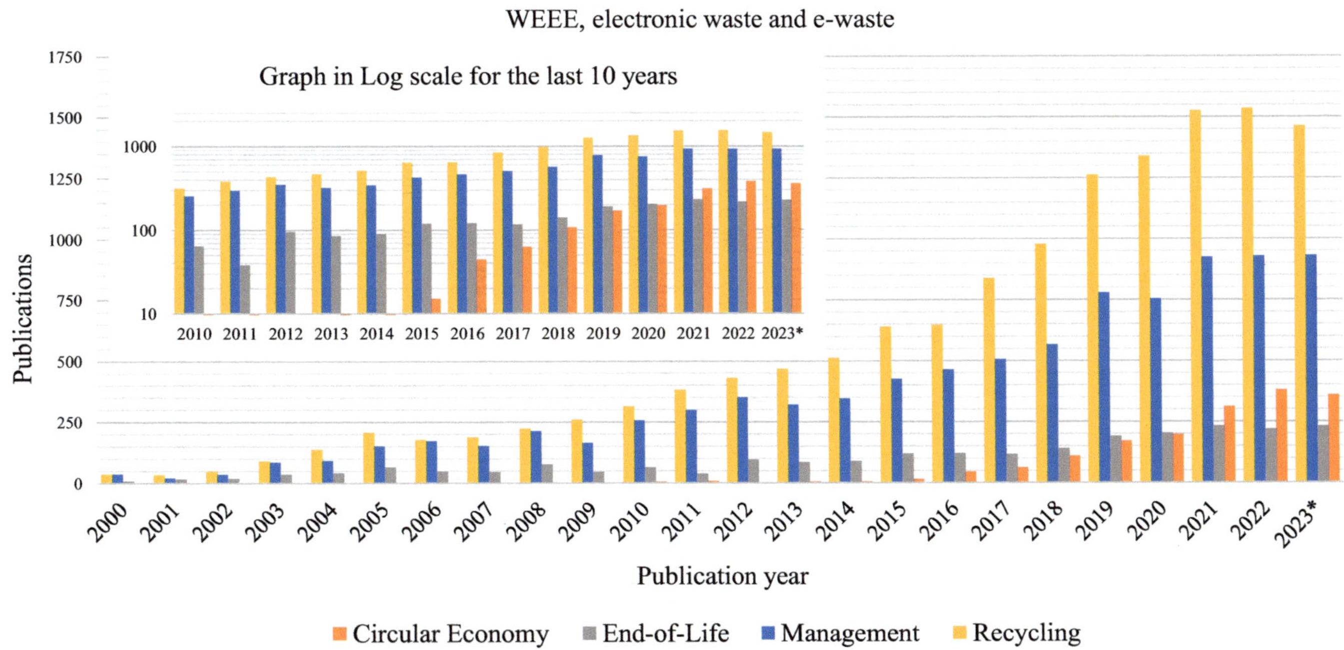

Fig. 6.9 Evolution of the number of published articles related to e-waste handling since 2000, with the search terms independently presented

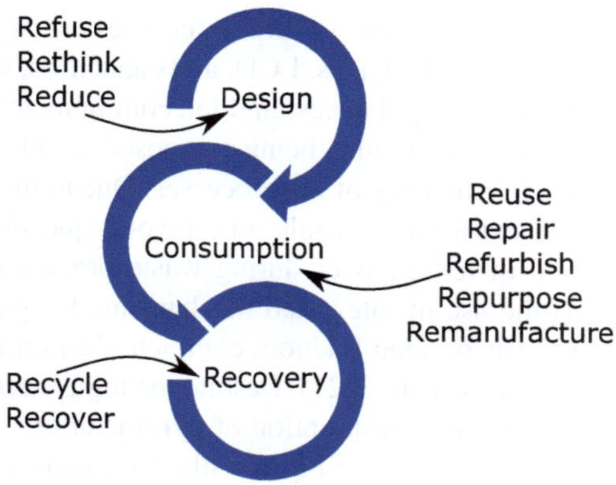

Fig. 6.10 EoL strategies to promote the circular economy of electronic products

- *Rethink* is an action adopted when the purpose of the product is expanded to encompass other functionalities. Cell phones that include many devices or applications, such as GPS, calculator, agenda, camera, for example;
- *Reduce* is the initiative adopted when the manufacturing process is planned to use fewer natural resources or raw materials (Kupfer et al., 2022).

In the consumption stage, the most adopted actions toward the promotion of the Circular Economy are:

- *Reuse* is the usage, for another person, of the discarded functional products;
- *Repair* before disposing of some malfunctioning products. Maintenance should be incorporated into this stage;
- *Refurbish* is the actualization, or restoration of obsolete items to keep them useful;
- *Remanufacture* and *repurpose* are similar. In both cases, parts, or the entire discarded product, are used for manufacturing a new product. The difference stands in the fact that in *Remanufacturing,* the new product has the same purpose, whereas in *repurpose*, the new product may have a different application (Oliveira et al., 2021).

After the extinction of the processes during the product's use phase, there are two ways to recover the valuable components of the WEEE. *Recycling* is the most important tool employed. During the recycling process, some energy may be retrieved by burning, in the so-called *recovering*. Recycling is essential and inevitable for the closing of the life cycle, after the attempts to apply the other mentioned EoL alternatives have been resolved.

Among the commonly adopted recycling techniques, we can mention the traditional pyrometallurgy, hydrometallurgy, and biometallurgy:

- Pyrometallurgy—This is a conventional technology applied in the ores handling. It has been successfully employed in the WEEE recycling. The process is based on the use of elevated temperatures to obtain and purify non-ferrous and precious metals. Pyrometallurgy comprises a set of techniques: smelting, incineration, combustion, pyrolysis, molten salt, sintering, and plasma (Makuza et al., 2021;

Wang et al., 2017). These processes are applied to the metallic components of cables, PCBs, CRTs, LCD, and capacitors, which are the parts of large and small household appliances, and telecommunications equipment (Ebin & Isik, 2016). PCBs, in particular, being composed mainly of flame retardants, cause a decrease in the efficiency of the processes. Due to the elevated temperature used to reduce and extract the metals, one pivotal question of pyrometallurgy is the generation of toxic gasses during waste processing. However, recent advances, such as the use of integrated mechanisms for gas treatment, have resulted in a non-volatile organic fraction, complete dust removal, and low volume of gas release (Makuza et al., 2021). Despite the high energy consumption, the carbon footprint and energy consumption of pyrometallurgy are lower than those resulting from primary mining (Ebin & Isik, 2016) and the elimination of impacts related to the disposal of WEEE makes this alternative a widely used technique by WEEE processing industries (Santos et al., 2021). One last issue related to pyrometallurgy is the uncontrolled burning, without safe precautions, that is performed in some countries to recover only the valuable metals. This approach releases the brominated flame retardant derivatives, such as dioxins and furans, and may cause harm to human health and the environment (Ebin & Isik, 2016).

- Hydrometallurgy—One of the most used processing techniques to extract and purify valuable metals from WEEE. It consists in the employment of the acidic or alkaline aqueous solutions to leach the metals of interest from ores or substrates. It is comprised two main steps: leaching and recovering. In the former, the metallic species are segregated from the substrate and remain into the solution. Afterward, some separation procedures such as precipitation, solvent extraction, or resin ion exchange are applied in the recovering stage to separate the metals from the undesired elements in the solution (Tunsu & Retegan, 2016). It is an extremely versatile approach that allows the use of strong or mild acidic solutions, with or without heating and stirring, and yet it can be executed in combination with other adjuvant techniques, such as ultrasound. The main concerning about the hydrometallurgical processes is the generation of the highly contaminated effluents, but with recent advances in the technology, a wide range of methods are available to purify the effluent streams (Sabia et al., 2022; Sohn, 2017) aiming to meet environmental regulations, and to recover useful reagents or materials. In this sense, Jia et al. (2020) reported an increase in the biohydrometallurgy followed by an obvious decrease in the proportion of studies using solvent extraction.
- Biometallurgy appears as an eco-friendly alternative in the recovery of the metallic fraction of the WEEE. This technique employs microorganisms with higher efficiency. However, the time consumption of this technique represents a source of concern. There are two main areas in the metallic recovering field bioleaching and biosorption. In the bioleaching, metals are removed from their ores or substrates by means of ferrous iron and sulfur oxidizing bacteria. The main microorganisms used are *Acidithiobacillus ferrooxidans* and *Acidithiobacillus thiooxidans*.

  The biosorption mechanism is the adoption of dead or inactive microbial biomass that attaches to metallic ions, present in aqueous solution and effluents, thus enhancing the concentration. It arises as a promising inexpensive technology with industrial scalability (Williams, 2016). In this scenario, biohydrometallurgy, which combines bioleaching and hydrometallurgy, emerges as an efficient and faster technique for metal extraction from WEEE (Kanaujia et al., 2021). In order to present an optimal activity, the microorganism requires the supply of electron donors that can also act as a source of carbon. However, they are sensitive to changes in the operational conditions such as pH and temperature, making it difficult to be controlled. The higher specificity and better performance at lower metallic concentrations, when compared to pyrometallurgical and chemical processes, point to the biometallurgy as a suitable alternative for the remediation of metallic contaminants in WEEE. The challenges related to this technique are the slow kinetic and large space required for industrial applications (Ramprasad et al., 2022).

Alternatively, we can also mention some advanced technologies like:

- Supercritical fluid technology is emerging as the most attractive technique for e-waste recycling. It exhibits expressive results in the recovering of valuable metals and, at same time, to decompose harmful compounds (Oliveira et al., 2021). When the temperature and pressure exceed the critical point of water ($T_c \geq 374.29$ °C, $P_c \geq 22.089$ MPa), the supercritical state is achieved, and then the physical properties of water, such as viscosity, ionic product, density, and dielectric constant, vary widely becoming similar to that of organic compounds, and under these conditions, water can dissolve non-polar organic compounds and also exhibit certain solubility of polar or ionic compounds (de Souza et al., 2022). Compared to conventional techniques, such as pyrolysis or acid leaching, the use of supercritical fluids leads to superior performance in terms of environmental aspects, kinetics, and reaction (Preetam et al., 2022). The most known supercritical fluids are water, alcohols, and carbon dioxide, but the last ones are in the beginning of their applicability, while supercritical water technology has already been commercialized by several corporations (de Souza et al., 2022). Being a neutral process, it does not demand the addition of harmful chemicals. Some concerns about the technology are regarded the corrosion, salt precipitation, clogging issues, and high operational costs, but the employment of green solvents such as supercritical fluids and ionic liquids whose environmental impact is minimum must be explored (Rajesh et al., 2022).
- Ionic liquids possess the ability to manipulate some physicochemical properties, such as viscosity, density, hydrophilicity, and solubility. This feature results from the combination of cations and anions or the addition of functional groups, the latter known as tunability (Francisco et al., 2013). Among its many applications, ionic liquids have great potential when used as solvents and leaching extractors in high value-added metal recovery processes. In this way, ionic liquids can replace critical steps in traditional WEEE treatment and recovery processes. This is particularly important in the pre-treatment in acidic media and the use of extractors, thus avoiding the use of aggressive solvents and the generation of toxic volatile organic compounds. Likewise, ionic liquids can be integrated with other advanced technologies in the search for more efficient processes to recover electronic waste.

- Membrane separation processes play a key role in removing toxic compounds from liquid and gaseous effluents and offer significant advantages (Shi et al., 2019), such as simple operation procedure; continuous work with low or variable flow; non-destructive method with easy scalability; high energy efficiency; and low potential for environmental impacts. The use of membrane technology has wide application in the removal of heavy metals from water and wastewater treatment processes. However, the use of membrane technology for the treatment of effluents generated from the processing of WEEE deserves further investigation.

Figure 6.11 summarizes the possibilities in the WEEE recycling process, showing the interconnection between the presented technologies.

Although a lot of effort has been dedicated to the development of techniques to increase the recovery rate of the valuable resources related to WEEE while minimizing their harmful effects, all the technologies need to be made available to those interested. Nowadays, we face two distinct realities: developed countries investing in the whole e-waste processing, from collection to final disposal, allowing the recovery of all materials with high added value, in a safe way for the worker and the environment. At the other extreme, developing countries to which immense amounts of waste are sent, sometimes illegally, which are dismantled in a rudimentary manner and burned in the open sky, without any protection for workers and the environment.

The circularity of the valuable components in the WEEE works beyond the environmental aspects. Recovering of critical elements can surpass the extraction obtained from natural resources. The recycling of WEEE allows substituting primary resources with secondary raw materials (SRMs) and producing a more secure and uninterrupted supply chain for scarce and critical materials (Huisman et al., 2017). This is acknowledged by the European Commission (EC) that introduced the Circular Economy Action Plan (CEAP) to enhance closing the loop of materials (Ellen MacArthur Foundation, 2022).

6.7 Final Considerations

EEE has specific characteristics, such as component compression, size, and volume reduction, or shortening of useful life cycles, among others, that make the EoL cycle of these products and components critical.

The advancement of WEEE recycling technologies, therefore, becomes essential for environmental sustainability on the planet and for the economic and technological sustainability of various sectors that are part of the electronic production chain. The adoption of the described strategies and technologies, individually or combined, aims the Circular Economy principles, shifting the product conception from cradle-to-grave to cradle-to-cradle, where materials and energy are returned to their respective useful life cycles. The opportunity of the harmful compounds' depletion besides the critical and valuable metals recovering converts a troublesome waste into an economically attractive and environmentally friendly niche.

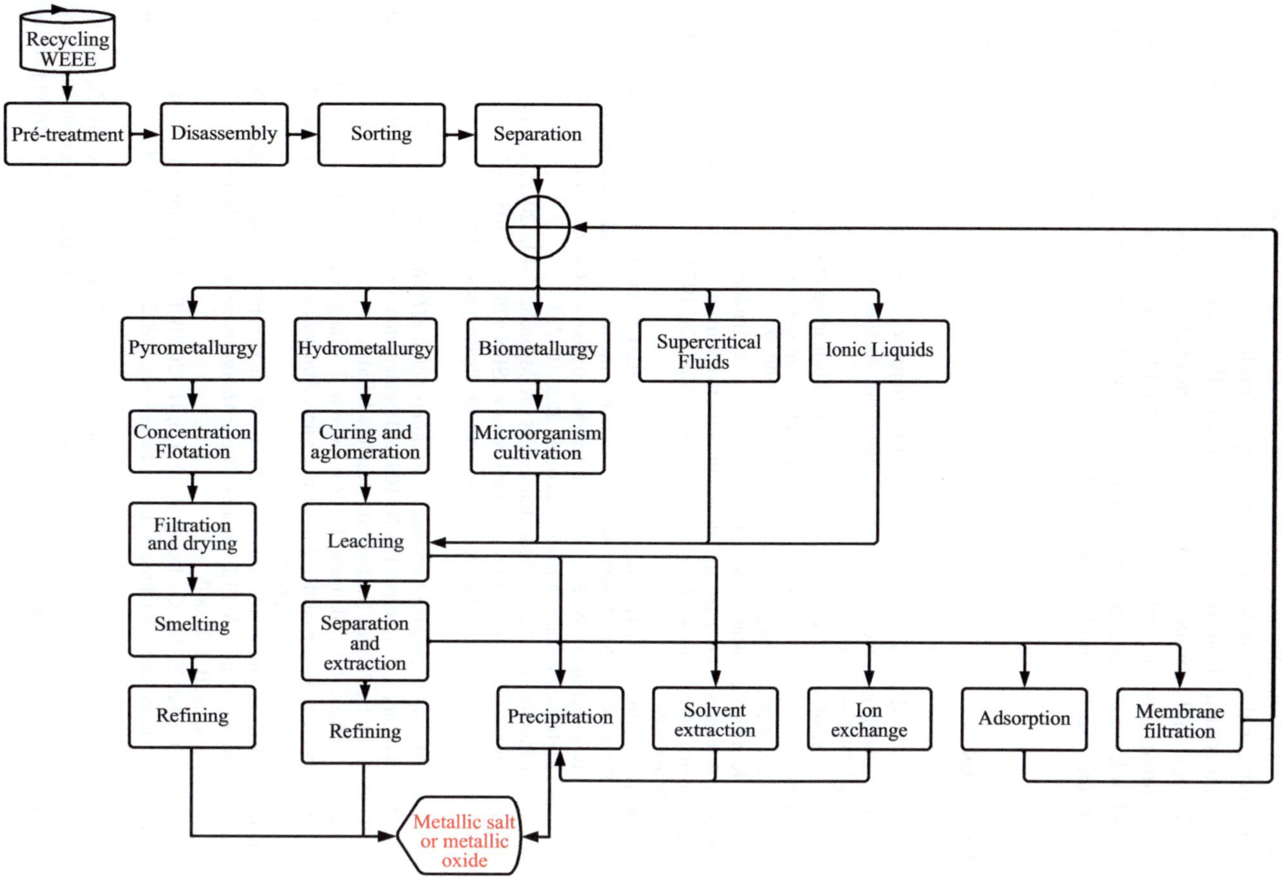

Fig. 6.11 WEEE recycling process and the interconnection between the presented technologies. *Source* Adapted from Ramprasad et al. (2022)

This chapter aimed to illustrate the relevance of the topic of WEEE recycling for the promotion of a Circular Economy and presents, in a summarized way, the compositions and functions of different elements that make up the EEE. In addition, this work introduces the main WEEE recycling technologies, and in summary, presents the potential contributions of each one, individually and in combination, to close the EEE life cycle and promote circularity in the whole electronics sector.

By summarizing the technologies, their benefits, and challenges, it is expected that this book chapter can guide future research on the topic of Circular Economy in the electronics sector and to result as a basis for technological advancement.

References

Amanor-Boadu, J. M., & Guiseppi-Elie, A. (2020). Improved performance of Li-ion polymer batteries through improved pulse charging algorithm. *Applied Sciences, 10*(3), 1–11. https://doi.org/10.3390/app10030895

Aminoff, A., & Sundqvist-Andberg, H. (2021). Constraints leading to system-level lock-ins—The case of electronic waste management in the circular economy. *Journal of Cleaner Production, 322*, 129029. https://doi.org/10.1016/j.jclepro.2021.129029

Ankit, S. L., Kumar, V., Tiwari, J., Sweta, R. S., Singh, J., & Bauddh, K. (2021). Electronic waste and their leachates impact on human health and environment: Global ecological threat and management. *Environmental Technology and Innovation, 24*, 102049. https://doi.org/10.1016/j.eti.2021.102049

Arduin, R. H., Mathieux, F., Huisman, J., Blengini, G. A., Charbuillet, C., Wagner, M., Baldé, C. P., & Perry, N. (2020). Novel indicators to better monitor the collection and recovery of (critical) raw materials in WEEE: Focus on screens. *Resources, Conservation, and Recycling, 157*, 104772. https://doi.org/10.1016/j.resconrec.2020.104772

Bacher, J., Yli-Rantala, E., Castell-Rüdenhausen, M., & Mroueh, U.-M. (2017). *Future trends in WEEE composition and treatment—A review report*. ISBN 978-952-7205-23-5.

Barak, M. (Ed.). (1980). *Electrochemical power sources: Primary and secondary batteries* (No. 1). IET.

Brasil. (2010, August 2). *Política Nacional de Resíduos Sólidos (PNRS)*. Lei N. 12305.

Brasil. (2022, January 12). Decreto Lei N. 10936. Regulamenta a Lei 12305, de 2 de agosto de 2010, que institui a Política Nacional de Resíduos Sólidos.

Binnemans, K., Jones, P. T., Blanpain, B., Van Gerven, T., Yang, Y., Walton, A., & Buchert, M. (2013). Recycling of rare earths: A critical review. *Journal of Cleaner Production, 51*, 1–22. https://doi.org/10.1016/j.jclepro.2012.12.037

Cardamone, G. F., Ardolino, F., & Arena, U. (2021). About the environmental sustainability of the European management of WEEE plastics. *Waste Management, 126*, 119–132. https://doi.org/10.1016/j.wasman.2021.02.040

Chai, B., Wei, Q., She, Y., Lu, G., Dang, Z., & Yin, H. (2020). Soil microplastic pollution in an e-waste dismantling zone of China. *Waste Management, 118*, 291–301. https://doi.org/10.1016/j.wasman.2020.08.048

de Souza, G. B. M., Bisinotto Pereira, M., Francisco dos Santos, L., Gonçalves Alonso, C., Jegatheesan, V., & Cardozo-Filho, L. (2022). Management of waste printed circuit boards via supercritical water technology. *Journal of Cleaner Production, 368*. https://doi.org/10.1016/j.jclepro.2022.133198

Directive 2011/65/EU of the European Parliament and of the Council of 8 June 2011 on the Restriction on the Use of Certain Hazardous Substances in electronic equipment (2011).

Directive 2012/19/EU of the European Parliament and of the Council of 4 June 2012 on electronic waste (WEEE) (2012).

Ebin, B., & Isik, M. I. (2016). Pyrometallurgical processes for the recovery of metals from WEEE. *WEEE Recycling*. https://doi.org/10.1016/B978-0-12-803363-0.00005-5

Ellen MacArthur Foundation Towards the Circular Economy. (2022). Available in. https://ellenmacarthurfoundation.org/topics/circular-economy-introduction/overview. Retrieved July 25, 2022

Forti, V., Baldé, C. P., Kuehr, R., & Bel, G. (2020). *The global E-waste monitor 2020: Quantities, flows, and the circular economy potential*. United Nations University (UNU)/United Nations Institute for Training and Research (UNITAR)—co-hosted SCYCLE Programme. https://collections.unu.edu/view/UNU:7737#.YmFDOI2v_Hs.mendeley. In International Telecommunication Union and International Solid Waste Association (International Solid Waste Association).

Francisco, M., van den Bruinhorst, A., & Kroon, M. C. (2013, March 11). Low-transition-temperature mixtures (LTTMs): A new generation of designer solvents. *Angewandte Chemie International Edition in English, 52*(11), 3074–3085. https://doi.org/10.1002/anie.201207548. https://doi.org/10.1002/anie.201207548

Frazzoli, C., Ruggieri, F., Battistini, B., Orisakwe, O. E., Igbo, J. K., & Bocca, B. (2022). E-WASTE threatens health: The scientific solution adopts the one health strategy. *Environmental Research, 212*(A)(PA), 113227. https://doi.org/10.1016/j.envres.2022.113227

Ghadimi, P., Wang, C., & Lim, M. K. (2019). Sustainable supply chain modeling and analysis: Past debate, present problems and future challenges. *Resources, Conservation and Recycling, 140*, 72–84. https://doi.org/10.1016/j.resconrec.2018.09.005

Goswami, P., Valsalan, N. V., & Dharani, G. (2020). First evidence of microplastics bioaccumulation by marine organisms in the Port Blair Bay. *Marine Pollution [Bulletin], 155*, 111163. https://doi.org/10.1016/j.marpolbul.2020.111163

Habib, H., Wagner, M., Baldé, C. P., Martínez, L. H., Huisman, J., & Dewulf, J. (2022). What gets measured gets managed—Does it? Uncovering the waste electrical and electronic equipment flows in the European Union. *Resources, Conservation and Recycling*, 181. https://doi.org/10.1016/j.resconrec.2022.106222

Huisman, J., Leroy, P., Tertre, F., Söderman, M. L., Chancerel, P., Cassard, D., Amund, N., Wäger, P., Kushnir, D., Rotter, V. S., Mählitz, P., Herreras, L., Emmerich, J., Hallberg, A., Habib, H., Wagner, M., & Downes, S. (2017, December 21). *ProSUM Project urban mine and mining wastes—Final report* (pp. 978–992). Print 808-9061-7 (electronic). ISBN: 978-92-808-9060-0.

Işıldar, A., Rene, E. R., van Hullebusch, E. D., & Lens, P. N. L. (2018). Electronic waste as a secondary source of critical metals: Management and recovery technologies. *Resources, Conservation and Recycling, 135*, 296–312. https://doi.org/10.1016/j.resconrec.2017.07.031

Jia, L. P., Huang, J. J., Ma, Z. L., Liu, X. H., Chen, X. Y., Li, J. T., He, L. H., & Zhao, Z. W. (2020). Research and development trends of hydrometallurgy: An overview based on Hydrometallurgy literature from 1975 to 2019. *Transactions of Nonferrous Metals Society of China (English ed), 30*(11), 3147–3160. https://doi.org/10.1016/S1003-6326(20)65450-4

Kanaujia, K., Trivedi, A., Upvan, K., & Hait, S. (2021). *Hybrid bioleaching: An emerging technique for extraction of critical metals from WEEE*. Elsevier, Inc.

Kupfer, R., Schilling, L., Spitzer, S., Zichner, M., & Gude, M. (2022). Neutral lightweight engineering: A holistic approach towards sustainability driven engineering. *Discover Sustainability, 3*(1). https://doi.org/10.1007/s43621-022-00084-9

Lahaye, J., Wetterwald, M. J., & Messiet, J. (1984). Electrolyte absorption capacity of conducting carbon blacks used in Leclanche type batteries. *Journal of Applied Electrochemistry, 14*(1), 117–122. https://doi.org/10.1007/BF00611268

Lahtela, V., Hamod, H., & Kärki, T. (2022). Assessment of critical factors in waste electrical and electronic equipment (WEEE) plastics on the recyclability: A case study in Finland. *Science of the Total Environment, 830*, 155627. https://doi.org/10.1016/j.scitotenv.2022.155627

Liu, X., Lu, X., Feng, Y., Zhang, L., & Yuan, Z. (2022). Recycled WEEE plastics in China: Generation trend and environmental impacts. *Resources, Conservation and Recycling, 177*, 105978. https://doi.org/10.1016/j.resconrec.2021.105978

Lu, Y., Yang, B., Gao, Y., & Xu, Z. (2022, October). An automatic sorting system for electronic components detached from waste printed circuit boards. *Waste Management, 137*, 1–8. https://doi.org/10.1016/j.wasman.2021.10.016

Makuza, B., Tian, Q., Guo, X., Chattopadhyay, K., & Yu, D. (2021). Pyrometallurgical options for recycling spent lithium-ion batteries: A comprehensive review. *Journal of Power Sources, 491*(February), 229622. https://doi.org/10.1016/j.jpowsour.2021.229622

Ministério do Meio Ambiente (MMA). (2019). *Sectoral agreement for the implementation of a Reverse Logistics System for household electrical and electronic products and their components.* Available in. https://www.mma.gov.br/images/Acordo%20Setorial/Acordo%20Setorial%20-640%20Eletroeletr%C3%B4nicos.pdf. Retrieved July 21, 2022

Mostaghel, S., & Samuelsson, C. (2010). Metallurgical use of glass fractions from waste electric and electronic equipment (WEEE). *Waste Management, 30*(1), 140–144. https://doi.org/10.1016/j.wasman.2009.09.025

NCSL, National Conference of State Legislatures. (2018). Available in. https://www.ncsl.org/research/environment-and-natural-resources/e-waste-recycling-legislation.aspx Retrieved July 26, 2022.

Oliveira, J. A., Lopes Silva, D. A., Puglieri, F. N., & Saavedra, Y. M. B. (2021). Life cycle engineering and management of products: Theroy and practice. *Springer International Publishing*. https://doi.org/10.1007/978-3-030-78044-9

Padoan, F. C. S. M., Altimari, P., & Pagnanelli, F. (2019). Recycling of end of life photovoltaic panels: A chemical prospective on process development. *Solar Energy, 177*, 746–761. https://doi.org/10.1016/j.solener.2018.12.003

Parajuly, K., Fitzpatrick, C., Muldoon, O., & Kuehr, R. (2020). Behavioral change for the circular economy: A review with focus on electronic waste management in the EU. *Resources, Conservation and Recycling, 6*(6), 100035. https://doi.org/10.1016/j.rcrx.2020.100035

Pelamatti, T., Cardelli, L. R., & Rios-Mendoza, L. M. (2022). The role of microplastics in bioaccumulation of pollutants. In T. Rocha-Santos, M. F. Costa & C. Mouneyrac (Eds.), *Handbook of microplastics in the environment*. Springer. https://doi.org/10.1007/978-3-030-39041-9_18

Peeters, J. R., Vanegas, P., Kellens, K., Wang, F., Huisman, J., Dewulf, W., & Duflou, J. R. (2015). Forecasting waste compositions: A case study on plastic waste of electronic display housings. *Waste Management, 46*, 28–39. https://doi.org/10.1016/j.wasman.2015.09.019

Preetam, A., Mishra, S., Naik, S. N., Pant, K. K., & Kumar, V. (2022). A sustainable approach for material and metal recovery from e-waste using subcritical to supercritical methanol. *Waste Management, 145*(April), 29–37. https://doi.org/10.1016/j.wasman.2022.04.011

Rajesh, R., Kanakadhurga, D., & Prabaharan, N. (2022). Electronic waste: A critical assessment on the unimaginable growing pollutant, legislations and environmental impacts. *Environmental Challenges, 7*, 100507. https://doi.org/10.1016/j.envc.2022.100507

Ramprasad, C., Gwenzi, W., Chaukura, N., Izyan Wan Azelee, N., Upamali Rajapaksha, A., Naushad, M., & Rangabhashiyam, S. (2022). Strategies and options for the sustainable recovery of rare earth elements from electrical and electronic waste. *Chemical Engineering Journal, 442*(P1), 135992. https://doi.org/10.1016/j.cej.2022.135992

Rodrigues, J. P., Duarte, A. C., & Santos-Echeandía, J. (2022). Interaction of microplastics with metal(oid)s in aquatic environments: What is done so far? *Journal of Hazardous Materials Advances, 6*, 100072. https://doi.org/10.1016/j.hazadv.2022.100072

Sabia, G., Tammaro, M., Cerchier, P., Salluzzo, A., & Brunelli, K. (2022). Treatment and management of the effluents generated by hydrometallurgical processes applied to End-of-Life Photovoltaic Panels. *Journal of Water Process Engineering, 47*, 102814. https://doi.org/10.1016/j.jwpe.2022.102814

Santos, M. Pd, Garde, I. A. A., Ronchini, C. M. B., Filho, L. C., Souza, G. B. M., Abbade, M. L. F., Regone, N. N., Jegatheesan, & Oliveira, J. A. (2021, January). A technology for recycling lithium-ion batteries promoting the circular economy: The RecycLib. *Resources, Conservation and Recycling, 175*. https://doi.org/10.1016/j.resconrec.2021.105863

Shi, Y., Huang, J., Zeng, G., Cheng, W., & Hu, J. (2019). Photocatalytic membrane in water purification: Is it stepping closer to be driven by visible light? *Journal of Membrane Science, 584*, 364–392. https://doi.org/10.1016/J.MEMSCI.2019.04.078

Shittu, O. S., Williams, I. D., & Shaw, P. J. (2021). Global E-waste management: Can WEEE make a difference? A review of e-waste trends, legislation, contemporary issues and future challenges. *Waste Management, 120*, 549–563. https://doi.org/10.1016/j.wasman.2020.10.016

Sohn, H. Y. (2017, May). *Hydrometallurgical principles white star. Reference module in materials science and materials engineering* (pp. 2245–2250). https://doi.org/10.1016/b978-0-12-803581-8.03589-x

Stenvall, E., Tostar, S., Boldizar, A., Foreman, M. R. S. J., & Möller, K. (2013). An analysis of the composition and metal contamination of plastics from waste electrical and electronic equipment (WEEE). *Waste Management, 33*(4), 915–922. https://doi.org/10.1016/j.wasman.2012.12.022

Tunsu, C., & Retegan, T. (2016). Hydrometallurgical processes for the recovery of metals from WEEE. *In WEEE Recycling*. https://doi.org/10.1016/B978-0-12-803363-0.00006-7

US EPA. (2022). *National strategy for electronics stewardship (NSES)*. https://www.epa.gov/smm-electronics/national-strategy-electronics-stewardship-nses. Retrieved July 21, 2022

Wang, H., Zhang, S., Li, B., Pan, D., Wu, Y., & Zuo, T. (2017). Recovery of waste printed circuit boards through pyrometallurgical processing: A review. *Resources, Conservation and Recycling, 126*, 209–218. https://doi.org/10.1016/j.resconrec.2017.08.001

Williams, I. D. (2016). Global metal reuse, and formal and informal recycling from electronic and other high-tech wastes. *Metal Sustainability*, 23–51. https://doi.org/10.1002/9781119009115.ch2

Wu, Y., Yang, L., Tian, X., Li, Y., & Zuo, T. (2020). Temporal and spatial analysis for end-of- life power batteries from electric vehicles in China. *Resources, Conservation and Recycling, 155*, 104651. https://doi.org/10.1016/j.resconrec.2019.104651

Chapter 7
Sustainable Development and Circular Economy

Maeen Md. Khairul Akter, Upama Nasrin Haq, and Mohammad Abbas Uddin

7.1 Introduction

The demand for natural resources is unsustainable and increasing at a rate that would require two planets by 2030 to meet the demand (Esposito et al., 2018). Every year, 90 billion tons of primary materials are extracted for plastic, of which only 9 percent are recycled (Schandl & Baynes, 2021; UNEP, 2019). The UN Sustainable Development Goals (SDGs) are considered a global blueprint to long-term sustainability. This chapter discusses the concept of sustainability and the circular economy approach to achieve Sustainable Development Goals.

7.2 Sustainable Development

7.2.1 Sustainability

Sustainability has coined significant attention from scholars, development practitioners, factory stakeholders, and government policymakers in the last few years. Its relationship with our existence is a potential basis for this global attraction,

M. Md. K. Akter
Department of Textile Engineering Management, Bangladesh University of Textiles, Dhaka, Bangladesh

U. N. Haq
Department of Apparel Engineering, Bangladesh University of Textiles, Dhaka, Bangladesh

M. A. Uddin (✉)
Department of Dyes and Chemical Engineering, Bangladesh University of Textiles, Dhaka, Bangladesh
e-mail: abbas.shiyak@dce.butex.edu.bd

S. A. Bandh and F. A. Malla (eds.), *Waste Management in the Circular Economy*,
https://doi.org/10.1007/978-3-031-42426-7_7

and its pervasiveness and tremendous popularity were unaffected by the obscurity of its multiple definitions. Even the concept got worldwide recognition in a way that other previous development concepts have not, it appears to be on the path of becoming the dominant development framework for a long period (Mensah, 2019). The concept of sustainability in the quest for sustainable development (SD) originated in 1987 by World Commission on Environment and Development (WCED) (Shi et al., 2019). Sustainability and the concept of sustainable development are consistently related; thus, both terms are used as synonyms, especially in academic and scientific contexts. Despite being used interchangeably, the terms "sustainability" and "sustainable development" have multiple theoretical definitions, prospects, and constraints. (Ruggerio, 2021). As an open universal concept, sustainability has 300 different definitions from various viewpoints (Geissdoerfer et al., 2017). Given the numerous theoretical streams guiding perspectives on the topic, extracting a complete definition by reviewing the literature on sustainability has proven almost impossible (Haidar, 2021). Table 7.1 summarizes the definitions of sustainability from multiple viewpoints found in the literature between 2015 and 2021.

Table 7.1 Sustainability definitions

Definition	Authors and year
Sustainability represents a paradigm shift away from traditional accounting practices which concentrate mainly on the financial (economic) aspects of a business, and includes environmental accounting, which places greater emphasis on examining the economic aspects of a business's main environmentally related activities	Haidar (2021)
The meaning of sustainability implies maintaining the capacity of ecological systems, to support and enhance the quality of social systems	Sakalasooriya (2021)
Sustainability can be seen as the capacity of a certain community to create and maintain communal existence through the management of the local natural resources in a way that assures the survival and interconnectedness of the members of both the community and the environment	Virtanen et al. (2020)
Sustainability is the concept to connote improving and sustaining a healthy economic, ecological, and social system for human development	Mensah (2019)
Sustainability is the combination of three things: people, planet, and profit	Razzaq et al. (2018)
Comprehensive definition of sustainability includes the following five constructs: (1) after a defined period of time, (2) the program, clinical intervention, and/or implementation strategies continue to be delivered and/or (3) individual behavior change (i.e., clinician, patient) is maintained; (4) the program and individual behavior change may evolve or adapt while (5) continuing to produce benefits for individuals/systems	Moore et al. (2017)
Sustainability may be defined as the capacity to maintain or improve the state and availability of desirable materials or conditions over the long term	Harrington (2016)
A company's delivery of long-term value in financial, social, environmental, and ethical terms	Haffar and Searcy (2017)

It is obvious from Table 7.1 that no matter from which perspective sustainability is defined, its foundational principles stay unchanged. Each concept emphasizes the interdependence and interconnectedness of either ecological systems and human growth, the environmental impact and healthy profit of people and systems, or the equilibrium of environmental capacity, monetary gain, and communal survival. Since the terminology was first introduced in the WCED report, it has been used as a baseline in environmental scientific studies. So, sustainability can be stated as a strategy for bringing social and economic transformation without jeopardizing environmental integrity. To define sustainability holistically, a three-pillar conception consisting of environmental, social, and economic factors or goals has been introduced, which is commonly represented by three intersecting circles with overall sustainability at the center. This three-pillar concept is also prevalent as Triple Bottom Line (TBL), which implies the balance between three pillars of sustainability. These three perspectives of sustainability are interconnected, and human behavior acts as a trigger for this relationship. People can affect the environment, economy, and society simultaneously by their actions. Suppose natural resources can be protected by human action. In that case, the environment will be conserved, the economy will prosper and be sustainable, and social life will be healthy because of harmony to protect human rights (Purvis et al., 2019; Ranjbari et al., 2021). Environmental problems like climate change, global warming, carbon dioxide emissions, air and water pollution, soil degradation, biodiversity loss, resource depletion, and excessive land usage cause threats to the Earth's life support systems. Human activities and modern living are pervasively responsible for these environmental issues.

Moreover, overpopulation causes excessive unemployment, unfavorable working conditions, social vulnerability, massive waste, and a depressed economy, which impedes meeting societal expectations. Financial and economic instability affects individual businesses and entire economies due to economic issues such as supply risk, complex ownership structures, deregulated markets, and inadequate incentive structures (Geissdoerfer et al., 2017). The interrelationship between these three pillars will likely satisfy societal demands while being economically and environmentally viable or socially and economically balanced if environmentally acceptable, Fig. 7.1.

Suppose an individual in a given spatial area becomes unemployed (economic), who could be likely to be poor (social). If one is jobless and lacks money, would be more engaged in practices to harm ecology, like cutting down trees for firewood for cooking meals and warming up his home (environment). As a result of his activities and those of others who cut down trees in his area, deforestation will cause fewer crops and minerals loss from the soil (environmental). Lacking dietary nutrients of the inhabitants will cause subpar performance, resulting in lower productivity (economic). If productivity declines or stagnates (economic), impoverished people will stay poor or become destitute (social), and the cycle will continue (Mensah, 2019).

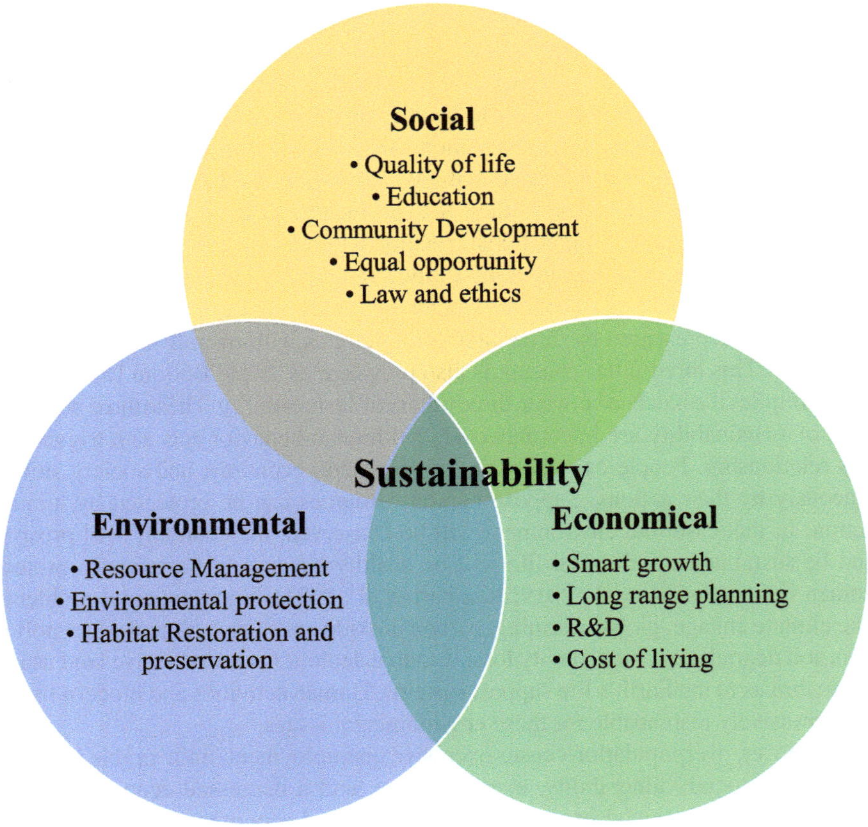

Fig. 7.1 Relationships among three pillars of sustainability (Mensah, 2019)

7.2.2 Classical Concept of Sustainable Development

The classical concept of SD that experienced three stages: the embryonic stage (before 1972), the molding stage (1972–1987), and the developing stage (1987–present) (Shi et al., 2019), includes the concepts of development, needs, and future generations. The development concept refers to socioeconomic development in accordance with ecological constraints. The needs concept highlights the necessity of resource redistribution to ensure the quality of life for all. The future generations' concept pursues the sustainable use of natural resources to ensure a healthy life for future generations (Klarin, 2018). The world's socioeconomic transformation was the earnest need to preserve ecological balance in the early twentieth century. The industrial revolution's blessing of overproduction led to the exploitation of natural resources and the release of massive amounts of waste and toxins into the environment. As a result of environmental damage and societal discontent brought on by human activity, economic growth was slow. In those circumstances, the idea of SD evolved as a savior in this unexpected situation and developed into a key plan of action for the global socioeconomic transition for environmental equilibrium. The International Union for the Nature Conservation has been addressing the harmony between ecology and economy for SD for the benefit of the global environment since 1951; however, most of the organizations in the twentieth century associated SD with human activities and environmental conservation (Izvercian & Ivascu, 2015). Therefore, it is clear that environmental concerns for the global ecological balance have been the main focus of SD since its establishment.

7.2.3 Modern Concept of Sustainable Development—The SDG

The modern concept of SD and the literature of sustainability in a global context emerged in the late twentieth century. Recent literature on sustainability and the underlying concepts appears to be focused on the diverse set of Sustainable Development Goals (SDGs) established by the United Nations (UN) in 2015. The SDGs describe the elements covered in the three sustainability pillars with precise elaboration (Purvis et al., 2019). The SDGs aspired to protect and preserve the environment while ensuring people's prosperity. According to the SDGs, society or system must be developed to properly meet the general population's socioeconomic needs and interests while minimizing the detrimental impacts that endanger the ecosystems and biodiversity (Ukko et al., 2019). 2030 Agenda for SDGs has enhanced and expanded the dialogue on sustainable development in the modern era. It has 17 goals and generally shows a shift from a concentration on ecological issues toward more general societal challenges like zero hunger, good health, quality education, gender equality, sustainable cities, communities, etc. The modern concept focuses on the sustainable development of societal issues to ensure human well-being, underlining poverty eradication, human-induced environmental damage prevention, and economic prosperity (Manioudis & Meramveliotakis, 2022). Table 7.2 shows the UN SDGs (Meschede, 2020; Smith, 2020).

7.2.4 Global Initiatives Toward Achieving Sustainable Development

The SDGs focus on the environment, seeking aspirational goals of alleviating inequality through socioeconomic transition in a sustainable manner. 169 targets and 232 quantifiable indicators have been introduced to achieve the SDGs by 2030. The goal outlines the expected changes or effects or the state of global affairs, the target defines how SDGs can be accomplished practically, and the indicator states how these targets can be measured. For instance, Goal 1, "End poverty in all its

Table 7.2 UN sustainable development goals

SDGs	Title
1	End poverty in all its forms everywhere
2	End hunger, achieve food security and improved nutrition, and promote sustainable agriculture
3	Ensure healthy lives and promote well-being for all at all ages
4	Ensure inclusive and equitable quality education and promote lifelong learning opportunities for all
5	Achieve gender equality and empower all women and girls
6	Ensure availability and sustainable management of water and sanitation for all
7	Ensure access to affordable, reliable, sustainable, and modern energy for all
8	Promote sustained, inclusive and sustainable economic growth, full and productive employment and decent work for all
9	Build resilient infrastructure, promote inclusive and sustainable industrialization, and foster innovation
10	Reduce inequality within and among countries
11	Make cities and human settlements inclusive, safe, resilient, and sustainable
12	Ensure sustainable consumption and production patterns
13	Take urgent action to combat climate change and its impacts
14	Conserve and sustainably use the oceans, seas, and marine resources for sustainable development
15	Protect, restore, and promote sustainable use of terrestrial ecosystems, sustainably manage forests, combat desertification, and halt and reverse land degradation and halt biodiversity loss
16	Promote peaceful and inclusive societies for sustainable development, provide access to justice for all and build effective, accountable, and inclusive institutions at all levels
17	Strengthen the means of implementation and revitalize the Global Partnership for Sustainable Development

forms worldwide," sounds very ambitious and unreachable, for which the UN has set five targets. One target is to eliminate severe poverty worldwide by 2030, which is currently applied to individuals making less than USD 1.25 daily, and the indicator of this example is "percentage of people living below the poverty level internationally." Goals, targets, and indicators for SDGs seem complicated but logical because, without measurable indicators, it is challenging to know the scope and type of activities needed to reach these ambitious goals (Smith, 2020). These goals cover SD's economic, social, and environmental facets in an unprecedented way. The SDGs combined the Rio+ approach for SD with the Millennium Development Goals (MDGs) focusing on human development, and significantly broadened the themes they covered and underlined the need for a change in legislative tactics. The developmental objectives are formulated by SDGs, and their targets are inherently interdependent and interlinked (Breuer et al., 2019). Most nations and societies support these SGDs according to their status of poverty, hunger, decent work, inequality, green cities, waste management, and infrastructure. Though the SDGs have been established in response to global demand, each country has a different priority for each goal domestically. Some developed countries like UK, USA, Germany, Italy, Spain, and Sweden have chosen circular strategies in manufacturing to facilitate the journey toward sustainability. The UK's textiles, upcycling, and food processing industries follow recycling, reusing, and reducing waste to lessen the negative environmental impact. Collaborating with global universities, institutions, research centers, and other stakeholders, companies in the UK seek sustainable solutions for their production line, which is linked to SDGs 17. By extending the product life cycle from the initial phase of material extraction to product disposal, manufacturing industries in the UK attempt to migrate circular operating systems for the socioeconomic growth and environmental preservation.

For example, a typical steel company in the UK recycles an average of 1.2 billion steel waste annually, while plastic businesses are researching alternatives to plastic as recycling plastic packaging in the UK is quite expensive (Galvão et al., 2022). Despite being a developing nation, Brazil excels at recycling, particularly in the paper, plastic, and aluminum sectors. Brazil has 858 recycling industries, including 22 aluminum centers, 27 paper recycling industries, and 809 plastic recycling enterprises. According to the "Brazilian Corporate Commitment to Recycling" in 2019, the Brazilian recycling industry sold 590 million BRL (about 114 million USD) and produced 1.28 million t/y of recycled goods. The amount of recycling varies in Brazil depending on the sort of waste gathered. For instance, due to its high value and ease of transportation, the nation recycles more than 90% of aluminum waste, yet only 1.28% of plastic waste is recycled yearly (de Almeida et al., 2021). The National Policy on Solid Waste—2010 NPSW's introduction of "Reverse Logistics" as a civil contract is another important Brazilian endeavor. Brazilian enterprises are mandated by law to ensure the proper reuse, recycling, treatment, and disposal of post-consumer packaging, which is apparently in line with SDGs 12 (Presidência da República, 2010).

In order to achieve SDG 13 and promote a green transformation, Norway specifically focuses on climate change and greenhouse gas emissions. The climate goals of Norway are to achieve a low-emission society by 2050 and a reduction in emissions of up to 55% by 2030. Norway seeks to promote a green economy, focusing on agriculture and fisheries as sectors with the potential to lead a green transformation in light of recent developments in global policy, such as the Paris Agreement (2015) and the United Nations' SDGs (2015). The agriculture and fishing industries in Norway strongly emphasize making sensible use of the available resources for the benefit of the local economy, people, and environment (Karlsson & Hovelsrud, 2021). By 2050, it is projected that global resource use will have increased by three times, and by 2025, cities will produce 2.2 billion tons of garbage with noticeable overuse of trash landfill. Global businesses explicitly strive to decrease waste to reduce negative effects on the environment for SD. Worldwide, manufacturing industries have prioritized waste reduction for environmental sustainability. Still, they do not have a cradle-to-cradle cycle yet, which would allow for the optimum elimination of waste through the design of materials, products, systems, and business models.

Some electrical companies charge a minimal recycling fee to their clients, which is then used to manage the recycling because recycling is more expensive than using virgin materials. Handling electronic waste safely and reducing global waste, the association of recycling companies promotes conservation energy for environmental sustainability (Jaeger & Upadhyay, 2020). Bangladesh, a developing country, has also started to take initiatives for the nation's long-term prosperity. One of the green initiatives is executing comprehensive policies and development strategies to ensure a steady energy supply by leveraging the nation's renewable energy sources to meet. It has implemented the fuel diversification advantage to expand its renewable energy sector and meet the nation's growing electricity needs. To diversify fuel supply and provide energy security, the government initiated a renewable energy policy in 2008 to replace fossil fuel and non-renewable energy with renewable sources (biomass, biofuel, biogas, hydropower, solar energy, wind power, hydrogen cell, geothermal, tide, and wave) that would meet local energy demands. Another significant initiative of Bangladeshi government was launching the "500 MW Solar Power Mission" to satisfy the rising energy consumption demand. The country has also taken initiatives to meet technological, socio-cultural, and environmental challenges for sustainable growth and development. In September 2016, Bangladesh ratified the Paris Agreement of the United Nations Framework Convention on Climate Change (UNFCC). Under which it is making necessary efforts to conserve the environment and keep the climate stable. The nation has committed to employing domestic resources to lessen its Green House Gas emissions by 5% by 2030, regardless of the Gross Domestic Product (GDP) of the nation, and also formulated a legal framework for addressing environmental issues, developing regulations, and policies to reduce Green House Gas emissions, and implementing a sustainable economic strategy to produce electricity aligned with SDGs 7 and 13 (Karim et al., 2019).

7.3 Circular Economy (CE)

7.3.1 Conceptual Definition

The concept of circular economy (CE) emerged as a defense against the linear economic model of take-make-dump which is proved to generate wastes in the amount exceeding the Earth's capacity. The concept is both new and old in a way that the classical and modern descriptions differ in content but promotes the same phenomena—"circularity." Again, the practical implication of CE is multifaceted in nature in different industries. In this chapter, the definitions and practical implications of CE are analyzed in order to make an understandable conceptualization of CE.

Though the concept of circularity evolved during the 80 s in Europe and 90 s in USA, research and practice of CE has gained a global momentum in the recent past (Stahel, 2020). Since 2003, increased number of research papers have been published on CE with an all-time high in 2018 (Del Río et al., 2021; Mendes, 2020). However, with time the complexity and the dimensions from which CE is approached is getting more involved (Del Río et al., 2021).

7.3.1.1 The Classical Concept

The classical concept of CE was concerned with the transformation from a linear industrial economy to a circular industrial economy where economic growth was measured with the quality and quantity of industrial stocks, i.e., assets and capital (Stahel, 2020). This concept stresses the idea of wealth (value) creation as a result of retention of the stock values, which evolved against resource consumption. Since then, CE has been defined and described from many aspects of material prevention and economic development (Nikolaou et al., 2021). An analysis of 114 definitions by Kirchherr et al. (2017) shows at least 17 dimensions of circularity with dimensions like reuse, recover, reduce, and recycle as the prominent ones.

In the linear economy, commodities produced in the industries, distributed through various channels to the consumers, are eventually disposed of as waste in landfills. The idea of circularity (Fig. 7.2) is to recover wastes and inject them into the economic system again, enabling the products and materials to be reused by the consumers, resold in the distribution channels, and recycled for further processing (production) of new products and commodities. Such a system as a whole defines what the circular economy is about. The prime objective was to increase the stock of materials, i.e., more stock of materials produced and consumed with less resource intake. Hence, the classical concept of circularity was involved with maximizing the time of materials/ products used in the economic system before it is passed down as disposal in landfills. The consideration of environmental and societal issues beyond the economic concern was obscure.

7.3.1.2 The Modern Concept

On the contrary, the modern concept of CE is evolving and promoting global economic development while solving ever-increasing environmental and societal challenges. Modern environmentalists also prefer to link CE with corporate strategy and the global supply chain as the global economy today is led by corporate firms linked within themselves in a complex supply chain, Fig. 7.3 (Maranesi & De, 2020).

Hence, CE today is not only a tool to maximize material stocks but also a proven method to solve the global environmental and societal problems if the corporate strategy complements it. European Union (EU) is an example of CE being implemented as a means for sustainable economic development (Grdic et al., 2020). At the macro-level EU has formulated CE-friendly policies bolstering and facilitating the corporate firms to impregnate CE in their company vision and as actions in their operational activities through innovative business models (Kirchherr et al., 2017). The firms at the micro-level practice different CE norms in their supply chain to help the

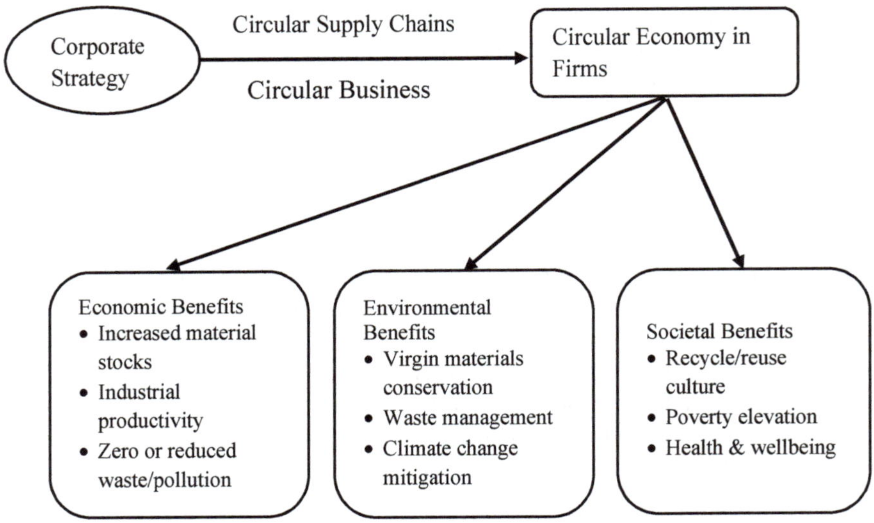

Fig. 7.2 Modern concept of circular economy

Fig. 7.3 Traditional process of sample cutting results noticeable waste

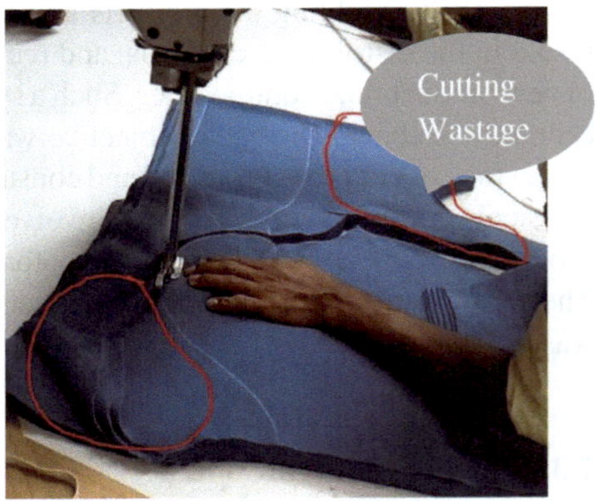

EU achieve the SDG goals set by the UN in a view to eradicating the global problems related to poverty, climate change, equity, and social well-being (Ellen Macarthur Foundation, 2022). Hence, in the modern world, the notion is to facilitate economic development without compromising its environmental and societal consequences. CE is regarded as one of the most effective tools to achieve it.

7.3.1.3 The Three CE Outlooks

Analysis of the modern literature on CE depicts that there are three outlooks to be considered when discussing CE, i.e., the perspective, the implementation (measurement) method, and the strategy (line of action) (Table 7.3).

CE can be explained from two perspectives. The closing cycle perspective, as in the classical concept, is concerned with transforming the linear economic model to a circular one through recycling, reusing, and recovering wastes, thus closing the material or energy loop (Aguilar Esteva et al., 2021). The notion is to design products with slow and close-end loops, enabling every material, component, and product to contribute maximum to the economy (Mestre & Cooper, 2017). A notable feature of closed-loop supply chain studies appears to be focusing more on monetary issues and lacking the integration of sustainability dynamics (MahmoumGonbadi et al., 2021). However, the system (holistic) perspectives are not only confined to the transition of linear models to circular models but consider the application of CE principles in a holistic way (Mangers et al., 2021). The holistic perspective also studies the implication of CE adoption in the industry and the long-term impacts on the environment and society to ensure net benefits to the global system (Iacovidou et al., 2021). According to the system approach, CE is more conceivable in the modern globalized world where every individual, industry, nation, and economic system is interconnected through information technology (Vanchukhina et al., 2019).

Measurement of the production and consumption pattern of an economy is a must for implementing CE. Life cycle assessment is an established method to measure CE's impacts, thus helping measure the environmental consequences of a specific product or materials (Peña et al., 2021). Most CE scholars agreed that the life cycle assessment is the most effective tool to evaluate options for CE solutions (Haupt & Zschokke, 2017). Though the use of LCA is prominent at the product level, several

Table 7.3 Circular economy outlook

Circular economy outlook		References
Perspective	Closing cycle perspective	Aguilar Esteva et al. (2021) and Mestre and Cooper (2017)
	System/holistic perspective	Iacovidou et al. (2021) and Mangers et al. (2021)
Implementation/measurement method	Life cycle assessment	Peña et al. (2021) and Haupt and Zschokke (2017)
	Material flow analysis	Gao et al. (2020) and Merli et al. (2018)
Strategy/line of action	Transform social and economic dynamics at macro-level	Kurniawan et al. (2022) and Zhu et al. (2019)
	Implement circular processes in the industries at the micro-level	Saavedra et al. (2018) and Ghisellini et al. (2016)

applications show its effective use at the societal level and make it a universal tool for measuring CE impacts (Roos Lindgreen et al., 2021). Another tool used in assessing CE, especially to identify values lost through waste in a complex manufacturing supply chain, is the material flow analysis (Khairul Akter et al., 2022). The tool is used extensively to evaluate CE impacts and frameworks in regional economies like China (Gao et al., 2020). In complex systems, material flow analysis is used to measure material balances. It can provide potentially beneficial information to firms looking to handle wastes economically, looking at waste as valuable input (Millette et al., 2020).

Strategy-wise, two approaches can be distinguished from literature to adopt CE. Simply put, one is at the macro-level and one at the micro-level. However, to ensure the diffusion of benefits to society and ensure a sustainable environment in the longer term, consideration of both macro- and micro-levels is essential (Maranesi & De Giovanni, 2020). China is a great example of how circularity is taken as a top-down national policy to be implemented by transforming the economy and society. They opted for digitization, a big driving force toward their economic transformation within the CE framework (Kurniawan et al., 2022). Digitization enabled smooth connectivity between the government and the private sector, where the emphasis is given to accountability for implementing CE. To regulate the private sector, the use of policy is comprehensive through various government agencies, especially regarding resources and production in light of CE (Zhu et al., 2019). This is a macro-level approach to transform the industries and comply with them to adopt CE principles with stringent accountability.

On the other hand, the EU is an example where CE adoption has been activated from the micro-level. Impregnation of concepts like industrial ecology, industrial symbiosis with company vision, and adopting CE principles at the firm level to achieve SDG goals is evident in the EU (Saavedra et al., 2018). Apart from the EU, in USA and Japan, a bottom-up approach is followed where the industry is at the forefront of CE drive with environmental and waste management strategies (Ghisellini et al., 2016). Adoption of cleaner production patterns at the firm level, and the involvement of consumers are the key features of this approach.

7.3.2 The Practical Implications

Definitions of CE are diverse and have been analyzed from different viewpoints in the literature. One aspect of CE that can be noted from the above discussion is that it is not an aim but rather a pathway to achieve the aim of sustainability. Thus, the practical implication of CE may differ in different industries though the target aim is similar, which can be explained by the theory of Gallie published in 1956 (Meers, 2021), in which he describes concepts like Cas essentially contested because there is agreement on the goals. Still, there is ambiguity on the methodology to reach the goal and disagreements on the units of analysis of the CE dynamism. The implication of CE in different tiers of the economy, i.e., the government, public sector, private sector, industry unit, and at the consumer level, is thus different. Economic implications are easier to measure as the monetary unit is widely accepted, but environmental and societal implications are debated due to the difficulty in quantifying them. Many governments and even the firms are using CO_2 emission reduction as a unit to depict their attempt to save the environment (Wang & Feng, 2018). This also represents how circular they are or to what extent they are adopting the circular principle. At the societal level, consumer awareness is vital. Consumers inherently desire a healthy lifestyle, however, understanding which one is a healthy lifestyle may not be clear to all (Zhang & Dong, 2020). Media and promotion are regarded as effective ways to increase consumer awareness. However, more awareness does not always lead to better actions. Tan et al. (2022) described this phenomenon as an intention-action gap and identified it as a major barrier to CE implementation at a societal level. Survey research at IKEA reveals that most consumers prefer environmentally friendly products, but only a few are ready to pay more (Biekert, 2021). As such, the implication of CE is a complex dynamic to put in a common framework. However, it can be said that the net result of implementing CE by different actors in different segments of the economy is to drive sustainability and make progress economy-wise, environment-wise, and society-wise.

7.4 Circular Economy as a Sustainable Development Tool

7.4.1 Case 1: "Waste"—A Fictional Term in the Apparel Industry of Bangladesh

The term "waste" is a myth in the apparel industries of Bangladesh. It's a common saying in clothing industries that "dust mixed in the apparel waste turn into money too," which means utilizing every single fabric used in the manufacturing, even the wastage of the bulk production or sample section. Material consumption, waste generation, waste disposal in bulk production, and traceability of apparel waste have recently received more attention from academics than in the past (Khairul Akter et al., 2022). Despite being enormous, the material waste produced in the product development (PD) division has not been given the same concern. This case study focuses on the classification of PD waste, its management, and the USD worth of a well-known Bangladeshi textile manufacturer. The case factory is located in Narayangonj, Dhaka, Bangladesh, a large-scale 100% export-oriented knit composite with 84.29 million annual revenue in the 2021–22 fiscal year. Field observations and open-ended interviews with three plant employees were used to gather the data. The head of the PD unit, the assistant manager of PD, and a senior merchandiser of the plant participated in the interviews. PD unit has 500 pieces daily, producing unavoidable cutting wastes, surplus fabric, and rejected samples addressed as product development or PD waste.

7.4.1.1 Cutting Wastes

Comparing sample cutting to bulk cutting, Fig. 7.1 shows that sample cutting is done piece by piece and is responsible for 25–35% of the cutting waste. Initially, large plastic drums are used for collecting cutting wastages. After filling the drums with waste, operators often wrap the waste in large poly bags and deliver it immediately to the factory's final waste collecting division. During observation, it appeared like that generated waste had disappeared from the sample section's floor in the length of an eye blink. The final destination of the PD waste is concealed and difficult to find in the factory. The team managing the PD wastes operates independently, leaving no data behind due to data secrecy. The waste collecting sector uses no computers, and all records are maintained manually. All these cutting wastes are eventually sold to local traders as "*Jhu*t" for a price ranging from 0.12 to 0.95 USD/kg, depending on the type, size, quality, and market demand of *Jhut*. It was found that this apparently "waste" was employed as a "resource" in the local economy by recycling, reusing, and re-selling as value-added products. So, even though the entire process is informal and limited to Bangladesh's local market, converting apparel waste into value-added products is a remarkable example of sustainable practice (Khairul Akter et al., 2022).

7.4.1.2 Surplus Fabric

Leftover fabric is the quantity of excess fabric purchased by the company to prevent shortages of fabric when additional samples are required due to rejection, the prerequisite for a minimum amount of sample dyeing machine, and other unavoidable flaws like crease marks, shrinkage, etc. Repetition of the same procedure would almost certainly result in missing the order because fabric booking for sample preparation is a lengthy process. Therefore, the factory booked 25–50% additional fabric when booking the material to secure the order. Although the manufacturer cannot avoid purchasing this additional fabric, which results in a substantial quantity of waste in the PD sector, the business uses this so-called waste for various purposes. For example, excess fabrics are stored in the PD section and used in the bulk if balance is possible. It is also used for product development of subsequent orders if fabric specifications are similar or to produce samples of repeat orders. Another excellent application of this surplus as the resource is used in the training school of the factory for skill development of the cutting and sewing operators. A small percentage of 10–15% extra fabric is used for training purposes. Suppose still there is a balance that the factory cannot utilize. In that case, it is stored in the PD section until shipment, and then after dispatching the order from the factory, surplus fabrics are sold as wastages.

7.4.1.3 Rejected Pieces

The percentage of waste from rejected samples is insignificant, accounting for less than 1% of all samples produced. Rejected items are sold as stock lots to local markets rather than being directly disposed of in a landfill or destroyed in another way.

7.4.1.4 Economic Value of Annual Cutting Waste of PD Section

The study found that depending on the fabric specifications and amount of cutting, the sample or PD section generates 80–100 kg of cutting wastes daily. The amount of cutting waste has been traced from 8.00 am to 6.00 pm to find empirical data. Three cutting waste drums weighing 34.24, 14.36, and 21.10 kg, respectively, were identified after five hours of surveillance (Fig. 7.4). After taking the weight of the drum into account, the cutting waste was weighed 54.7 kg. The following four hours (2.00–6.00) pm produced 29.3 kg of cutting trash. Out of the 250 kg of fabrics cut that day, 84 kg of cutting waste was generated, or 33.6% of cutting wastes produced on that particular day. The factory typically has 26 working days, and if we assume the cutting waste has a minimum selling price of 0.12 USD, the estimated annual revenue from just the PD section cutting waste is 84 × 26 × USD 0.12, or USD 3144.96~USD 3200.

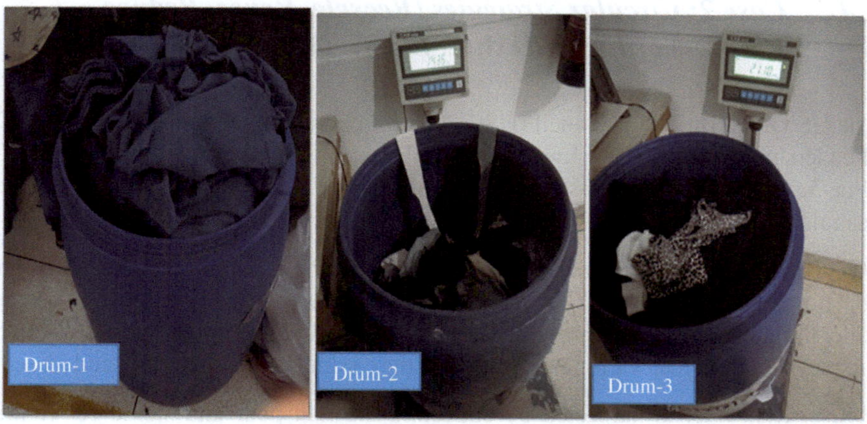

Fig. 7.4 Three drums full of cutting wastes (8.00 am–1.00 pm)

Fig. 7.5 Circular strategies (recycle, reuse, reduce, recover) of Envoy Textiles Ltd (ETL), Bangladesh, **a** LEED badge, **b** effluent treatment plant (ETP), **c** reservoir for rainwater harvesting, and **d** Ozone finishing machine

7.4.2 Case 2: Circular Strategies (Recycle, Reuse, Reduce, Recover) of Envoy Textiles Ltd. (ETL)

The US Green Building Council (USGBC) awarded Envoy Textiles Limited LEED (Leadership in Energy and Environmental Design) Platinum certification in 2016 (Fig. 5a). Envoy Textiles was the first ever to achieve LEED Platinum of any textile manufacturer in Bangladesh and denim manufacturer worldwide.

Recycling: Company's Effluent Treatment Plant (ETP) discharges only treated water to nature, and most of the ETP water is reused for other internal purposes, such as toilet flash, gardening, and fire hydrant (Fig. 5b). Hard wastes of the spinning section are recycled from ball warping, and 100% of the carding waste is recycled and reused.

Reuse: Rainwater is harvested by collecting the rain pour and stored for future use, rather than allowing it to run off. It provides an independent water supply for firefighting and other usages. An aesthetic lake (Fig. 5c), has been created, used for rainwater harvesting, water supply for fire hydrants, and fish farming. Cooling water from the finishing section machine is reused through process water optimization.

Reduce: Groundwater consumption is planned to be reduced to 50% by 2025. To do so, Zero Liquid Discharge will be implemented by recycling ETP-treated water using Reverse Osmosis (RO) technology. Using Ozone (O_3) technology, up to 10% of water is being saved in finishing processes (Fig. 5d). The use of caustic usage has been reduced by installing a Caustic Recovery Plant. The CO_2 reduction roadmap to 2028 is to reduce emissions by 60%.

Recover: Auto blow down, condensate recovery system, economizer, and exhaust-gas-boiler have been installed as a part of process optimization. The factory has a regular air leak detection system, steam trap management system, and heat recovery from stenter machines for energy optimization.

7.5 Conclusions

Sustainable development has been a key global policy for a sustainable planet and resources, and circular economy has been considered as a major commercial opportunity to achieve that goal. It could support the development of new industries and jobs and increase the efficient use of natural resources. Transitioning to a circular economy can create opportunities and benefits for society and the environment. Cooperation and innovation across the spectrum of government, industry, research and development, and consumption practices will be essential to prolong product life cycles, minimize waste, and implement other sustainable ecological practices.

References

Aguilar Esteva, L. C., Kasliwal, A., Kinzler, M. S., Kim, H. C., & Keoleian, G. A. (2021). Circular economy framework for automobiles: Closing energy and material loops. *Journal of Industrial Ecology, 25*(4), 877–889. https://doi.org/10.1111/JIEC.13088

Biekert, M. (2021). Greenwashing less dangerous than silence on climate. https://www.bloomberg.com/news/articles/2021-11-09/ikea-at-cop26-ceo-urges-transparency-action-on-climate-change. Bloomberg.

Breuer, A., Janetschek, H., & Malerba, D. (2019). Translating sustainable development goal (SDG) interdependencies into policy advice. *Sustainability, 11*(7), 2092. https://doi.org/10.3390/su11072092

de Almeida, R., de Souza, R. G., & Campos, J. C. (2021). Lessons and challenges for the recycling sector of Brazil from the pandemic outbreak of COVID-19. *Waste Disposal and Sustainable Energy, 3*(2), 145–154. https://doi.org/10.1007/s42768-021-00075-y

Ellen MacArthur Foundation. (2022). Transitioning to a circular economy. *In Transitioning to a Circular Economy*. https://doi.org/10.1596/37331

Esposito, M., Tse, T., & Soufani, K. (2018). Introducing a circular economy: New thinking with new managerial and policy implications. *California Management Review, 60*(3), 5–19. https://doi.org/10.1177/0008125618764691

Galvão, G. D. A., Evans, S., Ferrer, P. S. S., & de Carvalho, M. M. (2022). Circular business model: Breaking down barriers towards sustainable development. *Business Strategy and the Environment, 31*(4), 1504–1524. https://doi.org/10.1002/bse.2966

Gao, C., Gao, C., Song, K., & Fang, K. (2020). Pathways towards regional circular economy evaluated using material flow analysis and system dynamics. *Resources, Conservation and Recycling, 154*, 104527. https://doi.org/10.1016/J.RESCONREC.2019.104527

Geissdoerfer, M., Savaget, P., Bocken, N. M. P., & Hultink, E. J. (2017). The circular economy—A new sustainability paradigm? *Journal of Cleaner Production, 143*, 757–768. https://doi.org/10.1016/j.jclepro.2016.12.048

Ghisellini, P., Cialani, C., & Ulgiati, S. (2016). A review on circular economy: The expected transition to a balanced interplay of environmental and economic systems. *Journal of Cleaner Production, 114*, 11–32. https://doi.org/10.1016/j.jclepro.2015.09.007

Haffar, M., & Searcy, C. (2017). Classification of trade-offs encountered in the practice of corporate sustainability. *Journal of Business Ethics, 140*(3), 495–522. https://doi.org/10.1007/s10551-015-2678-1

Haidar, H. M. (2021). Historical perspectives and the emergence of the sustainability concept in organizations. *Open Journal of Business and Management, 09*(5), 2277–2298. https://doi.org/10.4236/ojbm.2021.95123

Harrington, L. M. B. (2016). Sustainability theory and conceptual considerations: A review of key ideas for sustainability, and the rural context. *Papers in Applied Geography, 2*(4), 365–382. https://doi.org/10.1080/23754931.2016.1239222

Haupt, M., & Zschokke, M. (2017). How can LCA support the circular economy?—63rd discussion forum on life cycle assessment, Zurich, Switzerland, November 30, 2016. *International Journal of Life Cycle Assessment, 22*(5), 832–837. https://doi.org/10.1007/s11367-017-1267-1

Iacovidou, E., Hahladakis, J. N., & Purnell, P. (2021). A systems thinking approach to understanding the challenges of achieving the circular economy. *Environmental Science and Pollution Research International, 28*(19), 24785–24806. https://doi.org/10.1007/s11356-020-11725-9

Izvercian, M., & Ivascu, L. (2015). Waste management in the context of sustainable development: Case study in Romania. *Procedia Economics and Finance, 26*(15), 717–721. https://doi.org/10.1016/S2212-5671(15)00825-4

Jaeger, B., & Upadhyay, A. (2020). Understanding barriers to circular economy: Cases from the manufacturing industry. *Journal of Enterprise Information Management, 33*(4), 729–745. https://doi.org/10.1108/JEIM-02-2019-0047

Klarin, T. (2018). The Concept of Sustainable Development: From its Beginning to the Contemporary Issues. *Zagreb International Review of Economics and Business, 21*(1), 67–94. https://doi.org/10.2478/zireb-2018-0005

Karim, M. E., Karim, R., Islam, M. T., Muhammad-Sukki, F., Bani, N. A., & Muhtazaruddin, M. N. (2019). Renewable energy for sustainable growth and development: An evaluation of law and policy of Bangladesh. *Sustainability, 11*(20). https://doi.org/10.3390/su11205774

Karlsson, M., & Hovelsrud, G. K. (2021). "Everyone comes with their own shade of green": Negotiating the meaning of transformation in Norway's agriculture and fisheries sectors. *Journal of Rural Studies, 81*, 259–268.

Khairul Akter, M. M., Haq, U. N., Islam, M. M., & Uddin, M. A. (2022). Textile-apparel manufacturing and material waste management in the circular economy: A conceptual model to achieve sustainable development goal (SDG) 12 for Bangladesh. *Cleaner Environmental Systems, 4*(January), 100070. https://doi.org/10.1016/j.cesys.2022.100070

Kirchherr, J., Reike, D., & Hekkert, M. (2017). Conceptualizing the circular economy: An analysis of 114 definitions. *In Resources, Conservation and Recycling, 127*. https://doi.org/10.1016/j.resconrec.2017.09.005

Kurniawan, T. A., Liang, X., O'Callaghan, E., Goh, H., Othman, M. H. D., Avtar, R., & Kusworo, T. D. (2022). Transformation of solid waste management in China: Moving towards sustainability through digitalization-based circular economy. *In Sustainability, MDPI, 14*(4), 2374. https://doi.org/10.3390/su14042374

MahmoumGonbadi, A., Genovese, A., & Sgalambro, A. (2021). Closed-loop supply chain design for the transition towards a circular economy: A systematic literature review of methods, applications and current gaps. *Journal of Cleaner Production, 323*, 129101. https://doi.org/10.1016/j.jclepro.2021.129101

Mangers, J., Minoufekr, M., Plapper, P., & Kolla, S. (2021). An innovative strategy allowing a holistic system change towards circular economy within supply-chains. *Energies, 14*(14), 4375. https://doi.org/10.3390/EN14144375
Manioudis, M., & Meramveliotakis, G. (2022). Broad strokes towards a grand theory in the analysis of sustainable development: A return to the classical political economy. *New Political Economy, 27*(5), 866–878. https://doi.org/10.1080/13563467.2022.2038114
Maranesi, C., & De Giovanni, P. (2020). Modern circular economy: Corporate strategy, supply chain, and industrial symbiosis. *Sustainability, 12*(22), 1–25. https://doi.org/10.3390/su12229383
Meers, J. (2021). 'Home' as an essentially contested concept and why this matters. *Housing Studies*, 1–18. https://doi.org/10.1080/02673037.2021.1893281
Mendes, I. (2020). The Circular Economy:An Ancient Term That Became Polysemic. ISEG – Departamento de Economia. https://www.repository.utl.pt/handle/10400.5/20883
Mensah, J. (2019). Sustainable development: Meaning, history, principles, pillars, and implications for human action: Literature review. *Cogent Social Sciences, 5*(1), 1–21. https://doi.org/10.1080/23311886.2019.1653531
Merli, R., Preziosi, M., & Acampora, A. (2018). How do scholars approach the circular economy? A systematic literature review. *Journal of Cleaner Production, 178*, 703–722. https://doi.org/10.1016/J.JCLEPRO.2017.12.112
Meschede, C. (2020). The sustainable development goals in scientific literature: A bibliometric overview at the meta-level. *Sustainability, 12*(11). https://doi.org/10.3390/su12114461
Mestre, A., & Cooper, T. (2017). Circular product design. A multiple loops life cycle design approach for the circular economy. *Design Journal, 20*(sup1), S1620–S1635. https://doi.org/10.1080/14606925.2017.1352686
Millette, S., Eiríkur Hull, C., & Williams, E. (2020). Business incubators as effective tools for driving circular economy. *Journal of Cleaner Production, 266*, 121999. https://doi.org/10.1016/j.jclepro.2020.121999
Moore, J. E., Mascarenhas, A., Bain, J., & Straus, S. E. (2017). Developing a comprehensive definition of sustainability. *Implementation Science, 12*(1), 110. https://doi.org/10.1186/s13012-017-0637-1
Nikolaou, I. E., Jones, N., & Stefanakis, A. (2021). Circular economy and sustainability: The past, the present and the future directions. *Circular Economy and Sustainability, 1*(1), 1–20. https://doi.org/10.1007/s43615-021-00030-3
Peña, C., Civit, B., Gallego-Schmid, A., Druckman, A., Pires, A. C., Weidema, B., Mieras, E., Wang, F., Fava, J., Canals, L. M., Cordella, M., Arbuckle, P., Valdivia, S., Fallaha, S., & Motta, W. (2021). Using life cycle assessment to achieve a circular economy. *International Journal of Life Cycle Assessment, 26*(2), 215–220. https://doi.org/10.1007/s11367-020-01856-z
Presidência da República (2010). Federal Law n. 12.305. Brazil: Brasília (DF). http://www.planalto.gov.br/ccivil_03/_ato2007-2010/2010/lei/l12305.htm. Accessed 18 August 2022.
Purvis, B., Mao, Y., & Robinson, D. (2019). Three pillars of sustainability: In search of conceptual origins. *Sustainability Science, 14*(3), 681–695. https://doi.org/10.1007/s11625-018-0627-5
Ranjbari, M., Shams Esfandabadi, Z., Zanetti, M. C., Scagnelli, S. D., Siebers, P. O., Aghbashlo, M., Peng, W., Quatraro, F., & Tabatabaei, M. (2021). Three pillars of sustainability in the wake of COVID-19: A systematic review and future research agenda for sustainable development. *Journal of Cleaner Production, 297*, 126660. https://doi.org/10.1016/j.jclepro.2021.126660
Razzaq, A., Ansari, N. Y., Razzaq, Z., & Awan, H. M. (2018). The impact of fashion involvement and pro-environmental attitude on sustainable clothing consumption: The moderating role of Islamic religiosity. *SAGE Open, 8*(2). https://doi.org/10.1177/2158244018774611
Río, Pablo Del, Christoph P. Kiefer, Javier Carrillo-Hermosilla, and Totti Könnölä. (2021). The Circular Economy Economic, Managerial and Policy Implications. *Revista Gestão & Sustentabilidade*. https://doi.org/10.1007/978-3-030-74792-3
Roos Lindgreen, E., Mondello, G., Salomone, R., Lanuzza, F., & Saija, G. (2021). Exploring the effectiveness of grey literature indicators and life cycle assessment in assessing circular economy

at the micro level: A comparative analysis. *International Journal of Life Cycle Assessment, 26*(11), 2171–2191. https://doi.org/10.1007/s11367-021-01972-4

Ruggerio, C. A. (2021). Sustainability and sustainable development: A review of principles and definitions. *Science of the Total Environment, 786*, 147481. https://doi.org/10.1016/j.scitotenv.2021.147481

Saavedra, Y. M. B., Iritani, D. R., Pavan, A. L. R., & Ometto, A. R. (2018). Theoretical contribution of industrial ecology to circular economy. *In Journal of Cleaner Production, 170.* https://doi.org/10.1016/j.jclepro.2017.09.260

Sakalasooriya, N. (2021). Conceptual analysis of sustainability and sustainable development. *Open Journal of Social Sciences, 09*(3), 396–414. https://doi.org/10.4236/jss.2021.93026

Schandl, H., K. S., & Baynes, T. M. (2021). *A circular economy road map for plastics, tyres, glass and paper in Australia.* CSIRO.

Shi, L., Han, L., Yang, F., & Gao, L. (2019). The evolution of sustainable development theory: Types, goals, and research prospects. *Sustainability, 11*(24), 1–16. https://doi.org/10.3390/su11247158

Smith, M. J. (2020). Sustainable development goals: Genuine global change requires genuine measures of efficacy. *Journal of Maps, 16*(2), (i)–(iii). https://doi.org/10.1080/17445647.2020.1749386

Stahel, W. R. (2020). History of the circular economy. The historic development of circularity and the circular economy. In The circular economy in the European Union (pp. 7–19). Springer. https://doi.org/10.1007/978-3-030-50239-3_2

Sverko Grdic, Z., Krstinic Nizic, M., & Rudan, E. (2020). Circular economy concept in the context of economic development in EU countries. *Sustainability, 12*(7), 3060. https://doi.org/10.3390/su12073060

Tan, J., Tan, F. J., & Ramakrishna, S. (2022). Transitioning to a circular economy: A systematic review of its drivers and barriers. *Sustainability, 14*(3), 1757. https://doi.org/10.3390/su14031757

Ukko, J., Saunila, M., Rantala, T., & Havukainen, J. (2019). Sustainable development: Implications and definition for open sustainability. *Sustainable Development, 27*(3), 321–336. https://doi.org/10.1002/sd.1904

UN Environmental Program. (2019). Beat plastic pollution. https://www.unep.org/interactives/beat-plastic-pollution/. United Nations Environment Programme.

Vanchukhina, L. I., Leybert, T. B., Khalikova, E. A., Rudneva, Y. R., & Rogacheva, A. M. (2019). Industry 4.0 and closed-loop economy in the context of solving the global problems of modern times. In *Studies in systems, decision and control* (p. 169) Springer International Publishing. https://doi.org/10.1007/978-3-319-94310-7_4

Virtanen, P. K., Siragusa, L., & Guttorm, H. (2020). Introduction: Toward more inclusive definitions of sustainability. *Current Opinion in Environmental Sustainability, 43*, 77–82. https://doi.org/10.1016/j.cosust.2020.04.003

Wang, M., & Feng, C. (2018). Using an extended logarithmic mean Divisia index approach to assess the roles of economic factors on industrial CO_2 emissions of China. *Energy Economics, 76*, 101–114. https://doi.org/10.1016/j.eneco.2018.10.008

Zhang, X., & Dong, F. (2020). Why do consumers make green purchase decisions? Insights from a systematic review. *International Journal of Environmental Research and Public Health, 17*(18), 6607. https://doi.org/10.3390/ijerph17186607

Zhu, J., Fan, C., Shi, H., & Shi, L. (2019). Efforts for a circular economy in China: A comprehensive review of policies. *Journal of Industrial Ecology, 23*(1), 110–118. https://doi.org/10.1111/jiec.12754

Chapter 8
Integrated Waste Management and Circular Economy

Malonga Hazemba

8.1 Introduction

Waste management continues to be a public health and environmental issue worldwide and requires that the conversation moves beyond the physical scope of waste generation, collection, transportation, and disposal. Globally, a combined estimated 7 to 10 billion tonnes of waste is generated every year constituting Municipal Solid Waste (MSW), Commercial and Industrial waste (C&I), and Construction and Demolition waste (C&D). MSW covers about 2.01 billion tonnes of this combined total and is projected to reach 2.59 billion tonnes by 2030, and 3.40 billion tonnes by 2050 (Silpa et al., 2018; UNEP & ISWA, 2015).

As countries continue to urbanise and their populations continue to grow, citizens demand more goods and services, which leads to a further increase in trade of imported goods, and ultimately an increase in waste generated (Silpa et al., 2018). It thus falls on countries to manage to their waste in an environmentally sound manner by encouraging collaboration among different stakeholders and skillsets, opportunity recognition and benefit realisation (UNEP, 2016).

The Waste Management Hierarchy (WMH) represents the priority preference with which countries should aspire to manage their waste. Although portrayed in a number of ways the general order usually constitutes, in preferential order, prevention, minimisation, re-use, recycling, recovery, treatment, and disposal. Integrated waste management (IWM) applies itself to this hierarchy by seeking to integrate these components into societal systems and aiming to manage waste. However, the term has also been used to refer to integrating the physical elements of waste management, the stakeholders, and all strategic aspects to managing waste (UNEP & ISWA, 2015).

M. Hazemba (✉)
Emerging Themes Unit, Zambia Institute for Policy Analysis and Research, Corner of John Mbita and Nationalist Roads, 10101 Lusaka, Zambia
e-mail: m.hazemba@uq.net.au

S. A. Bandh and F. A. Malla (eds.), *Waste Management in the Circular Economy*,
https://doi.org/10.1007/978-3-031-42426-7_8

There has also been a push to realise the economic value of waste as a resource as a way of waste prevention and has led to talks of transitioning economics from linear models to circular ones. A circular economy seeks to promote waste elimination in the design of products, prolonged material and product circulation, and environmental regeneration. The emphasis is on keeping materials and products in use for as long as possible before sending them to recycling, or allowing them to enter the biological cycle.

This chapter starts by looking at the Waste Management Hierarchy (WMH) and how it relates to integrated waste management (IWM). It then explores the elements of Integrated Sustainable Waste Management (ISWM) and their importance. The focus then returns to waste prevention and how a transition to a circular economy can facilitate ambitions to move up the Waste Management Hierarchy. The chapter then closes on the role that policy plays in this transition and provides some examples as well.

8.2 The Waste Management Hierarchy

A common point of reference in exploring and designing options for a waste management system is the Waste Management Hierarchy (WMH), with the aim being to systematically move up the hierarchy and prevent waste altogether (UNEP, 2016; UNEP & ISWA, 2015; UNEP & UNITAR, 2013). A number of versions of the hierarchy exist, but generally all of them fall into the following categories from most preferred to least preferred: prevention, reduction, recycling, recovery, treatment, and disposal.

8.2.1 Prevention

The ultimate target in a waste management system is that no waste is generated at all. This requires a changed approach in the way products are designed so as to make them last longer, prevent the generation of waste, and ultimately reduce the demand for resources. Therefore, it requires that production and commercial stakeholders be engaged in shifting to a path where the design, material choice, production and marketing of goods can be used to push this agenda. The adoption of cleaner production methods can also be used towards achieving this goal (UNEP, 2016).

8.2.2 Reduction

Reduction looks at limiting the amount of waste generated through actions like re-use, maintenance, and repair. At the core of re-use is the design of products themselves. For end-users to keep on using a product, it must be designed in manner that makes it easy to use again and again. This also calls for businesses to take up practices based on cleaner production and to design products for durability to increase their re-use (UNEP, 2016).

8.2.3 Recycling

Recycling involves the collection, sorting, processing, and transformation of materials into other useful products, thereby saving considerable material and energy resources (UNEP, 2016). This also involves organic waste which is usually comprised of food and agricultural wastes. The biodegradable nature of organic wastes makes them viable for composting and anaerobic digestion and therefore presents an opportunity to divert them away from landfills (UNEP & UNITAR, 2013). A core aspect of recycling is the separation of materials at the source. This enables the efficient capture of recyclable materials before they end up mixed with other waste and become more difficult and expensive to retrieve. Separation requires that all stakeholders, including the designers and producers of goods, waste management service providers, formal and informal sector recyclers, and end-users and consumers of goods, cooperate at all levels of the waste management stream for it to be successful (Lawrence et al., 2020; UNEP & UNITAR, 2013).

8.2.4 Recovery

Recovery can be split into materials recovery and energy recovery. In material recovery, discarded products are broken down and sorted to retrieve useful materials that can then be re-used or recycled. This process also requires a level of separation at source before disposal for easier operation (UNEP, 2016). Energy recovery is whereby waste undergoes combustion or incineration to capture the energy produced. It requires that the composition of the waste is controlled to avoid contamination and toxic emissions. The common methods of energy recovery are combustion, pyrolysis, and gasification. Methane from anaerobic digestion can also be used for energy recovery.

8.2.5 *Treatment*

Treatment of waste through physical, biological, or chemical processes can be used to reduce its volume or character so that it can have little to no effect on the environment once disposed (UNEP, 2016). The operation of facilities that carry out treatment should also be monitored to ensure that any health or environmental impacts from the facilities' operations can be minimised (UNEP, 2016).

8.2.6 *Disposal*

The option of last resort is disposal and is for waste that cannot be processed through the preceding steps of the hierarchy. The purpose of disposal is to provide for the final, safe storage, or release of unwanted materials into the environment at the end of their life cycle (UNEP, 2016). Disposal involves the use of landfills, incineration without energy recovery, and controlled and uncontrolled dumpsites. Landfills are engineered facilities designed to prevent the release of pollutants into soil, water, and air. Incineration without energy recovery is the mere burning of waste for purpose of getting rid of it. Controlled dumpsites are disposal sites operated and managed with some form of registration procedures and control on incoming waste. Uncontrolled dumpsites are unmonitored disposal sites and are deemed unacceptable (UNEP & UNITAR, 2013). The pricing of environmentally sound disposal methods can be used to encourage waste generators to minimise their waste (UNEP, 2016).

8.3 Integrated Waste Management Versus Integrated Sustainable Waste Management

Integrated waste management has mostly been used to refer to the integration of the technical aspects of waste management into a complete system that is designed to protect the health of humans and the environment, while realising the efficient use of resources (Wilson et al., 2013; Wong, 2022). However, the success of any waste management system is dependent on meeting several other factors for it to be successful. The factors generally include the physical elements, the stakeholders involved, and the strategic aspects in and around the system (UNEP & ISWA, 2015).

The physical elements include the aspects of infrastructure and equipment needed to facilitate the management of waste from its generation to its final disposal. This includes factors like storage, collection, and transport, as well as facilities for recycling, treatment, and disposal.

The waste management system also involves a number of players starting from the producers of a product, through to the service providers, and the service users or eventual waste generators. It also involves all the players around the system that ensure its sustainability. This includes governments, municipal authorities, civil society, and non-governmental organisations and international agencies.

The strategic aspects include those factors that can influence the success or failure of a waste management system. These can be political, institutional, economic, financial, or technical. These aspects may also be health-related, social, or environmental among others. It is also said that such systems should factor in legal responsibilities (Batista et al., 2021).

8.4 Elements of Integrated Sustainable Waste Management

All the factors mentioned in the section influence waste management systems, but differ among countries worldwide (Batista et al., 2021). Thus, it is important to understand how they are interconnected and to strike a balance among them for the purpose of informed decision-making (Torkayesh et al., 2022). However, given that the focus of the Waste Management Hierarchy is more on the prevention, processing and disposal of waste, the UN-HABITAT programme developed an analytical framework for solid waste management that incorporates all these factors and creates an Integrated Sustainable Waste Management (ISWM) system (UN-HABITAT, 2010). They grouped these factors under 'physical elements' and 'governance aspects'.

8.4.1 Physical Elements of ISWM

The physical elements of ISWM focus on the social aspects of public health, environmental aspects of environmental protection, and the economic aspects of resource management (3Rs of reduce, re-use, and recycle). All these are needed to make an ISWM system work and remain sustainable (Satori et al., 2018).

8.4.1.1 Public Health

The public health factors revolve around maintaining clean, healthy, and habitable areas through the provision of well-functioning waste collection services. This is important as failure to properly manage the post-consumer or post-user end of the materials cycle could have a negative impact on human and environmental health (UN-HABITAT, 2010).

8.4.1.2 Environmental Protection

The environmental factors look at protecting the environment from uncontrolled dumping of waste through the provision of appropriate waste treatment and disposal services. It aims to put an end to uncontrolled disposal, while increasing the environmental standards of disposal facilities (UN-HABITAT, 2010). It also requires the use of locally appropriate and adaptable technologies and equipment.

8.4.1.3 Resource Management

Resource management focuses on realising the value in waste through the following types of actions: reducing waste generation quantities; prolonging the use of products through re-use, repair, refurbishment, or remanufacture; recycling those materials that can be reintroduced into value chains through extraction or recovery, and in returning the nutrients embedded in organic waste back to the soil through activities like composting and anaerobic digestion (UN-HABITAT, 2010).

8.4.2 *Governance Aspects of ISWM*

The governance aspects of ISWM include the inclusivity of stakeholders, financial sustainability, and having sound institutions and proactive policies.

8.4.2.1 Inclusivity

The well-functioning of an ISWM is dependent on the participation and contribution of service users who are waste generators, service providers who include local authorities and private actors, and service enablers who include the national government. This means that all stakeholders need to be involved in the planning, implementation, and monitoring of waste management systems.

8.4.2.2 Financial Sustainability

Financial sustainability focuses on ensuring that a waste management system can provide an adequate service for the cost it demands, is affordable for the service users, transparent in its costs, and is well-financed through either consistent user charges or government expenditure.

8.4.2.3 Sound Institutions and Proactive Policies

For a waste management system to operate effectively, it needs a strong, well-functioning, and transparent institutional framework that promotes good governance of waste. This will help ensure that services are adequate, effectively capacitated, and accountable, with every stakeholder understanding their role. This may involve having a national policy framework and well-equipped local institutions (UNEP & ISWA, 2015). It may also involve the use of a competitive, transparent and accountable private sector contracted by the municipal authority to carry out services. It should also take into consideration the rolls of micro-, small-, and medium-sized enterprises and that of the informal sector who may already be engaged in such activities and are dependent on them for their livelihoods (UN-HABITAT, 2010).

8.5 Waste Prevention as the Ultimate Goal

As has been earlier established, the prevention of waste is the ultimate goal of the Waste Management Hierarchy and thereby should influence the design of any waste management system. Waste prevention can occur from the moment a raw material is extracted to make a product to the time that product reaches its end-of-use, and therefore can be targeted at almost all stages along the life cycle of the product. According to Cobo et al. (2018), waste prevention reduces the quantity of waste, the negative impacts of waste generation, and the harmful substances that are possibly embedded in materials and products. It may include activities such as the restriction of planned obsolescence of electronic goods, lowering the weight of products, and designing products in a manner that allows for them to be easily disassembled (Cobo et al. 2018). As such, there are two fundamental aspects, these being quantitative and qualitative prevention, under which waste prevention can be understood.

8.5.1 Quantitative Prevention

Quantitative prevention measures are those that aim to reduce the amount of materials or products that end up as waste. Measures that can be taken to achieve prevention on the production side include avoiding unnecessary consumption, using less virgin material in product design and production processes, designing products which are durable and easy to disassemble, and allowing and purchasing fruits and vegetables that are not of perfect size or shape (UNEP & ISWA, 2015). On the consumer side, the aim would be to prolong the use of a product by re-using it, or having it repaired or refurbished (UNEP & ISWA, 2015).

8.5.2 Qualitative Prevention

The aim of qualitative prevention is to improve the environmental performance of products and production processes by eliminating the use of specific hazardous substances in materials or products that end up as waste (UNEP & ISWA, 2015).

Waste prevention, if done right, can lead to a reduction in the demand of resources and improved management or reduction in the use of hazardous materials. Therefore, it is paramount that waste prevention is well-understood by the public and within waste management sector itself (UNEP & UNITAR, 2013) and should lead to better decision-making on both the production and consumption ends of a product life cycle.

8.6 Circular Economy

This chapter has established that preference should be given to waste management solutions at the higher end of the Waste Management Hierarchy with waste prevention being the ultimate goal. It has also been established that for a waste management system to function sustainably it must take into account all the stakeholders involved and the strategic aspects within and around the system. However, the dominant economic model favours the disposal of products. With high rates of urbanisation and population growth, and problems of industrialisation and the lifestyles of people (Satori et al., 2018), this will not be sustainable as the earth's resources cannot cater to so many people and so much of development. As well, the number of areas that can be used for disposal are only so much and will not be able to sustain the waste generated and disposed of from this economic way of doing things (Cobo et al., 2018). Another issue is the costs linked with waste management and how these factors can add increasing weight on municipal budgets (Batista et al., 2021). It is therefore imperative that focus be shifted higher up the Waste Management Hierarchy and waste management systems be designed to support this shift. This is where a transition to a circular economy comes into play.

A circular economy is one in which the concept of waste prevention leans heavily towards designing out waste early in the life cycle of products so that they may be kept in use and re-used for as long as possible. It promotes the use of environmentally sound materials and products that can be used and re-used for as long as possible. Finally, it allows for a reduction in the extraction of new raw materials and the return of useful nutrients back to the nature, thereby giving natural resources a chance to regrow. As illustrated by the Ellen MacArthur Foundation in the butterfly diagram, the circular economy is designed around two distinct cycles: the biological cycle and the technical cycle (Fig. 8.1).

8.6.1 The Biological Cycle

Materials in the biological cycle are those that can be consumed or derived from natural products and are essentially biodegradable. This cycle contains processes capable of returning nutrients back to the soil which in turn can help regenerate natural

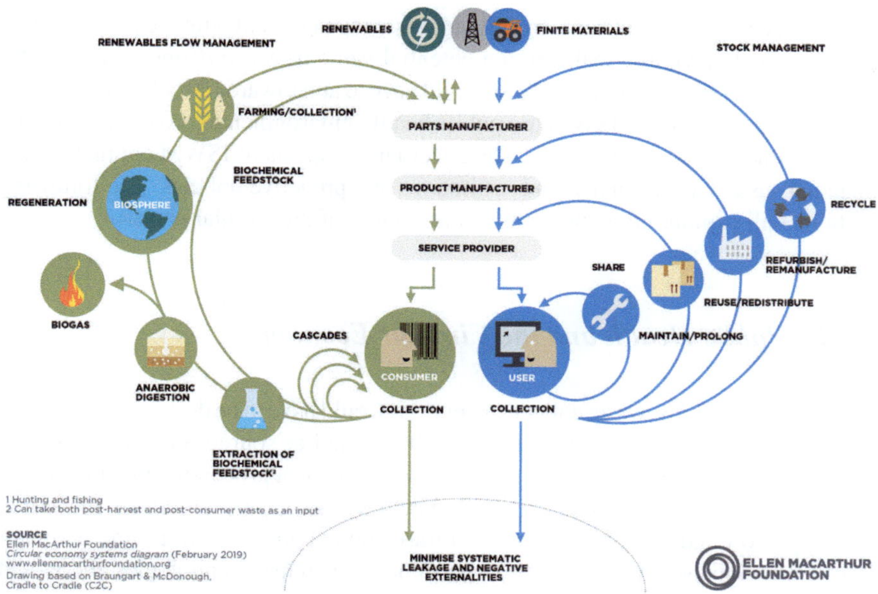

Fig. 8.1 Circular economy butterfly diagram. Source: EMF (2019b)

resources. Activities such as composting and anaerobic digestion can participate in this cycle.

8.6.2 The Technical Cycle

Materials and products in the technical cycle are those that are designed to be used and not consumed. The aim is to keep them in use for as long as possible to avoid them becoming waste. The preferred activities involve sharing, re-use, maintenance, and repair of products. These options are dependent on products being designed for durability. Beyond these, the next preferred options are refurbishment and remanufacture using re-usable materials. These options are better served when products are designed to be easily disassembled. The last option is that of recycling which aims to keep materials in use by reprocessing them into something new.

8.7 ISWM and Circular Economy

In a circular economy, an Integrated Sustainable Waste Management (ISWM) system would have to be designed to support the biological and technical cycles with the aim of moving towards waste prevention. Such a system would also be able to produce materials, energy, and nutrients by linking waste management with the processing of resources and inevitably realising the potential value in waste (Cobo, et al., 2018). Therefore, each element of the ISWM would have to aim towards circularity. To recap, the physical elements of ISWM are public health, environmental protection, and the 3Rs of reduce, re-use and recycle. The governance aspects of ISWM are inclusivity, financial sustainability, and sound institutions and proactive policies. The following sections will explore these elements in the context of the circular economy.

8.7.1 Public Health and the Circular Economy

In a circular economy, waste collection services would have to be designed as resource collection services, with resources preferably sorted at source. For the biological cycle, this would see organic waste being taken to its own processing plant to enable it return to nature. The technical cycle would see materials sorted for recycling, or dropped off and collected for repair, refurbishment and remanufacture, and later resold. This kind of arrangement is dependent on available technologies and would therefore have to be designed in the context of the area of interest.

8.7.2 Environmental Protection and the Circular Economy

A circular economy seeks to avoid the need for disposal. However, there are certain elements that eventually reach beyond usable standards and may need to be safely released back into the environment. In the biological cycle, this can be done through composting and anaerobic digestion of organic waste. In the technical cycle, this brings in the issue of designing products using environmentally safe materials and chemicals so that they do not cause harm to the environment when disposal is required.

8.7.3 Resource Management in the Circular Economy

Resource management seeks to realise the economic value of waste through the acts of reducing waste generation; prolonging the use of products by re-using, repairing, refurbishing and remanufacturing them; recycling those materials that cannot be re-used; and returning nutrients in organic waste back to nature through activities like composting. These activities are the hallmarks of a circular economy and are core to realising the value in materials and products that would otherwise end up as waste.

8.7.4 Inclusivity and the Circular Economy

A transition to a circular economy requires the participation of all stakeholders if it is to succeed. Efforts cannot be left to public authorities alone, but require that all stakeholders be involved in the planning, implementation, and monitoring of the waste management system. The waste generators and those who represent them would need to be made aware what a circular economy is, what it means for their waste management system, and the role they would have to play to make it successful. Waste management service providers would have facilitate or create links with institutions that provide re-use, repair, refurbishment, remanufacturing, composting or anaerobic digestion services, and work with public authorities and waste generators in areas such as the sorting of waste at source, or segregation after collection.

8.7.5 Financial Sustainability and the Circular Economy

The creation of separate resource streams, especially for recycling, would mean an inevitable increase in costs for collection and administration (UNEP & ISWA, 2015). Therefore, having a regular stream of resources and investments to finance such systems is vital. Stakeholders would have to share costs for the waste management system to be successful. This could involve the use of the following instruments among others: public–private partnerships; micro-financed micro-, small-, and medium-sized enterprises, Community-Based Organisations (CBOs) and informal sector services; innovative financing instruments; and Extended Producer Responsibility (EPR) charges (UNEP & ISWA, 2015). Inter-municipal cooperation is another option that has been suggested where municipal authorities move from managing waste independently to working with each other as a way of improving performance and lowering costs (Ferreira et al., 2022).

8.7.6 Sound Institutions and Proactive Policies

A strong institutional framework, a transparent organisational structure for financial and service management, and partnerships with stakeholders like the private sector are essential to a well-functioning ISWM. These will require sufficient capacities to prepare and enforce legislation and also facilitate collaboration with stakeholders. It may also require that the roles of legislator and regulator be made separate for effective and credible enforcement (UNEP & ISWA, 2015).

Having a strong and well-informed circular economy policy would do well in guiding institutions towards the successful implementation of waste management systems that feed into the circular economy. The following section will explore such policies and how they can provide for such a system.

8.8 ISWM and Circular Economy Policy

Waste prevention can occur at almost any stage of the life cycle of a product and may require specific interventions to ensure success. This may include the targeting of policies at specific stakeholders along the supply chain of a product; the extraction, production and transportation stage may be addressed by policies aimed at sustainable production; the consumption stage may be targeted by policies aimed at sustainable consumption; and the post-consumer and waste collection phase may be targeted by specific waste management policies (UNEP & ISWA, 2015).

A variety of policy instruments exist that can be used to ensure an effective integrated waste management system. Notable among these are direct regulation, economic instruments, and social instruments (UNEP & ISWA, 2015). The use of multiple instruments aims to address the broad nature of waste management systems and how different sectors require different interventions towards meeting sustainable consumption and production goals.

8.8.1 Direct Regulation

The use of legislation, including its consistent enforcement, is what constitutes direct regulation. In waste management, this includes having laws and regulations aimed at protecting society's access to public health and a clean environment. Such laws would have to be clear, coherent, and enforced in a fair and consistent manner. In promoting waste prevention, they would also have to address waste reduction and resource recovery as a means of addressing waste management costs and realising the value of recovered resources (UNEP & ISWA, 2015).

8.8.2 Economic Instruments

Economic instruments engage the use of market-based incentives and disincentives to influence the behaviours and practices of stakeholders. Such instruments may include pay-as-you-throw (PAYT) charges aimed at encouraging segregation at source; land-fill, or incineration taxes aimed at limiting their use; fiscal benefits for investments in waste management; and the use of extended producer responsibility (EPR) charges aimed at making producers and importers accountable for the products they release to markets (UNEP & ISWA, 2015).

8.8.3 Social Instruments

Communication, raising awareness, and interactions among stakeholders are necessary when trying to get buy-in for sustainable waste management systems. It goes beyond the provision of information by national or local governments and includes the encouragement and engagement of stakeholders, with the national or local government leading by example through actions such as the use of economically viable and socially acceptable green procurement measures (UNEP, 2016; UNEP & ISWA, 2015).

8.8.4 Circular Economy Policy

Circular economy policy is better driven at national level and aimed towards applying the more preferable options of the Waste Management Hierarchy (UNEP, 2016). It helps to facilitate processes that optimise the performance of other stakeholders. General steps that apply include initiation, development of a circular economy plan, and assignment of national and local government roles (UNEP, 2016).

8.8.4.1 Initiation

Initiation of the development of a circular economy may require the use of either a national strategy statement or legislation. The use of legislation would require the development of a plan that provides mechanisms for action.

8.8.4.2 Circular Economy Development Plan

Adopting a typical plan outline, the circular economy development plan systematically sets out the basic criteria and actions for the government. It should be preferably preceded by the development of guidelines for the development of the plan and take a whole-of-government approach for it to be acceptable. The plan should also have a regular review process given that it seeks to change the way current systems operate.

8.8.4.3 Assignment of National and Local Government Roles

The involvement of all stakeholders is important in a circular economy as roles and costs need to be shared among them. The implementation of a circular economy also requires that the plan is well-communicated to all stakeholders. The national government could take lead on this with a nationwide approach, which could then be followed up by local approaches to get the general public and businesses involved.

8.8.5 *Limitations of Technology*

Apart from the above steps the use of technology whose feasibility is economical and application is socially acceptable is required. This is because countries operate at different levels of development with differing drivers and therefore require technologies that best apply to their conditions. The ultimate goal should be to minimise waste with technologies that are easy to adopt and adapt to, are environmentally sound, and promote operations at the most preferred end of the Waste Management Hierarchy (UNEP, 2016).

8.8.6 *Importance of Data*

Also vital to policy is the need for data on waste management systems. This is essential to knowing if the system is functioning optimally. This data can be gathered by incorporating monitoring and evaluation into waste management systems and maintaining engagement with stakeholders on whether the system is meeting its targets (UNEP & ISWA, 2015).

8.9 Circular Economy Policies and Strategies

The following are the examples of enacted policies, legislation, and instituted strategies aimed at transitioning to a circular economy.

8.9.1 *The Basic Act for the Establishment of a Sound Material-Cycle Society in Japan (2000)*

The Japanese Government enacted a law that enabled them to make a plan that would allow them to comprehensively and systematically promote policies and measures for the establishment of a circular economy. The plan would have basic principles on policies and measures, outline the policies and measures that the government should make, and have other matters necessary to promote the comprehensive and systematic establishment of a circular economy (UNEP, 2016; Government of Japan, 2000). The act also calls for measures including the sharing roles and costs among stakeholders, the taking on of a whole-of-government approach, improved sensitisation on circular economy, and for the plan to be reviewed every five years or so.

8.9.2 *Act to Promote a Circular Economy and Safeguard the Environmentally-Compatible Management of Waste (Circular Economy Act) in Germany (2012)*

In 2012, the German Government enacted into law an act for the purpose of promoting a circular economy to conserve natural resources and ensure that the generation and management of waste did not harm human health and the environment (Government of Germany, 2012). The act obligates waste producers and holders to minimise their waste and prioritises waste minimisation ahead of disposal. In the likelihood of disposal being the only option, the act allows for it to be done in a manner that best ensures that the health of humans and the environment is protected. The act also calls for the use of waste management methods that are technically possible and economically reasonable and further provides guidelines to determining the technology that best applies to costs, benefits, and principles (UNEP, 2016).

8.9.3 *Circular Economy (Waste Reduction and Recycling) Act of the State of Victoria (2021)*

The state of Victoria in Australia enacted into law, in 2021, an act seeking to introduce a circular economy within the state, with the purpose of maximising the continued use of products and waste material across their life cycle, while taking into account their impact on the environment (State Government of Victoria, 2021). The act provides for the assignment of roles with a central recycling body as head, and municipal authorities as recycling service providers. It also calls for best practice in procurement and management of contracts pertaining to waste and recycling services, and further provides service standards. Finally, it establishes a container deposit scheme, provides for data collection and reporting, and creates a system of penalties and other enforcement measures.

8.9.4 *Western Cape Industrial Symbiosis Programme*

A prominent example of the application of circular economy in waste management is the Western Cape Industrial Symbiosis Programme (WISP) in South Africa (Ellen MacArthur Foundation, 2020). Industrial symbiosis involves the coming together of separate industries in an exchange of by-products to competitively take advantage of their value (Cobo, et al., 2018). The city of Cape Town in the Western Cape was generating between five to six thousand tonnes of waste per day, with 87% of this ending up at landfills. However, the City's three landfilling sites were getting close to their maximum and this required immediate action. The WISP linked up businesses whose by-product waste could be used in one another's operations and led to the diversion of 23.3 tonnes of waste from landfills in its pilot year.

8.9.5 Computer Reconditioning Centre in Belo Horizonte

The Computer Reconditioning Centre (CRC) in the city of Belo Horizonte in Brazil is another example of waste management, particularly electronic waste (e-waste) (Ellen MacArthur Foundation, 2019a). To meet demands for digital inclusion in the Belo Horizonte area, the city decided to take advantage of the waste landscape of electronic and electrical equipment by setting up a computer reconditioning centre where private and public institutions would donate equipment, which if suitable would be remanufactured and offered to registered digital inclusion institutions. Any unsuitable parts or materials would be auctioned off to other registered companies who would send them to recycling facilities.

8.10 Summary

This chapter has looked at how waste management continues to be a worldwide public health and environmental issue. Thus, it has sought to explain the role of integrated waste management and circular economy in creating sustainable waste management systems.

It looked at the Waste Management Hierarchy and how it is designed to promote waste management with a preferred order of prevention, reduction, recycling, recovery, treatment, and finally disposal.

The chapter then focused on the concept of integrated waste management and how it shifted to Integrated Sustainable Waste Management, due to the recognition that waste management goes beyond the physical elements and should consider stakeholders and other strategic aspects within and around the system.

The chapter then looked at the physical elements and governance aspects of an Integrated Sustainable Waste Management system and what they entail. The physical elements focused on public health, environmental protection, and resource management through the 3R (reduce, re-use, and recycle) approach. The governance aspects constituted inclusivity, financial sustainability, and sound institutions and proactive policies.

The focus then returned to waste prevention and how it can be applied at almost any level of the life cycle of a product. Waste prevention was then addressed under quantitative and qualitative aspects. Quantitative aspects aim to reduce the amount of materials or products that become waste, while qualitative aspects aim to eliminate the use of specific hazardous substances in materials and products likely to end up as waste so as to improve their environmental performance.

The chapter then looked at the circular economy and why a transition to it is required. It described a circular economy as one where waste is eliminated in the design of products, materials and products are kept in use for longer, and natural resources are allowed to regenerate, with nutrients returning to nature. The focus then moved to the biological and technical cycles of the circular economy and the activities that are considered under these. The section then concluded by considering the circular economy under the elements of the Integrated Sustainable Waste Management system.

The chapter then shifted to look at policy and the role it plays in Integrated Sustainable Waste Management and the circular economy. The section looked at Direct Regulation, Economic Instruments, and Social Instruments, including how they are applied to waste management systems. It then looked at circular economy policy and the requirements that allow for its creation. These include initiation, the creation of a Circular Economy Development Plan, and the assignment of national and local government roles. This section further considered the limitations of technology and the importance of data.

The chapter then concluded by looking at examples of various legislations and plans aimed at transitioning to or creating a circular economy. These included national and state laws, as well as city strategies.

References

Batista, M., Goyannes Gusmão Caiado, R., Gonçalves Quelhas, O. L., Brito Alves Lima, G., Leal Filho, W., & Rocha Yparraguirre, I. T. (2021). A framework for sustainable and integrated municipal solid waste management: Barriers and critical factors to developing countries. *Journal of Cleaner Production, 312*. https://doi.org/10.1016/j.jclepro.2021.127516

Cobo, S., Dominguez-Ramos, A., & Irabien, A. (2018). From linear to circular integrated waste management systems: A review of methodological approaches. *Resources, Conservation and Recycling, 135*, 279–295. https://doi.org/10.1016/j.resconrec.2017.08.003

Ellen MacArthur Foundation. (2019a). *Computer Reconditioning Centre*. s.l.: Ellen MacArthur Foundation.

Ellen MacArthur Foundation. (2019b). Circular economy systems diagram. https://www.ellenmacarthurfoundation.org/

Ellen MacArthur Foundation. (2020). *Africa's First Industrial Symbiosis Programme*. s.l.: Ellen MacArthur Foundation.

Government of Japan. (2000). *The basic act for establishing a sound material-cycle society (act No. 110 of 2000)*. In s.l.:s.n.

Government of Germany. (2012). Act to promote circular economy and safeguard the environmentally compatible management of waste. In *Act reorganising the law on closed cycle management and waste*. s.l.:s.n.

Lawrence, K., Cooper, V., & Kissoon, P. (2020). Sustaining voluntary recycling programmes in a country transitioning to an integrated solid waste management system. *Journal of Environmental Management, 257*, 109966. https://doi.org/10.1016/j.jenvman.2019.109966

Satori, M. (2018). Review of the influencing factors of integrated waste management. *International Journal of Geomate, 15*(48), 34–40. https://doi.org/10.21660/2018.48.55459

Silpa, K., Yao, L., Bhada-Tata, P., & Woerden, F. V. (2018). *What a Waste 2.0: A global snapshot of solid waste management to 2050*. World Bank.

State Government of Victoria. (2021). *Circular economy (waste reduction and recycling) act*. In s.l.:s.n.

Torkayesh, A. E., Rajaeifar, M. A., Rostom, M., Malmir, B., Yazdani, M., Suh, S., & Heidrich, O. (2022). Integrating life cycle assessment and multi criteria decision making for sustainable waste management: Key issues and recommendations for future studies. *Renewable and Sustainable Energy Reviews, 168*. https://doi.org/10.1016/j.rser.2022.112819

UN. (2010). *Solid waste management in the World's cities: Water and sanitation in the World's cities*. Habitat. London; Washington D.C.: Earthscan.

UN Environmental Program, & United Nations Institute for Training and Research. (2013). *Guidelines for national waste management strategies: Moving from challenges to opportunities*. s.l.:United Nations Environment Programme.

UN Environmental Program, & International Solid Waste Association. (2015). *Global waste management outlook*. s.l.: United Nations Environment Programme.

UN Environmental Program. (2016). *Guidelines for framework legislation for integrated waste management*, s.l.: s.n.

Villalba Ferreira, M. V., Dijkstra, G., Scholten, P., & Sucozhañay, D. (2022). The effectiveness of inter-municipal cooperation for integrated sustainable waste management: A case study in Ecuador. *Waste Management, 150*, 208–217. https://doi.org/10.1016/j.wasman.2022.07.008

Wilson, D. C., Velis, C. A., & Rodic, L. (2013). Integrated sustainable waste management in developing countries. *Proceedings of the Institution of Civil Engineers—Waste and Resource Management, 166*(2), 52–68. https://doi.org/10.1680/warm.12.00005

Wong, M. H. (2022). Integrated sustainable waste management in densely populated cities: The case of Hong Kong. *Sustainable Horizons, 2*. https://doi.org/10.1016/j.horiz.2022.100014

Chapter 9
Value Creation from Waste Through Remanufacturing: Understanding Barriers from the Perspective of Business Model Dimensions

Sayed Mohammad Ayati, Jukka Majava, Pasi Rönkkö, and Ehsan Shekarian

9.1 Introduction

The circular economy (CE) narrative and effective transition from a linear economic model is central to waste management and end-of-life (EoL) product treatments. In the CE, growth comes from within by increasing the retainable values from existing economic structures, products, and materials (Webster, 2017). Regenerative and restorative approaches are applied in the CE to keep materials, components, and products in use at their highest utility and value in the long term (Webster, 2017), enabling organizations to retain values of biodegradable waste and technical EoL products (Batista et al., 2018). The regenerative approach extracts biodegradable materials from waste and strives to build cooperation between multiple actors. In the restorative approach, organizations retain values from EoL products that involve reducing, reusing, remanufacturing, refurbishing, and recycling. An important aspect of implementing a restorative value chain is the cooperation of supply chains that take back materials from end-use points to where an organization can retain values (Fernando et al., 2022). Thus, implementing any recovery approaches is hinged on the business model (BM) capabilities and related supply chain operations (Murray et al., 2017).

The recovery approach can keep products in a tighter loop, closest to the original products, serve best values, or in the outer loop, recover less values (Shekarian, 2020). Retainable values are classified into two groups: material value and functional

S. M. Ayati (✉) · J. Majava · P. Rönkkö · E. Shekarian
Industrial Engineering and Management Research Unit, Faculty of Technology, University of Oulu, Oulu, Finland
e-mail: Sayed.Ayati@oulu.fi

E. Shekarian
European Shared Service Center, Rockwell Automation, Rotterdam, The Netherlands

S. A. Bandh and F. A. Malla (eds.), *Waste Management in the Circular Economy*,
https://doi.org/10.1007/978-3-031-42426-7_9

value. The material value is associated in the literature with the extraction of raw materials to produce a product. However, the functional value retains the value added to a product to realize its functionality. According to the inertia principle, recycling is the least preferable approach among various recovery approaches to diminish environmental costs over time and maintain products in the markets for as long as possible (Stahel, 2019). Recycling destroys the functional values of EoL products and disintegrates products to recover the material values (Shekarian et al., 2021). In contrast, remanufacturing retains the highest possible functional values, bringing the product or its component back to its useful life, and making it like new (Kumar et al., 2007).

Remanufacturing is a multi-process that consists of various operations to requalify an EoL product, such as inspecting, cleaning, disassembly, part processing, reassembly, and testing, to meet the desired product standards (Sakao & Sundin, 2019). The literature maintains that remanufacturing, in addition to the reduction of landfilling and mitigating raw material consumption (Peng et al., 2019), diminishes the amount of energy used for production by removing the need for raw materials (Jiang et al., 2019), which is cost-saving for both manufacturers and customers (Xiao et al., 2021). Although remanufacturing builds systematic benefits and recovers the highest amount of the EoL product's functional values, manufacturers are less willing to implement this approach (Masi et al., 2018).

The literature (e.g., de Lima et al., 2021) asserts that several barriers prevent organizations from implementing remanufacturing, such as a lack of the take-back process from consumers, significant divergencies in the stream of products, and labor-intensive processes. In addition, many papers (e.g., Sakao & Sundin, 2019) argue that a remarkable part of the barriers that organizations have encountered in implementing remanufacturing is related to the necessity of rethinking supply chains, and therefore, the way values are created and delivered through their BMs. In adopting remanufacturing, the organization must realize the ways to create, deliver, and capture values by retaining values from EoL products. The value chain in remanufacturing is complicated as its entities are not static but dynamic and require embracing the entire value network, considering different elements and their relations in the system. Effective remanufacturing thinking demonstrates waste increases in the take-back process and services in the absence of any planning for facilitating incentive-driven downstream management ideas. Particularly, waste becomes more problematic when EoL products are not remanufacturable or the reverse logistics solution is not ecological.

Remanufacturing is still at an early stage of implementation, and many organizations are undergoing the exploratory stage of development. To progress past this stage and facilitate the implementation of remanufacturing, it is essential to identify barriers and develop effective methods of overcoming them. Although the authors of previous studies (e.g., Kirchherr et al., 2018) have investigated barriers to the CE implementations, such as life cycle extensions in BMs, Ayati et al. (2022) assert that only a few articles investigated the barriers related to remanufacturing implementation in-depth. Notably, the emphasis on identifying barriers can provide a context for elucidating BM modification and remanufacturing integration with the supply chain.

To date, few articles focus on developing discussions that would address remanufacturing waste of the heavy vehicle industry (HVI) (Xiao et al., 2018). There is also evidence that although some organizations remanufacture HVI waste, their performance falls short of the CE concept (Barreiro-Gen and Lozano, 2020; Oncioiu et al., 2018). The HVI industry is rarely subject to EoL waste management regulations in contrast to the regulations that address waste management of the automotive industry (Directive, 2000). Thus, in the lack of supportive or proactive regulations, other incentives, such as economic, drive organizations to retain value from HVI waste. Considering substantial remaining material and functional values in the EoL HVI products (Ridley et al., 2019), from the economic and environmental perspective, there is a real need to take such EoL products back to the market (Rönkkö et al., 2021).

Therefore, this study is designed to reveal barriers to remanufacturing implementation in HVI. As part of the research, the barriers related to BMs were analyzed in conjunction with a thematic analysis that established a meaningful interrelationship between BM dimensions and barriers. In this regard, identifying and analyzing the more affected dimensions provided a comprehensive conceptual understanding for modifying BM dimensions to create values from remanufacturing. To deepen discussion for creating values from remanufacturing, the following research questions were formulated:

1. What barriers hinder remanufacturing from the perspective of a BM?
2. How can organizations modify the BM dimensions to create value from waste in remanufacturing?

Section 9.2 describes the remanufacturing implementation, BMs, and barriers. Next, Sect. 9.3 explains the methodology this research deployed in the case study analysis. Analyzing the results in Sect. 9.4, we distinguish specific barriers to implementing remanufacturing. Section 9.5 discusses BM modifications required to create values from remanufacturing. Finally, Sect. 9.6 concludes the research.

9.2 Literature Review

9.2.1 Remanufacturing

Remanufacturing is an approach that involves putting an EoL product across a manufacturing process to maintain the functional values of its worn components and produce a like new product (Vogtlander et al., 2017). Implementing CE goals and mitigating industrial waste in accordance with the triple-bottom-line sustainability are the key characteristics that have received much attention in the literature

(Shekarian et al., 2023; Jensen et al., 2019; Korhonen et al., 2018). Although remanufacturing trend implies that many firms intend to employ remanufacturing extensively, this operation needs to be overhauled (Zhang et al., 2019). Substantial investment emphasizes the need for remanufacturing feasibility assessment compared to other recovery methods. On the other hand, studies cite various strategic reasons for exploring remanufacturing, including cost reductions for manufacturers, lower prices for customers, diminished consumption of virgin materials, emissions leakage, higher skilled labor jobs, and the development of industrial technology (Saidani et al., 2020). Various industries, including automotive, aerospace, electrical and electronic equipment (EEE), machinery, and medical equipment, have used remanufacturing in such situations (Akano et al., 2021). According to the literature (e.g., Andersen et al., 2022), it is critical to find effective take-back operations and design production planning to capitalize on remanufacturing in addition to manufacturing engineering.

The remanufacturing operation is based on a take-back system that gathers and returns EoL products to organizations to maintain values (Andersen et al., 2022). As an external process, the take-back process necessitates collaboration among all stakeholders (Ali et al., 2022) to alleviate the uncertain quality and quantity of returned products that affect the inventory, limiting the production plan and profitability (Vogtlander et al., 2017). Moreover, organizations must utilize reverse logistics (RL) to shift EoL items from the collection to the manufacturing site (Mishra et al., 2022). For many years, organizations have been using RL as a structure to manage waste by returning EoL products and extending the use of materials in more than one cycle. RL is a logistical technique that reduces waste formation by returning products to maintain value points (Acerbi & Taisch, 2020).

The remanufacturing process begins after the EoL items are returned to the production site. Many returned products do not match the minimum scope for requalification and cannot be remanufactured. Product designers can create remanufacturable standards by evaluating the remanufacturability of EoL products and developing remanufacturing criteria (Omwando et al., 2018). Standardization is an approach for determining remanufacturability based on time, quality, and cost factors (Sakao & Sundin, 2019). Remanufacturability depends on the product's longevity, functionality, retainable value, component interchangeability, cost of retaining values, technical stability for requalifying, and consumer desire to purchase remanufactured goods (Zhang et al., 2021). It is a complex technique due to the variances in the state of returned items, product type, mechanical conditions, product structure, and degree of damage. As a result, organizations evaluate remanufacturability based on its technical characteristics, economics, and environmental benefits (Dominguez et al., 2021).

Disassembly, cleaning, inspections, mending, and reassembly are technical characteristics. Disassembly means breaking a mechanical product down into its parts. Cleaning removes filth through chemical, mechanical, and physical processes, resulting in a required level of cleanliness. Inspection verifies the product's performance and components to verify warranty eligibility (Saidani et al., 2020). Repairing entails recovering faulty or out-of-place parts and putting them back to work to restore them to their original function. Reassembly is the process of merging new spare parts, recovered parts, and new components to create a remanufactured product. During all steps, operators consider consumer willingness for products which can compel them to downgrade a medium- or low-quality product and send them to recycling centers rather than repair and reassemble them. In addition to technical evaluation, producers reassess economic aspects to decide whether remanufacturing is worthwhile (Rönkkö et al., 2021). Economic feasibility is cost prediction for remanufacturing that evaluates the cost of requalifying EoL components compared to the cost of new parts. This assessment partially happens after scrapping a product to evaluate and characterize the EoL product's ability and potential for remanufacturing. This examination is a labor-intensive operation that has an impact on economic viability. According to the literature, manufacturers must implement a comprehensive decision model for operating effective disassembly activities to deal with waste products (Goltsos et al., 2019; Jiang et al., 2019). Although various papers have undertaken economic analyses, the remanufacturing process is still in its early stages, and the technical steps require further investigation.

Furthermore, numerous papers state that judging remanufacturing based on its technical and economic benefits is not in line with the growing waste and environmental concerns (Cong et al., 2019; Xiao et al., 2018). Remanufacturing is largely recognized as an environmentally favorable process compared to new component manufacturing for several reasons, including reduced use of raw material, energy use, and emission leakages (Saidani et al., 2020). However, the authors emphasize the role of regulations in accelerating remanufacturing implementation and improving environmental advantage. By standardizing developments and national plans, policymakers improve organizations' willingness to remanufacture (Xiao et al., 2018).

9.2.2 Business Model

A business model (BM) is a logic that governs how an organization designs its business to create, deliver, and capture value (Geissdoerfer et al., 2018). A generally accepted BM framework has three to four dimensions. Table 9.1 shows the definitions of dimensions, sub-dimensions, and meanings for BM dimensions (Lüdeke-Freund et al., 2019). BMs were initially implemented to communicate complex company ideas to potential investors (Vermunt et al., 2019), but it has eventually grown into a strategic asset in a competitive economy. Apart from growing competitiveness and earnings, the literature also highlights the significance of reducing the negative effects of the linear BM on the environment (Majava & Hyvärinen, 2022; Masi et al., 2017). Although any change in the architecture of the BM mechanism complicates interactions among supply chain actors (Homrich et al., 2018), organizations must transition to alternative models, such as the sustainable business model (SBM), product-service systems (PSS), or circular business model (CBM). SBM generates long-term value for diverse stakeholders through collaborative and stakeholder-centered management (Geissdoerfer et al., 2020). PSS is a subset of SBM that merges tangible products

Table 9.1 BM dimensions definitions

Dimension	Sub-dimension	Description
Value proposition	• Products • Services	• The organization's main objective in offering products and services needs to gain sufficient revenue compared to costs. It is required to ensure long-term capacity to obtain three bottom lines of sustainability
Value creation	• Target customers • Value creation process	• It is essential to develop a value network with motivated stakeholders and contribute to economic viability, environmental benefits, social concerns, and preparation for long-term challenges in business associations
Value delivery	• Partners and stakeholders • Value delivery process	
Value capture	• Revenues • Costs	• Capturing economic benefits and society's well-being and preserving natural resources in the short and long terms

with intangible services to meet customers' needs (de Sousa et al., 2019; Yang et al., 2017). CBM is similar to SBM and PSS in that it uses the functional and material values of EoL products by implementing circular ideas. Circular notions, such as extending, cascading, purifying, and slowing resources in loops, reduce waste and emission leakage.

Moreover, stakeholders' connections and collaborations improve the efficiency of a BM's circularity features from different perspectives, such as consumer values, public and private societal advantages, and economic and environmental benefits (de Sousa et al., 2019). So, it has been proposed that the configuration of BM dimensions must be circular (Nascimento et al., 2019). A value proposition shows that a company sells products and services to generate income to cover fixed and variable expenditures. Characterizing value propositions with circularity necessitates considerations, such as eco-design and modular architectural product design, to promote take-back and maintain the value of EoL products during the disassembly and quality evaluation phases (Lüdeke-Freund et al., 2019). Creating and delivering value necessitates a shift in the value network of stakeholders to seek economic viability, environmental benefits, social welfare, and long-term business success. Organizations that want to incorporate value capture and circularity must reconsider both natural resource preservation and societal considerations in the short and long terms (de Sousa et al., 2019; Mont et al., 2017).

9.2.3 Remanufacturing Barriers

In contrast to the manufacturing process in which demand and supply are clearly identified, barriers, such as poor quality, availability of EoL products, and customers' willingness, can hinder value creation in remanufacturing and reverse supply chains.

From the perspective of recovery approaches, Ayati et al. (2022) identified that the most frequent barriers affecting the prospect of remanufacturing were related to economics, finance, regulations, and culture. These barriers occur once the original equipment manufacturer (OEM), third party, or subsidiaries establish a take-back process, remanufacturing process, or even once they fulfill the customer's needs. For instance, the cost of remanufacturing determines the costs of initiating remanufacturing, whereas continuation may not be feasible, or a lack of technologies involved in sorting and collecting slows down production, affecting the inventory cost (for detail, see Ayati et al., 2022). Therefore, we examined the barriers in the case company based on the definition of barriers in Ayati et al.

9.3 Methodology

The research approach for this study is the qualitative single case study (Yin, 2011). A single case study is selected to provide a greater depth of exploration at the early stage of adopting remanufacturing in a heavy vehicle company's BM and identifying barriers related to remanufacturing (Karlsson, 2010). The data comprise a triangulation, combining various methods used to study the phenomenon (Voss et al., 2002), including questionnaires, document content analysis, archival research, and workshops.

9.3.1 Industrial Case Company

The case company is a Finnish manufacturer of special heavy vehicles with headquarters in Finland and subsidiaries in other countries. The case company has operated reusing and reconditioning for ten years. Their business idea is to increase the retaining value of EoL products and waste to improve the quality of components. In the recovering center, the company disassembles products, evaluates their quality, and sorts components into two groups, recoverable products and products that cannot be requalified. The sorting process assesses the recoverable components' quality based on three categories: good, very good, and good as new. The case company recognized the additional potential that can retain values from EoL products and waste in terms of remanufacturing.

Further, they looked into functional upgrading of subsidiaries according to customer needs, and initiating remanufacturing is considered a complementary strategy to increase their market share and profits in other locations. The challenges of implementing remanufacturing and upgrading subsidiaries functions are related to the take-back process, retaining values from EoL products, and delivering these remanufactured products to customers. The selection of this manufacturer is made according to collaboration through a research project with the university and the case study. During this project, we analyzed the organizational documents and archives, including archival information, records, and Excel sheets of the reusing rate of components.

9.3.2 Research Data Collection and Sample

In the case company, the qualitative data were collected through questionnaires and two workshops on the current status of the manufacturer in adopting other recovery approaches, such as reusing and maintenance. The units of observation and analysis were the manufacturer and their location. The questionnaires were sent to the organization's members of the higher hierarchy level to involve knowledgeable specialists with different managerial positions, including supply chain manager, after-sales manager, service manager, parts manager, spares manager, and logistics manager. The questionnaires' objective was to gain preliminary insight into adopting heavy vehicle remanufacturing. In formulating questions, the researchers ensured that the language of the questionnaire was consistent within all sectors. The questionnaires were distributed through the Webropol tool, and the access links were emailed to the respondents. Deploying a web-based questionnaire tool helped automatic data inputting and increased data reliability. It was crucial for all managers to recognize different terminology, such as remanufacturing or core acquisition, between sectors to obtain a consistent interpretation of the questions (Flick, 2018). First, the researchers reviewed the literature to obtain a set of questions that contextualized barriers related to recovering approach, the best way to adopt remanufacturing, and exploring new facets of adopting remanufacturing in different locations. Next, a set of basic rules in designing questions was followed to establish a well-formatted questionnaire that reflected the rules of courtesy, presentation, and readability (Ignatow & Mihalcea, 2018). Moreover, the questions were designed by mixing items belonging to different measures, employing different lengths of questions and reversal questions.

9.3.3 Data Analysis

The thematic analysis was deployed to analyze the questionnaires and interviews. Braun and Clarke (2006) highlight the thematic approach to analyzing qualitative data for themes or patterns. This approach identifies, analyzes, and reports patterns involving important aspects of the data related to research questions and outlines a level of patterned response or meaning within the data set. In six steps, the thematic analysis looks for patterns of meaning and issues of potential interests in the data, including a triangulation, such as questionnaires. Additionally, two workshops provided a better understanding of the results behind the responses. The researchers discussed with the representative of the case study organization in workshops to look at the obtained data and determine the general ideas, credibility, and the use of the information. The data analysis step entails transcribing the data to textual coding and looking for emerging categories (Creswell & Creswell, 2017), such as reverse logistics, core availability, and customer intentions.

9.4 Results

The data analysis allowed the researchers to organize the patterns of barriers into three themes: take-back process, remanufacturing process, and market of the remanufactured product. Figure 9.1 shows a map of barrier themes and their associated sub-barriers. For each theme, we provided a table to describe barriers from the subsidiary's perspective.

9.4.1 *Take-Back Process*

The take-back process includes the collection, movement, and transformation of EoL components and products from downstream customers to upstream, where the remanufacturing process begins. In this pattern (Table 9.2), barriers include product acquisition and reverse logistics from any perspective (economics, environmental, or social). At the step of the product acquisition, the quality data and other information are not transferred upstream. The questionnaire results show that the first step of the take-back process, assessing the quality of cores, the core acquisition deposit system, regulation of taking products back to headquarters, and environmental and economic feasibility of taking them back to the headquarters are the challenges for adopting remanufacturing at the headquarters' location. Furthermore, subsidiaries that have a deposit system or can manage core collections could not assess the quality of the cores; therefore, they must be transported to remanufacturing locations.

In addition, subsidiaries mentioned other reasons that impact data monitoring: they could not provide information sources for remanufacturability. Thus, subsidiaries

Fig. 9.1 Thematic classification of barriers

Table 9.2 Theme 1: Take-back process

Barrier	Description
Ability to assess the quality	Due to a lack of instructions, trainings, and measurement tools, subsidies cannot evaluate the economic and technical feasibility of EoL products and waste
Lack of deposit system for collecting cores	Some subsidiaries do not have collection points and deposit systems or warehouses. Some subsidiaries require training and development of collection points
Regulation hinders taking cores back to headquarters	Some subsidiaries mentioned that due to regulations, take-back to headquarters takes time, and sometimes, they could not transport EoL products to the headquarters
Environmental and economic concerns for re-transportation (take-back) to headquarters	Economically and environmentally, transporting cores is not feasible for headquarters. Re-logistics costs include the cost of labor, freight, delay in delivery, and bureaucratic costs for subsidiaries. In addition, headquarters must pay tax for receiving core

cannot assess the current quality to evaluate the cost of remanufacturing and expected economic benefits for the headquarters.

9.4.2 Remanufacturing Process

The remanufacturing process consists of any action performed to retain values from EoL components, aiming to rebuild or requalify EoL products to the new or like new state. Before reassembling and testing the product, operations may include disassembly, inspection, cleaning, replacing, and restoring all missing, defective, worn, or broken parts. The special physical characteristics of components and products make the disassembly process complex (Sakao & Sundin, 2019), which is also reflected in material handling, particularly when a subsidiary is not well equipped. The barriers noted by the subsidiaries concerning this theme are presented in Table 9.3.

9.4.3 Market

The demand in the HVI is very sensitive to market uncertainties. When analyzing the market theme data, the remanufactured product's price was mentioned several times. The case company stated that the product's price was high and providing a lower

Table 9.3 Theme 2: Remanufacturing process

Barrier	Description
Difficulty in disassembly	Some subsidiaries asserted they could not disassemble some products due to their specific or complicated designs
Labor intensive	Working with complicated products and a variety of EoL products requires large personnel
Time scheduling	Different uncertainties, including take-back streams and divergencies of EoL product streams, have influenced production planning in terms of assigning labor and scheduling a smooth workflow
Mature technology or equipment for some components	A lack of mature technology for disassembling and material handling, ergonomic tools, and equipment makes the process difficult and time-consuming
Skills and information	Subsidiaries mentioned that they needed a communication method to exchange knowledge and instruction with the headquarters. In the absence of sufficient training, some levels of the remanufacturing process were not feasible
Lack of standards and norms	Some subsidiaries stated they needed standards and norms; otherwise, for any step of remanufacturing, they had to contact the headquarters, which was not technically and economically feasible. They also mentioned that standards and norms were particularly crucial when an EoL component was remanufactured for the first time
Cost of remanufacturing	Some subsidiaries asserted that operating remanufacturing for engines and some components was pricy and less economical than using new components. Moreover, sending parts to headquarters was not feasible

price with a guarantee and satisfactory quality was an advantage. Moreover, the headquarters looked for opportunities to remanufacture products that limited suppliers or few manufacturers could produce. They classified the products and components that subsidiaries and third parties could not remanufacture under the groups of products that were valuable for initiating remanufacturing. Table 9.4 presents the barriers related to the market theme.

9.5 Discussion

This study investigated how a heavy vehicle manufacturer can create value from EoL products through remanufacturing. It is revealed that the company under study encounters various barriers in the implementation of remanufacturing. The barriers limit remanufacturing ideas in three aspects: (1) when collecting and taking EoL products back is hindered; (2) when remanufacturing EoL products, and (3) when forwarding remanufactured products to customers and selling them. The findings of

Table 9.4 Theme 3: Market

Barrier	Description
Cannibalization	One subsidiary felt that selling remanufactured products caused cannibalization; the rest of the subsidiaries and the headquarters did not notice this risk
Competition between third party and manufacturer	Some subsidiaries mentioned that competition influenced consumers' perceived values. Other subsidiaries found that third parties were well equipped for remanufacturing; therefore, it was a collaboration rather than competition
Market location	Various locations of manufacturers, headquarters, and customers affected collaboration between supply chain actors, creating uncertainty in the supply of remanufactured products and taking EoL products back
Structure of authorized and unauthorized markets (the price of remanufactured product)	The price of new components in some regions is lower than remanufactured ones. Some retailers can also collect EoL products and remanufacture them at lower prices
Tax and regulations	In some regions of subsidiaries, manufacturers are taxed with no tax deduction for remanufactured products

this study suggest that the current case company's BM should be modified to qualify in their circularity features.

From the value proposition viewpoint, this modification should ensure the production of durable remanufactured products with a feasible economy of scale and environmental and social benefits. In terms of the product, modification emphasizes the product's architecture, which poses challenges given the absence of technology. To modify this sub-dimension, a product needs to meet reusability characteristics. A modular architecture is a design with reusability that splits components into common and conceptual and helps eliminate redundant processes for evaluating the quality of, disassembling, and requalifying EoL products (Lindkvist Haziri & Sundin, 2020). Modifying the value proposition involves services. Having difficulty assessing and identifying the performance of EoL products suggests that the nexus between information and products is essential. Enhancing the aftermarket services enables organizations to achieve a better trade-off between information and products during their use-life (Gatenholm et al., 2021), facilitating the take-back process at the end of their use.

Modification of value creation and delivery requires adjustment of stakeholders' views. This alignment is a way to create more than only economic value (Jayarathna et al., 2022). Some barriers indicate an absence of integration between management systems that prevents collaboration between different actors at collecting points and manufacturing sites and fosters a lack of trust between entities. The case study designated barriers related to the market and take-back process that hindered value creation and delivery in remanufacturing. The alignment and integration of organizational capabilities, particularly when the case company strives to upgrade their subsidiary functions, are crucial to overcoming the barriers. More specifically, organization limitations emerge at the subsidiary level such as the take-back process, material handling of EoL products at the manufacturing site, marketing, finance, and customer service. In this regard, we propose that aligning value creation and value delivery should rely on creating a portfolio that integrates several firms (Breuer & Lüdeke-Freund, 2017) and improving the interrelationships between departments of a single firm. Several articles assert that outsourcing remanufacturing activities enhances overcoming barriers (e.g., Misni & Lee, 2017; Pushpamali et al., 2021). Further, considering the limitations related to shipping EoL products to the headquarters, outsourcing improves reverse logistics and effectively configures it taking the supply chain into account.

From the value-capture dimension, carrying out all steps of remanufacturing by the headquarters and subsidiaries is limited due to several financial crisis limitations, such as lack of investments, labor costs, and inflations. Value-capture modifications address strategic decisions, decreasing costs, and increasing revenue. These modifications comprehend the interrelationships between adopting the take-back process, remanufacturing process, and the market. Reverse logistics is a cost factor that is impacted by the current barriers. In addition, the labor cost in the absence of proper training is very high, so improving the training system and changing the facility layout to facilitate material handling can further decrease the cost of remanufacturing at the manufacturing site.

It should be noted that remanufacturing is not typically the core of a business (Vogtlander et al., 2017); rather, remanufacturing is used by organizations as part of their complementary strategy to achieve strong market diffusion and expand their market share. Hence, focusing on cost-saving while operating remanufacturing is imperative for the organization. In addition to cost-saving, organizational responsibility is also tied to environmental operating remanufacturing. During the take-back process, they must follow various laws to abate emissions while retaining the value of their EoL products (Singhal et al., 2020). Accordingly, we claim that developing a circular supply chain (CSC) (Batista et al., 2018) encompasses most barriers as it boosts circular thinking by focusing more on creating values, saving potential in energy and raw material consumption, and opportunities to promote environmentally friendly products. CSC also integrates a closed-loop supply chain with an open-loop of the take-back system (Farooque et al., 2019) to implement effective reverse logistics. This approach allows the organization to adapt its management system for long-term partnerships. By initiating CSC, the organization develops trust and collaboration, offering products as a service, ensuring recourse efficiency and, consequently, enhancing its profitability (Liu et al., 2018; Ramsheva et al., 2019).

Although many organizations contend that remanufacturing increases the risk of cannibalization, remanufacturing durable products is a key opportunity to expand the market (Vogtlander et al., 2017). Consequently, remanufacturing HVI products at a lower price does not compromise the manufacturer's brand image; on the contrary, it bolsters its profitability. Therefore, organizations must modify and develop a value network with actors and partners that remain highly responsive to economic, environmental, and social concerns.

9.6 Conclusions

This paper identifies and analyzes barriers through three themes: the take-back process, the remanufacturing process, and the market. These themes indicate the patterns of barriers that are recolonized by studying a case company in HVI. The identified barriers explore implementing a remanufacturing approach as a complementary business strategy, revealing the need of the case company to modify the current BM model. The value proposition can be improved by considering modular architecture in product design. Moreover, findings indicate that the company needs to adjust their stakeholders' views, refine the interrelationships with organizations, and develop outsourcing to modify the value creation and value delivery dimensions. Finally, to modify value capture, the case company can focus on improving the cost-driving elements, such as reverse logistics. We conclude that focusing on CSC can significantly impact the implementation of remanufacturing, and HVI remanufacturing can extend the market and become a profitable business for the case company.

References

Acerbi, F., & Taisch, M. (2020). A literature review on circular economy adoption in the manufacturing sector. *Journal of Cleaner Production, 273*, 123086. https://doi.org/10.1016/j.jclepro.2020.123086

Akano, D. I., Ijomah, W., & Windmill, J. (2021). Stakeholder considerations in remanufacturability decision-making: Findings from a systematic literature review. *Journal of Cleaner Production, 298*, 126709. https://doi.org/10.1016/j.jclepro.2021.126709

Ali, F., Haapasalo, H., Tampio, K. P., & Haapasalo, H. (2022). Analysing the challenges in stakeholder relationship management in the healthcare process: A social network perspective. *International Journal of Networking and Virtual Organisations, 26*(1/2), 125–156. https://doi.org/10.1504/IJNVO.2022.121934

Andersen, A.-L., Brunoe, T. D., Bockholt, M. T., Napoleone, A., Hemdrup Kristensen, J., Colli, M., Vejrum Wæhrens, B., & Nielsen, K. (2022). Changeable closed-loop manufacturing systems: Challenges in product take-back and evaluation of reconfigurable solutions. *International Journal of Production Research*, 1–20. https://doi.org/10.1080/00207543.2021.2017504

Ayati, S. M., Shekarian, E., Majava, J., & Wæhrens, B. V. (2022). Toward a circular supply chain: Understanding barriers from the perspective of recovery approaches. *Journal of Cleaner Production, 359*, 131775. https://doi.org/10.1016/j.jclepro.2022.131775

Barreiro-Gen, M., & Lozano, R. (2020). How circular is the circular economy? Analysing the implementation of circular economy in organisations. *Business Strategy and the Environment, 29*(8), 3484–3494. https://doi.org/10.1002/bse.2590

Batista, L., Bourlakis, M., Smart, P., & Maull, R. (2018). In search of a circular supply chain archetype—A content-analysis-based literature review. *Production Planning and Control, 29*(6), 438–451. https://doi.org/10.1080/09537287.2017.1343502

Braun, V., & Clarke, V. (2006). Using thematic analysis in psychology. *Qualitative Research in Psychology, 3*(2), 77–101. https://doi.org/10.1191/1478088706qp063oa

Breuer, H., & Lüdeke-Freund, F. (2017). Values-based network and business model innovation. *International Journal of Innovation Management, 21*(3), 1750028. https://doi.org/10.1142/S1363919617500281

Cong, L., Zhao, F., & Sutherland, J. W. (2019). A design method to improve end-of-use product value recovery for circular economy. *Journal of Mechanical Design, 141*(4). https://doi.org/10.1115/1.4041574

Creswell, J. W., & Creswell, J. D. (2017). *Research design: Qualitative, quantitative, and mixed methods approaches*. Sage publications.

de Lima, F. A., Seuring, S., & Sauer, P. C. (2021). A systematic literature review exploring uncertainty management and sustainability outcomes in circular supply chains. *International Journal of Production Research*, 1–34.

Diaz Lopez, F. J. D., Bastein, T., & Tukker, A. (2019). Business model innovation for resource-efficiency, circularity and cleaner production: What 143 cases tell us. *Ecological Economics, 155*, 20–35. https://doi.org/10.1016/j.ecolecon.2018.03.009

Directive, E.U. (2000). Directive 53/EC of the European Parliament and of the Council of 18 September 2000 on end-of life vehicles. *Official Journal of the European Communities, L Ser. 21*, 34–42.

Dominguez, R., Cannella, S., & Framinan, J. M. (2021). Remanufacturing configuration in complex supply chains. *Omega, 101*, 102268. https://doi.org/10.1016/j.omega.2020.102268

de Sousa, L., Jabbour, A. B., Rojas Luiz, J. V., Rojas Luiz, O., Jabbour, C. J. C., Ndubisi, N. O., Caldeira de Oliveira, J. H., & Junior, F. H. (2019). Circular economy business models and operations management. *Journal of Cleaner Production, 235*, 1525–1539. https://doi.org/10.1016/j.jclepro.2019.06.349

Farooque, M., Zhang, A., Thürer, M., Qu, T., & Huisingh, D. (2019). Circular supply chain management: A definition and structured literature review. *Journal of Cleaner Production, 228*, 882–900. https://doi.org/10.1016/j.jclepro.2019.04.303

Fernando, Y., Shaharudin, M. S., & Abideen, A. Z. (2022). Circular economy-based reverse logistics: Dynamic interplay between sustainable resource commitment and financial performance. *European Journal of Management and Business Economics*. https://doi.org/10.1108/EJMBE-08-2020-0254

Flick, U. (2018). *The SAGE handbook of qualitative data collection.*

Gatenholm, G., Halldórsson, Á., & Bäckstrand, J. (2021). Enhanced circularity in aftermarkets: Logistics tradeoffs. *International Journal of Physical Distribution and Logistics Management, 51*(9), 999–1021. https://doi.org/10.1108/IJPDLM-11-2020-0367

Geissdoerfer, M., Morioka, S. N., de Carvalho, M. M., & Evans, S. (2018). Business models and supply chains for the circular economy. *Journal of Cleaner Production, 190*, 712–721. https://doi.org/10.1016/j.jclepro.2018.04.159

Geissdoerfer, M., Pieroni, M. P. P., Pigosso, D. C. A., & Soufani, K. (2020). Circular business models: A review. *Journal of Cleaner Production, 277*, 123741. https://doi.org/10.1016/j.jclepro.2020.123741

Goltsos, T. E., Ponte, B., Wang, S., Liu, Y., Naim, M. M., & Syntetos, A. A. (2019). The boomerang returns? Accounting for the impact of uncertainties on the dynamics of remanufacturing systems. *International Journal of Production Research, 57*(23), 7361–7394. https://doi.org/10.1080/00207543.2018.1510191

Homrich, A. S., Galvão, G., Abadia, L. G., & Carvalho, M. M. (2018). The circular economy umbrella: Trends and gaps on integrating pathways. *Journal of Cleaner Production, 175*, 525–543. https://doi.org/10.1016/j.jclepro.2017.11.064

Ignatow, G., & Mihalcea, R. (2018). *An introduction to text mining: Research design, data collection, and analysis.*

Jayarathna, C. P., Agdas, D., & Dawes, L. (2022). Exploring sustainable logistics practices toward a circular economy: A value creation perspective. Bus. Strateg. *Environmentalist.*

Jensen, J. P., Prendeville, S. M., Bocken, N. M. P., & Peck, D. (2019). Creating sustainable value through remanufacturing: Three industry cases. *Journal of Cleaner Production, 218*, 304–314. https://doi.org/10.1016/j.jclepro.2019.01.301

Jiang, Z., Ding, Z., Zhang, H., Cai, W., & Liu, Y. (2019). Data-driven ecological performance evaluation for remanufacturing process. *Energy Conversion and Management, 198*, 111844. https://doi.org/10.1016/j.enconman.2019.111844

Karlsson, C. (2010). *Researching operations management*. Routledge.

Kirchherr, J., Piscicelli, L., Bour, R., Kostense-Smit, E., Muller, J., Huibrechtse-Truijens, A., & Hekkert, M. (2018). Barriers to the circular economy: Evidence from the European Union (EU). *Ecological Economics, 150*, 264–272. https://doi.org/10.1016/j.ecolecon.2018.04.028

Korhonen, J., Honkasalo, A., & Seppälä, J. (2018). Circular economy: The concept and its limitations. *Ecological Economics, 143*, 37–46. https://doi.org/10.1016/j.ecolecon.2017.06.041

Kumar, V., Shirodkar, P. S., Camelio, J. A., & Sutherland, J. W. (2007). Value flow characterization during product lifecycle to assist in recovery decisions. *International Journal of Production Research, 45*(18–19), 4555–4572.

Lindkvist Haziri, L., & Sundin, E. (2020). Supporting design for remanufacturing-A framework for implementing information feedback from remanufacturing to product design. *Journal of Remanufacturing, 10*(1), 57–76. https://doi.org/10.1007/s13243-019-00074-7

Liu, B., Chen, D., Zhou, W., Nasr, N., Wang, T., Hu, S., & Zhu, B. (2018). The effect of remanufacturing and direct reuse on resource productivity of China's automotive production. *Journal of Cleaner Production, 194*, 309–317. https://doi.org/10.1016/j.jclepro.2018.05.119

Lüdeke-Freund, F., Gold, S., & Bocken, N. M. P. (2019). A review and typology of circular economy business model patterns. *Journal of Industrial Ecology, 23*(1), 36–61. https://doi.org/10.1111/jiec.12763

Majava, J., & Hyvärinen, K. (2022). Crowdsourcing-based business model for online customer service: A case study. *International Journal of Value Chain Management, 13*(1), 33–46. https://doi.org/10.1504/IJVCM.2022.122161

Masi, D., Day, S. H., & Godsell, J. (2017). Supply chain configurations in the circular economy: A systematic literature review. *Sustainability, 9*(9), 1602. https://doi.org/10.3390/su9091602

Masi, D., Kumar, V., Garza-Reyes, J. A., & Godsell, J. (2018). Towards a more circular economy: Exploring the awareness, practices, and barriers from a focal firm perspective. *Production Planning and Control, 29*(6), 539–550. https://doi.org/10.1080/09537287.2018.1449246

Mishra, A., Dutta, P., Jayasankar, S., Jain, P., & Mathiyazhagan, K. (2022). A review of reverse logistics and closed-loop supply chains in the perspective of circular economy. *Benchmarking.* https://doi.org/10.1108/BIJ-11-2021-0669

Misni, F., & Lee, L. S. (2017). A review on strategic, tactical and operational decision planning in reverse logistics of green supply chain network design. *Journal of Computer and Communications, 05*(8), 83–104. https://doi.org/10.4236/jcc.2017.58007

Mont, O., Plepys, A., Whalen, K., & Nußholz, J. L. K. (2017). *Business model innovation for a circular economy: Drivers and barriers for the Swedish industry—The voice of REES companies.*

Murray, A., Skene, K., & Haynes, K. (2017). The circular economy: An interdisciplinary exploration of the concept and application in a global context. *Journal of Business Ethics, 140*(3), 369–380. https://doi.org/10.1007/s10551-015-2693-2

Nascimento, D. L. M., Alencastro, V., Quelhas, O. L. G., Caiado, R. G. G., Garza-Reyes, J. A., Rocha-Lona, L. R., & Tortorella, G. (2019). Exploring Industry 4.0 technologies to enable circular economy practices in a manufacturing context A business model proposal. *Journal of Manufacturing Technology Management, 30*(3), 607–627. https://doi.org/10.1108/JMTM-03-2018-0071

Omwando, T. A., Otieno, W. A., Farahani, S., & Ross, A. D. (2018). A bi-level fuzzy analytical decision support tool for assessing product remanufacturability. *Journal of Cleaner Production, 174*, 1534–1549. https://doi.org/10.1016/j.jclepro.2017.11.050

Oncioiu, I., Căpuşneanu, S., Türkeş, M. C., Topor, D. I., Constantin, D. M. O., Marin-Pantelescu, A., & Ştefan Hint, M. S. (2018). The sustainability of Romanian SMEs and their involvement in the circular economy. *Sustainability, 10*(8). https://doi.org/10.3390/su10082761

Peng, S., Li, T., Zhao, J., Guo, Y., Lv, S., Tan, G. Z., & Zhang, H. (2019). Petri net-based scheduling strategy and energy modeling for the cylinder block remanufacturing under uncertainty. *Robotics and Computer-Integrated Manufacturing, 58*, 208–219. https://doi.org/10.1016/j.rcim.2019.03.004

Pushpamali, N. N. C., Agdas, D., Rose, T. M., & Yigitcanlar, T. (2021). Stakeholder perception of reverse logistics practices on supply chain performance. *Business Strategy and the Environment, 30*(1), 60–70. https://doi.org/10.1002/bse.2609

Ramsheva, Y. K., Prosman, E. J., & Wæhrens, B. V. (2019). Dare to make investments in industrial symbiosis? A conceptual framework and research agenda for developing trust. *Journal of Cleaner Production, 223*, 989–997. https://doi.org/10.1016/j.jclepro.2019.03.180

Ridley, S. J., Ijomah, W. L., & Corney, J. R. (2019). Improving the efficiency of remanufacture through enhanced preprocessing inspection–a comprehensive study of over 2000 engines at Caterpillar remanufacturing, UK. *Production Planning and Control, 30*(4), 259–270. https://doi.org/10.1080/09537287.2018.1471750

Rönkkö, P., Ayati, S. M., & Majava, J. (2021). Remanufacturing in the heavy vehicle industry—Case study of a Finnish machine manufacturer. *Sustainability, 13*(19), 11120. https://doi.org/10.3390/su131911120

Saidani, M., Yannou, B., Leroy, Y., & Cluzel, F. (2020). Dismantling, remanufacturing and recovering heavy vehicles in a circular economy—Technico-economic and organisational lessons learnt from an industrial pilot study. *Resources, Conservation and Recycling, 156*, 104684. https://doi.org/10.1016/j.resconrec.2020.104684

Sakao, T., & Sundin, E. (2019). How to improve remanufacturing?—A systematic analysis of practices and theories. *Journal of Manufacturing Science and Engineering, 141*(2). https://doi.org/10.1115/1.4041746

Shekarian, E. (2020). A review of factors affecting closed-loop supply chain models. *Journal of Cleaner Production, 253*, 119823. https://doi.org/10.1016/j.jclepro.2019.119823

Shekarian, E., Marandi, A., & Majava, J. (2021). Dual-channel remanufacturing closed-loop supply chains under carbon footprint and collection competition. *Sustainable Production and Consumption, 28*, 1050–1075.

Shekarian, E., Prashar, A., Majava, J., Khan, I. S., Ayati, S. M., & Sillanpää, I. (2023). Sustainable supply chains in the heavy vehicle and equipment industry: A multiple-case study of four manufacturers. *Benchmarking: An International Journal.*

Singhal, D., Tripathy, S., & Jena, S. K. (2020). Remanufacturing for the circular economy: Study and evaluation of critical factors. *Resources, Conservation and Recycling, 156*, 104681. https://doi.org/10.1016/j.resconrec.2020.104681

Stahel, W. R. (2019). *The circular economy: A user's guide*. Routledge.

Vermunt, D. A., Negro, S. O., Verweij, P. A., Kuppens, D. V., & Hekkert, M. P. (2019). Exploring barriers to implementing different circular business models. *Journal of Cleaner Production, 222*, 891–902. https://doi.org/10.1016/j.jclepro.2019.03.052

Vogtlander, J. G., Scheepens, A. E., Bocken, N. M. P., & Peck, D. (2017). Combined analyses of costs, market value and eco-costs in circular business models: Eco-efficient value creation in remanufacturing. *Journal of Remanufacturing, 7*(1), 1–17. https://doi.org/10.1007/s13243-017-0031-9

Voss, C., Tsikriktsis, N., & Frohlich, M. (2002). Case research in operations management. *International Journal of Operations and Production Management, 22*(2), 195–219. https://doi.org/10.1108/01443570210414329

Webster, K. (2017). *The circular economy: A wealth of flows*. Ellen MacArthur Foundation Publishing.

Xiao, L., Liu, W., Guo, Q., Gao, L., Zhang, G., & Chen, X. (2018). Comparative life cycle assessment of manufactured and remanufactured loading machines in China. *Resources, Conservation and Recycling, 131*, 225–234. https://doi.org/10.1016/j.resconrec.2017.12.021

Xiao, L., Wu, Z.-Y., Zhang, S.-L., Chen, Z.-S., & Govindan, K. (2021). Product design and pricing strategies in a closed-loop supply chain with patent protection. *Kybernetes*.

Yang, M., Evans, S., Vladimirova, D., & Rana, P. (2017). Value uncaptured perspective for sustainable business model innovation. *Journal of Cleaner Production, 140*, 1794–1804.

Yin, R. K. (2011). *Applications of case study research. sage*.

Zhang, X., Ao, X., Cai, W., Jiang, Z., & Zhang, H. (2019). A sustainability evaluation method integrating the energy, economic and environment in remanufacturing systems. *Journal of Cleaner Production, 239*, 118100. https://doi.org/10.1016/j.jclepro.2019.118100

Zhang, X., Tang, Y., Zhang, H., Jiang, Z., & Cai, W. (2021). Remanufacturability evaluation of end-of-life products considering technology, economy and environment: A review. *Science of the Total Environment, 764*, 142922. https://doi.org/10.1016/j.scitotenv.2020.142922

Chapter 10
Waste Management Sector's Contribution to Meeting Sustainable Development Goals: A Social Life Cycle Assessment Approach

Novlloyd E. Celeste and Anthony B. Halog

10.1 Introduction: Establishing Link Between SWM and SDGs

Waste is the leading contributor to environmental pollution and public health problems. The problems are exacerbated by the challenges of solid waste management (SWM) associated with the primary aspects of the waste sector (e.g., waste generation and inadequate waste collection, transport, treatment, and disposal processes) (Das et al., 2019). Extensive research has shown that developing countries face a difficulty in reducing and managing waste due to their sociocultural, technical, financial, organizational, and legal-political barriers and population growth (Dhokhikah & Trihadiningrum, 2012). Additionally, a growing body of literature recognizes that insufficient infrastructure, weak strategic planning, registration, staff capacity, information systems, engagement with programs; unorganized waste management and fee collection systems, poor workmanship, lack of knowledge of the construction of waste management, unsustainable methods (Hassan et al., 2012; Yukalang, Clarke & Ross, 2017; Iyamu, Anda & Ho, 2020); no separation at source (Dhokhikah & Trihadiningrum, 2012; Hondo, Arthur, & Gamaralalage, 2020), complicated and inadequate collection processes, open dumped landfill, and no control of gas emissions and leachate in landfill (Dhokhikah & Trihadiningrum, 2012; Jain, 2017; Das et al., 2019); lack of financial support, environmental pollution awareness, social inclusion,

N. E. Celeste (✉)
Northwest Samar State University, Calbayog City, Samar, Philippines
e-mail: novlloyd.celeste@nwssu.edu.ph

A. B. Halog
Faculty of Science, School of Environment, The University of Queensland, Brisbane, QLD 4072, Australia
e-mail: a.halog@uq.edu.au

S. A. Bandh and F. A. Malla (eds.), *Waste Management in the Circular Economy*,
https://doi.org/10.1007/978-3-031-42426-7_10

technical knowledge; lack of managerial efficiency (Ferronato et al., 2021) are some of the major issues being faced in the waste sector.

Effective waste management is essential to curb the climate change-related impacts and subsequently achieve sustainable development. Environmental and economic benefits are not always the bottom line of such a sustainable approach, but social impacts also need to be analyzed for their contribution to the sustainable life concept (Yıldız-Geyhan et al., 2017). Hence, waste management needs to be done holistically to address the triple dimensions of sustainable development such as society, economy, and environment. In addition, policies and regulations, stakeholder participation, and social impacts are important aspects of sustainable solid waste management (SSWM) for properly managing, processing, and disposing of garbage (Tsai et al., 2021).

Gleaning on the benefits of waste management to sustainable development cannot be undermined, as (Fatimah et al., 2020) the use of ICT and IoT improves the efficiency and effectiveness of the waste management system and contributes to the SDGs such as Good health, and well-being (SDG 3); Clean water and sanitation (SDG 6); Decent Work and Economic Growth (SDG 8); Responsible Consumption and Production (SDG 12) and Climate Action (SDG 13). Hondo, Arthur, and Gamaralalage (2020) also argued that effective household SWM practices are vital in achieving SDGs 11, 12, and 14, which can also help protect the environment, especially at the municipal and local levels. Evidences suggest that the SDGs sub-indicators are related to SWM, as Rodić and Wilson (2017), reported 12 SDGs sub-indicators and targets are related to SWM while Sharma et al., (2021) considered 34 sub-indicators of SDGs to be related to SWM. Both authors have itemized these goals and indicators, but it is not categorized as to what affects the social dimensions per se.

As discussed above, existing research recognizes the critical role-played by the effective waste management in addressing SDGs 6 and 11 and other SDGs; however, there is a lack of data supporting its contribution (Elder, 2020). Still there is no mechanism to determine waste management activities in relation to the achievement of SDG goals. Thus, a great deal of work is to be done in this area in which this chapter delves.

10.1.1 Overview of Waste Management: ASEAN Context

Waste management is a major area of interest in ASEAN nations where it has an average of 1.1383 kg/capita/day of MSW generation, with Singapore having the highest (3.763 kg/cap/day) and Myanmar having the lowest (0.53). The other countries like Brunei Darussalam, Malaysia, Thailand, Vietnam, Indonesia, Philippines, and Cambodia have a per capita MSW generation of 1.4, 1.17, 1.05, 0.84, 0.70, 0.69, and 0.55 kg/cap/day, respectively (Jain, 2017).

Indonesia has the highest MSW generated among ASEAN countries. Households are the largest waste generator in the country. For small cities, the rate of MSW generation is 0.64 kg/capita/day, whereas metropolitan cities produce 0.75 kg/capita/day (Wibisono, Firdausi, & Kusuma, 2020) influenced by the households standard of living. In terms of waste management in the household level, Indonesian people particularly in Jakarta majority do not sort their waste and do not practice segregation. This behavior can be observed in their household waste wherein biowaste, inorganic waste, hazardous waste, and bulky waste are mixed.

Furthermore, the highest waste generated per capita is Singapore followed by Brunei Darussalam. Here most of the MSW are composed of organic waste, plastic, paper, glass, and metal. In terms of hazardous waste, Thailand ranks first, followed by the Philippines, Malaysia, and Singapore while for e-waste generation, Indonesia ranks first, followed by Thailand and Malaysia (Jain, 2017). Cambodia is lagging by instituting policies, programs, and projects at national and local levels, encouraging both producers and consumers to reduce the waste through greening production, greening lifestyle, and sustainable consumption to reduce MSW generation while Lao PDR has already designed policies, and Myanmar and Vietnam have no plans, strategies, and projects to address MSW reduction (Jain, 2017). Vietnam on the other hand has set policies and standards however its implementation needs to be improved with the participation of stakeholders to implement effectively (Verma et al., 2016). Likewise, the implementation of such policies would be effective if appropriate sanctions will be applied in relation to socialization or privatization for the classification, collection, transportation, and treatment of solid waste (Tsai et al., 2020).

The rapid increase in population and emerging waste streams continue to be a challenge in ASEAN. Hence, ASEAN countries need to establish link between effective waste management toward SDGs which to date is not yet being explored (Jain, 2017).

Most of the current literatures show the numerous ways that waste is managed in ASEAN member countries which are common to all. For instance, waste management practices in Indonesia start with household source segregation before it ends in landfill. Another way is household composting, and household then goes to temporary storage, composting, landfill gas to energy (Aprilia, Tezuka & Spaargaren, 2012). In Asian countries, solid waste is done in the process of collection, transport, and final disposal (Dhokhikah & Trihadiningrum, 2012), where final disposal is commonly open dump and landfill (Dhokhikah & Trihadiningrum, 2012; Hondo, Arthur & Gamaralalage, 2020). Treatment of waste such as waste sorting is usually done in waste sorting and storage facilities by waste transporters (Aprilia, Tezuka, & Spaargaren, 2012) or waste collectors (Fatimah et al., 2020).

To manage waste effectively, technologies and activities like composting, source segregation and community-based material recovery facility, Tempat Pengolahan Sampah Terpadu (TPST), intermediate waste facilities, waste banks, and 3Rs strategy (Budihardjo et al., 2022; Fatimah et al., 2020) have been introduced to households. There is a growing body of literature that suggests that one of the most important measures that are lacking in the developing countries is waste separation in which if done at source could reduce the amount of waste disposed in landfill sites. Thus, households and changing human behavior are the key to improve waste management (Hondo, Arthur, & Gamaralalage, 2020).

10.2 Methodology

10.2.1 Systematic Review of Literature

In evaluating the relationship or linkage of Social LCA indicators, viz. SDG indicators, the study relied on the published literature in the ScienceDirect database. To find relevant studies, the Preferred Reporting Items for Systematic Reviews and Meta-Analyses (PRISMA) checklist (Page et al., 2021) was used with the limit of the latest five years bearing an up-to-date literature. However, there are instances where other databases and journals were also included to substantiate the discussion. Keywords used included ASEAN to display only literatures related to ASEAN countries. Around 1800 articles appeared in search results using social life cycle assessment, municipal solid waste management and municipal solid waste management social life cycle assessment Asia. The keyword Social LCA and sustainable development goals had 261 articles; however, only 49 articles were deemed relevant to the study.

One of the goals of Social LCA is to highlight and improve social conditions and socioeconomic performance associated with a product throughout its life. In this sense, identifying social hotspots is necessary. According to Ashby et al. (2019), social hotspots are where the stakeholders and its life cycle phases of product development (e.g., materials, manufacture, distribution, and use) interact and may potentially cause damage in any phases. To put it simply, hotspots are the quality of practice below a certain threshold. In relation to this, Table 10.1 presents the Granta Social audit tool used best practices from a 1–100 rating scale and the performance characterization based on the LCA-SDG screening from Life Cycle Initiative and UNEP

Table 10.1 Description of performance characterization

Granta social audit results	Performance characterization	Description
80–100	+2	Ideal performance: beneficial output achieved and reported
60–79	+1	Progress beyond compliance is made and monitored
40–59	0	In compliance with local laws or aligned with international standards
20–39	−1	The non-compliant situation, but actions to improve have been taken
0–19	−2	No data or non-compliant situation; no action taken

(Weidema et al., 2020). Threshold of 50% from social audit represents zero (0) for the scale in performance characterization which means that "In compliance with local laws or aligned with international standards."

10.3 Linking Social Life Cycle Assessment and SDG

Linking solid waste management to SDGs is possible by using market-based sustainable solutions in managing solid waste under a circular economy approach (Sharma et al., 2021). While there are many frameworks proposed to achieve sustainable development, incorporating circular economy outperformed all other strategies regarding the delivery on SDG indicators in the low future waste growth scenario by 2030 (Fuldauer et al., 2019).

Since SWM is a crosscutting issue that transcends the scope of sustainable development (Rodić & Wilson, 2017), measuring the contribution of the waste management sector to SDG is a giant step toward achieving SDGs which includes participatory approaches involving local stakeholders (Fuldauer et al., 2019; Mathe, 2014).

Social LCA is valuable enough to measure social impacts, though the method needs to be improved to provide reliable measures in calculating a product's contribution to the SDGs (Herrera Almanza & Corona, 2020).

Social life cycle analysis (Social LCA) ((Herrera Almanza & Corona 2020) can help managerial decision making (Huertas-Valdivia et al., 2020) and can be used as a tool for sustainable management in illegal waste dumping in municipal services (Santos et al., 2019). It combines environmental and socioeconomic assessments, contributing to the complete assessment of products and services within sustainable development framework. Hence, Social LCA studies are often presented as case studies focusing on agriculture, bioenergy, and chemical products (Huertas-Valdivia et al., 2020), manufacturing, energy, waste management, and tourism (Petti et al., 2018).

Several authors have proposed an approach to assess waste management performance in developing countries (Ibáñez-Forés et al., 2019); as an extension of the S-LCA indicators. However, it did not tackle how it would contribute to the SDGs.

At present, tackling SDGs and achieving its goals could take a long time, even beyond 2030, making it more difficult with the onset of the COVID-19 pandemic (Halkos & Gkampoura, 2021). Moreover, using Social LCA to be aligned with SDGs requires strenuous documentation and analysis since SDGs are conceptualized from a country or regional level while Life Cycle Analysis (LCA) is on the by-product level (Wulf et al., 2018) as evident from many Social LCA published studies (Padilla-Rivera et al., 2019). Recently, Social LCA is being applied in bioeconomy and biobased products. Only two authors have worked on the waste sector, particularly municipal solid waste management (MSWM) and packaging waste collection system (Ramos Huarachi et al. 2020). In the field of circular economy, only a few studies address the contribution to SDGs using LCA, such as in Palm Oil Industry (Haryati et al., 2021) and the textile sector (Herrera Almanza & Corona, 2020), while other authors have explored Social LCA per se.

Based on the study of Wulf et al. (2018), Table 10.2 shows the relevant equivalent of SDG indicators according to the Social LCA indicators using the Product Social Impact Life Cycle Assessment (PSILCA) database. According to the study, out of 54 indicators from the PSILCA database, only 32 social indicators were aligned and characterized the SDGs (Wulf et al., 2018). In the Life Cycle Social Assessment (LCSA) indicators from the PSILCA database, drinking water coverage matched with SDG sub-indicator 6.1.1 (Proportion of population using safely managed to drink water services), and Sanitation coverage matched with 6.2.1 (Proportion of population using (a) safely managed sanitation services and (b) a hand-washing facility with soap and water). In the waste management sector, drinking water and sanitation coverage contribute to SDGs 6 and 11 though not explicitly (Elder, 2020).

Interestingly, Public Health Expenditure per capita and Press freedom, which are included in the Granta social audit tool with reference from UNEP-SETAC (Ashby et al., 2019), were not present in the assessment of Wulf et al. (2018). Thus, the LCSA indicator from PSILCA as assigned to SDGs appears to have lacked public health expenditure and press freedom.

On the other hand, from the Granta Social Audit Tool, the drinking water coverage, sanitation coverage, active involvement of enterprises in corruption and bribery, and presence of EMS are not explicitly stated indicators in the PSILCA database (Wulf et al., 2018). The two software tools may have pros and cons; nevertheless, they can provide us insights into social impact assessments, particularly at the regional level since both used guidelines from UNEP-SETAC.

A closer look at the review shows that Social LCA categories such as workers, local community, society, and value chain have only 17 SDG indicators matched with the PSILCA and Granta Social Audit tool database. It is important to understand that these indicators only point out social indicators which could be used to gauge waste management contribution in SDGs.

On the one hand, SDG 11.6.1 "Proportion of municipal solid waste collected and managed in controlled facilities out of total municipal waste generated, by cities," is not reflected in the list of indicators used by the two databases because it is not a social indicator but an environmental indicator.

10.3.1 Contribution of Waste Management in SDGs

The contribution of waste management in SDGs is categorized according to SLCA indicators presented previously. The vast majority of Social LCA studies included in the review found out that better performing social indicators are "Working hours and/or weekly rest" and "Safe and healthy living conditions (access to material resources)," followed by "Customer/citizen satisfaction. "The worst-performing social indicators are "Labor regulation" and "Fair salary," followed by "Social characteristics of population" and "Legal employment with social benefits/security"

Table 10.2 Comparison of PSILCA database and Granta social audit tool, viz. SDG sub-indicators

SDG sub-indicator	PSILCA database sub-indicators	Granta social audit	Social LCA stakeholders category
1.3.1 Proportion of population covered by social protection floors/systems, by sex, distinguishing children, unemployed persons, older persons, persons with disabilities, pregnant women, newborns, work-injury victims, and the poor and vulnerable	Social security expenditure	✓	Workers
1.a.2 Proportion of total government spending on essential services (education, health, and social protection)	Public expenditure on education	✓	Local community
3.8.2 Proportion of population with large household expenditures on health as a share of total household expenditure or income	Health expenditure, out of pocket	✓	Local community
4.6.1 Proportion of population in each age group achieving at least a fixed level of proficiency in functional (a) literacy and (b) numeracy skills, by sex	Illiteracy rate total, Youth illiteracy rate total	✓	Local Community
6.1.1 Proportion of population using safely managed drinking water services	Drinking water coverage	No specific indicator found	Local community
6.2.1 Proportion of population using (a) safely managed sanitation services and (b) a hand-washing facility with soap and water	Sanitation coverage	No specific indicator found	Local community
6.5.1 Degree of integrated water resources management	Presence of certified environmental management systems	No specific indicator found	Society
8.5.1 Average hourly earnings of employees, by sex, age, occupation, and persons with disabilities	Minimum wage per month, Sector average wage per month	✓	Workers
8.5.2 Unemployment rate, by sex, age, and persons with disabilities	The unemployment rate in a country	✓	Local community
8.7.1 Proportion and number of children (aged 5–17 years) engaged in child labor, by sex and age	Children in employment total	✓	Workers
8.8.1 Fatal and non-fatal occupational injuries per 100,000 workers, by sex and migrant status	Accident rate at workplace, Fatal accidents rate at workplace	✓	Workers

(continued)

Table 10.2 (continued)

SDG sub-indicator	PSILCA database sub-indicators	Granta social audit	Social LCA stakeholders category
8.8.2 Level of national compliance with labor rights (freedom of association and collective bargaining) based on International Labour Organization (ILO) textual sources and national legislation, by sex and migrant status	Right of Association, Right of collective bargaining, Right to Strike	✓	Workers
10.2.1 Proportion of people living below 50 percent of median income, by sex, age, and persons with disabilities	Living wage per month, Minimum wage per month	✓	Workers
10.3.1 Proportion of population reporting/ having personally felt discriminated against or harassed in the previous 12 months based on a ground of discrimination prohibited under international human rights law	Human rights issues faced by Indigenous people	✓	Value chain
16.2.2 number of victims of human trafficking per 100,000 population, by sex, age, and form of exploitation	Tier placement refers to trafficking in persons	✓	Workers
16.5.1 Proportion of persons who had at least one contact with a public official and who paid a bribe to a public official or were asked for a bribe by those public officials during the previous 12 months	Corruption Index or country	✓	Consumers
16.5.2 Proportion of businesses that had at least one contact with a public official and that paid a bribe to a public official or were asked for a bribe by those public officials during the previous 12 months	Active involvement of enterprises in corruption and bribery	Not specified in social LCA	Consumers

(Ibáñez-Forés et al., 2019). It is clear from previous research undertaken that some enhancements of SWM sector and 3Rs eventually help provide a strong contribution to the health and living conditions of those who lack better services and also provide decent jobs particularly in developing countries (Rodić & Wilson, 2017).

A number of studies have shown that the solid waste collection (SWC) as part of the SWM, largely contributes to SDG 3: Well-being and good health; SDG 11: Sustainable cities; SDG 8: Decent work and economic development and SDG 1: No poverty (Hannan et al., 2020). While, SWM contributes to SDG1: No poverty, poverty is likewise considered a barrier in achieving several other SDGs as contended by (Leal Filho et al., 2021) which is possible to include those that are related to SWM.

It should be noted here that according to Elsheekh et al. (2021), SDGs 11 and 13 followed by SDGs 8, 9, and 12 are in the lead that are affected by SWM plans and programs. Studies also show that achieving SDGs is being supported by some companies as is the case with Indonesia. Accordingly, companies in Indonesia support the following SDGs:

(1) Sustainable cities and communities,
(2) Good health and well-being,
(3) Decent work and economic growth,
(4) Responsible consumption and production, and
(5) Quality education (CSR) (Gunawan et al., 2020).

According to Jain (2017), Singapore and Malaysia have achieved significant increase in recycling rate of recyclables (e.g., plastic, paper, metal, etc.), by introducing policies and measures, and by setting up financial mechanisms and institutional frameworks involving relevant stakeholders (e.g., producers, consumers, recycling industry, users of recycled materials, etc.) and development of modern recycling industry.

Additionally, by institutionalizing policies, programs, and projects at national and local levels and motivating both producers and consumers to reduce the waste through greening production, greening lifestyle, and sustainable consumption would result in a reduction in MSW generation (Jain, 2017).

There should be a focus on the informal sector in waste management to help achieve SDGs. Informal sector has challenges in waste management and is constantly facing negative socioeconomic working conditions (Sharma et al., 2021). Other than this, policy and regulations, stakeholder participation, and social impacts are important aspects of sustainable solid waste management (SSWM) in particular for managing, processing and disposing of garbage (Tsai et al., 2021).

In Vietnam, there is a need to improve its household waste separation to reduce the waste in landfills. Policy standards for SWM may be enhanced despite its limited budget and thus require more stakeholder implementation for effective implementation (Verma et al., 2016). In 2018, Vietnam has processed 86.50% of municipal solid waste collected and managed in controlled facilities out of total municipal waste generated (https://data.aseanstats.org/about).

10.3.2 A Tale of Two Countries: Indonesia and Philippines SWM

In this section, we provide examples of different types of stakeholder's in Social LCA which are contributory to the SDGs achievement in relation to waste management. We considered Indonesia and the Philippines as the areas of study because these two countries have submitted two Voluntary National Reviews in the SDG Report among ten ASEAN countries (Elder, 2020). The figure below shows the case studies of these two countries (Fig. 10.1).

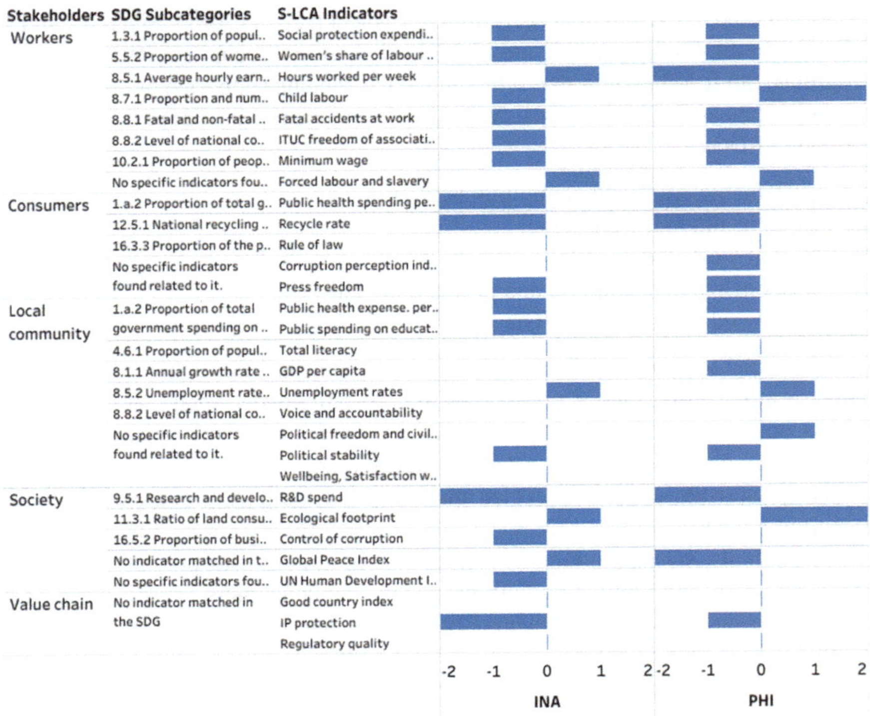

Fig. 10.1 Waste contribution in relation to SDGs using social LCA

One of the most significant findings in this chapter was that Indonesia is in a non-compliant situation, but actions to improve have been taken (−1) in consumers category and value chain. Only three Social LCA indicators appear to have progress beyond compliance made and monitored (+1), while four indicators have no data, or no action taken at all. On the other hand, Philippines has an Ideal performance: beneficial output achieved and reported in two indicators (ecological footprint and child labor) while force labor and slavery, political freedom and unemployment rates have progress beyond compliance (+1).

To top it all, waste management is influential in addressing SDGs in respect to the following sub-indicators 8.5.1 (Average hourly earnings of employees, by sex, age, occupation, and persons with disabilities), 8.5.2 (Unemployment rate, by sex, age, and persons with disabilities), 11.3.1 (Ratio of land consumption rate to population growth rate). In the Philippines, SWM may help address SDG 8.7.1 (Proportion and number of children aged 5–17 years engaged in child labor, by sex and age) and achieve its goal in addition to the ones previously mentioned.

In this country-level social audit, it appears that forced labor and slavery, hours worked per week, political freedom and civil rights, well-being and satisfaction in life, voice and accountability, unemployment rates, ecological footprint, global peace index, regulatory and good country index are positive indicators. In other words, in

SWM sector, Indonesia and Philippines may positively address SDG because of its current average hourly earnings per employee. Eventually, this will increase unemployment rate. Accordingly, collecting waste in the dumpsite can generate good income among waste pickers, but their economic status depends on Indonesia's economic growth (Sasaki et al., 2014).

This is likewise true to Ecological footprint being directly related to ratio of land consumption rate to the population growth rate. However, in terms of child labor, there is a need for Indonesia to double time in complying with international, national, and local policies and standards with regard to such indicator because of its non-compliance though it has already had some actions taken to address it as compared with Philippines.

10.3.2.1 Workers Category

The category has eight sub-indicators and of these indicators, forced labor and slavery and hours worked per week is the indicator that has progressed beyond compliance and is being monitored. Rest of the indicators are non-compliant in the situation. For instance, child labor having a non-compliant situation should be considered in the informal and even the formal waste management system because several pupils and dropouts joined in scavenging (Sasaki et al., 2014). Eliminating child and forced labor and improving the health safety and working conditions can improve the social performance of informal collection (Yıldız-Geyhan et al., 2017). Adapting the recommendation of other studies in the Asian context, working hours for all workers, child labor, and health issues were seen as the consequences of degraded waste management service. Thus, the integration of informal waste workers into the formal waste management system was seen as an essential factor in enhancing the life of the scavengers and recycling rate in the case of Kabul City (Azimi, Dente, & Hashimoto, 2020). Hence, public participation in policy-making is critical to recognizing waste collection organizations and rewarding their services to reveal their full potential and drive progress toward the SDGs (Gutberlet, 2021).

In the Philippines, workers in the SW are usually operating informally. However, they play a more significant role in implementing SWM (Domingo, Joy, & Manejar, 2021), like in Vietnam, where informal sectors refer to private micro-businesses working in secondary material collection and recovery services (Tong et al., 2021). However, informal sectors dominate waste management, which is hard to categorize because of the lack of data among informal sectors. Thus, it is worth noting that informal sectors should participate in the stakeholder category (Dickella, 2016). Regarding women's participation, the reduction of informal sector activity would impact women and make them vulnerable because they represent a more considerable proportion in the informal sector (Scheinberg, Simpson, & Gupt, 2010).

It is noticeable that the informal sector is playing a significant role in waste management in developing countries. Contrary to the stereotype that informal sectors pollute the environment, they were instead able to recycle waste. Through scrap sorting and trading activities, waste pickers are able to reduce the burden on the exploitation of non-renewable resources, and help to reduce environmental pollution (Tong et al., 2021). Thus, the informal sector should be increasingly strengthened because they helps put the waste back into the industry for recycling (Tong et al., 2021). The inclusion of the informal sector can be considered a viable way to improve the recycling rate and reduce the waste inflow into final disposal sites in developing countries due to low technological requirements and economic investments. However, further investigations and efforts should be implemented to understand the most appropriate strategy for its involvement (Ferronato & Torretta, 2019).

10.3.2.2 Consumers Category

In terms of consumers' category, both countries have no data, or no action taken on recycling rate and public health spending per capita. Press freedom is non-compliant but has actions taken. While for corruption perception index, only Indonesia has "In compliance with local laws or aligned with international standards. Meanwhile, both countries also are in compliance with local laws or aligned with international standards with the rule of law." Based on the study of (Sasaki et al., 2014), implementing waste management initiatives will have to consider several policy issues because laws and regulations were not strictly implemented. This is manifested in the solid waste management in Makassar City, Indonesia, where there is the unsatisfactory performance of the solid waste management authority in its implementation, and households received relatively inadequate support from authorities (Permana et al., 2015). While in the Philippines, despite that it has no data available, there is a technology implemented called refused-derived fuel (RDF) to lessen the solid waste in the landfill. This technology converts energy from non-recyclable materials to combustible waste material used for traditional alternative fuels such as coal. This is one of the strategies of the government as waste diversion goals (Sapuay, 2016). However, this is not implemented in all cities and municipalities in the Philippines. Likewise, in Indonesia technology on waste disposal is still the usual composting, incineration, sanitary landfill, open dump, and open burning. Though they have recycling to address the increasing problems of plastic waste, recycling rate is still at 50% below (Jain, 2017).

10.3.2.3 Local Community Category

Of the nine indicators, only one indicator has progressed beyond compliance: unemployment rates, while public spending on education, HDI (%GDP), public health expense per capita, and political stability, which are non-compliant but actions to improve have been taken. In Indonesia, Public health must be improved because the living conditions of waste pickers in the informal sector are detrimental to health and could be extremely dangerous because of hospital waste and polluted water (Sasaki et al., 2014). This is also true in the Philippines where many waste pickers are in the informal sectors and are vulnerable to health and hazard risks because no law protects them from unregulated work. Among these are women, children, and the elderly who have no other means of livelihood (Paul et al., 2012).

Meanwhile, GDP per capita, political freedom and civil rights, total literacy, well-being, satisfaction with life, voice, and accountability follow local laws or are aligned with international standards. Utilizing an advance intelligent waste management system can help Indonesia achieve at least five sustainable development goals such as good health and well-being (SDG 3); Clean water and sanitation (SDG 6); Decent Work and Economic Growth (SDG 8); Responsible Consumption, and Production (SDG 12) and Climate Action (SDG 13) (Fatimah et al., 2020).

Accordingly, having many wastes pickers part of the informal sector is challenging and requires more support policies for the informal waste sector. In addition, a formal collection system is better than other informal collection systems. However, it is less socially acceptable than informal ones (Yıldız-Geyhan et al., 2017). Policies that may reinforce the formalization and improve the informal sector since they play a significant role in "zero waste" initiatives and are vital to mainstreaming the circular approach in waste management. Doing so can help to achieve several SDG targets such as target 12.5 and other social objectives on reducing inequality (SDG10), decent work (SDG 8), and improved health and well-being (SDG 5) (Asian Development Bank, 2020).

10.3.2.4 Society Category

In Indonesia, the Ecological footprint and global peace index have progressed beyond compliance. While in the Philippines, ecological footprint seems to have an ideal performance. This may be due to the fact that in the Philippines, almost all regions have complied with the submission of the Ten-Year solid waste management plan (Sapuay, 2016) with a percentage of above 80–90%, although only 51% compliance of local government units (LGUs) implementing solid waste management act (Asian Development Bank, 2020). It is not clear what level and aspect of the waste management activity the LGUs are complying with, nor the rate of compliance showed how the RA 9003 is being implemented at the local level (Sapuay, 2016). According to (Pagunsan & Shimada, 2014), LGUs with higher income and good environmental governance can obtain better performance. On the other hand, budget constraints can significantly impact solid waste services in cost efficiency and economic effectiveness (Pagunsan & Shimada, 2014).

However, the UN Human development index and control of corruption are non-compliant, but actions to improve have been taken. Research and development have no data or action taken. This category has no direct relationship with SDG indicators and may be added in the future.

10.3.2.5 Value Chain Category

Integration of circular economy (CE) in waste management in Indonesia can potentially lead the country to be at par with Germany, where the CE paradigm is applicable (Kurniawan et al., 2021). Indonesia shows to follow local laws about regulatory quality and has a good country index. However, Intellectual property protection has no data available based on the grant social audit.

The Philippines shows to follow local laws concerning regulatory quality and has a good country index. However, Intellectual property protection has no data available based on the Granta social audit. Specific to the waste management sector, SDG 11.6.1 Proportion of municipal solid waste collected and managed in controlled facilities out of total municipal waste generated by cities. In Social LCA tools, in both PSILCA and Granta did not match with the sub-indicator. Since waste management is dealing with the collection of MSW in cities, this indicator should be addressed not directly. Moreover, this indicator only applies to waste generation and management, not to the welfare of the waste recyclers. Hence, tackling the contribution of waste management in this respect is insufficient since sustainability covers all aspects of a person.

One of the significant problems in waste management, particularly e-waste, is the lack of official data in e-waste generation, and the quantity of waste generated (Celestial, 2010). The significant challenges in the management of e-waste are the unavailability of accurate estimates of quantities of total e-waste generation, low level of awareness among the consumers about the health hazards and environmental impact of improper e-waste disposal, and lack of proper legislation and government policy on e-waste management (Alam, 2016).

10.4 Conclusions and Recommendations

The study presents the indicators from local and national levels that contribute to the SDGs using the Social LCA framework. In addition, in classifying and characterizing Social LCA indicators about SDGs, most ASEAN countries have not achieved the ideal performance. It is observed that only countries with high income achieved ideal performance like Singapore and Malaysia. On the one hand, low-income countries have neither no data available nor action was taken in some of the indicators in LCA-SDG screening. In terms of Social LCA contribution to SDGs, Indonesia's results show that six indicators are non-compliant in the worker's category. While in the Philippines, five indicators were non-compliant, and one has no data available based on the Granta social audit tool. Notably, the worker's category has more indicators that have non-compliant, and no action was taken based on the assessment. Hence, this category should be further investigated when making policies regarding solid waste management in ASEAN countries and other developing countries. Moreover, the value chain category, well-being, satisfaction with life, and global peace index have no indicators matched specifically to SDG. Furthermore, it is recommended that recyclers and the government should take seriously the implementation of solid waste management. In similar manner, independent informal waste pickers' organization should be aware of their role in curbing waste pollution and their rights to have a good working condition. Policymakers can propose and implement such policy to help informal waste pickers protect their individual rights and provide means.

References

Alam, Z. (2016). The assessment of the of E-waste management generated from cellular phones, laptops, and personal computers in the Philippines. *Manila Journal of Science, 9*(October), 27–42.

Aprilia, A., Tezuka, T., & Spaargaren, G. (2012). Household solid waste management in Jakarta, Indonesia: A socio-economic evaluation. *Intech, 32*, 137–144. http://www.intechopen.com/books/trends-in-telecommunications-technologies/gps-total-electron-content-tec-prediction-at-ionosphere-layer-over-the-equatorial-region%0AInTec

Ashby, M., Brechbühl, E., Vakhitova, T., & Vallejo, A. (2019). Social life-cycle assessment and social impact audit tool: A white paper. *In ANSYS Granta Design*. https://doi.org/10.1038/d41586-019-02139-w.2

Asian Development Bank. (2020). Supporting implementation of environment- related sustainable development goals in Asia and the pacific (Philippine subproject) (February).

Azimi, A. N., Dente, S. M. R., & Hashimoto, S. (2020). Social life-cycle assessment of household-waste management system in Kabul city. *Sustainability, 12*(8). https://doi.org/10.3390/SU12083217

Budihardjo, M. A., Ardiansyah, S. Y., & Ramadan, B. S. (2022). Community-driven material recovery facility (CdMRF) for sustainable economic incentives of waste management: Evidence from Semarang City, Indonesia. *Habitat International, 119*, 102488. https://doi.org/10.1016/j.habitatint.2021.102488

Celestial, R. G. A. (2010). E-waste management in the Philippines. In *E-waste management in the Philippines*. https://doi.org/10.13140/RG.2.2.17965.74728

Das, S., Lee, S.-H., Kumar, P., Kim, K. H., Lee, S. S., & Bhattacharya, S. S. (2019). Solid waste management: Scope and the challenge of sustainability. *Journal of Cleaner Production, 228*, 658–678. https://doi.org/10.1016/j.jclepro.2019.04.323

Dhokhikah, Y. and Trihadiningrum, Y. (2012). Solid waste management in asian developing countries: Challenges and opportunities. *Journal of Applied Environmental and Biological Sciences, 2*(7), pp. 329–335. Available at: https://www.textroad.com

Dickella, G. P. J. (2016). Development of E-waste management policy at the local level: A case study of Cebu City, Philippines. http://pub.iges.or.jp/modules/envirolib/view.php?docid=5754. Institute for Global Environmental Strategies (Institute for Global Environmental Strategies). Retrieved from Institute for Global Environmental Strategies (IGES), Japan website.

Domingo, S. N., Joy, A., & Manejar, A. (2021). *An analysis of regulatory policies on solid waste management in the Philippines: Ways forward* (pp. 2021–2023). Discussion Paper Series. https://www.pids.gov.ph

Elder, M. (2020). *Assessment of ASEAN countries' concrete SDG implementation efforts: Policies and Bugdets reported in their 2016–2020 voluntary national reviews (VNRs).*

Elsheekh, K. M., Kamel, R. R., Elsherif, D. M., & Shalaby, A. M. (2021). Achieving sustainable development goals from the perspective of solid waste management plans. *Journal of Engineering and Applied Science, 68*(1), 1–15. https://doi.org/10.1186/s44147-021-00009-9

Fatimah, Y. A., Govindan, K., Murniningsih, R., & Setiawan, A. (2020). Industry 4.0 based sustainable circular economy approach for smart waste management system to achieve sustainable development goals: A case study of Indonesia. *Journal of Cleaner Production, 269*, 122263. https://doi.org/10.1016/j.jclepro.2020.122263

Ferronato, N., & Torretta, V. (2019). Waste mismanagement in developing countries: A review of global issues. *International Journal of Environmental Research and Public Health, 16*(6). https://doi.org/10.3390/ijerph16061060

Ferronato, N. et al. (2020). Assessment of municipal solid waste collection in Bolivia: Perspectives for avoiding uncontrolled disposal and boosting waste recycling options. *Resources, Conservation and Recycling, 167*. https://doi.org/10.1016/j.resconrec.2020.105234

Fuldauer, L. I., Ives, M. C., Adshead, D., Thacker, S., & Hall, J. W. (2019). Participatory planning of the future of waste management in small island developing states to deliver on the sustainable development goals. *Journal of Cleaner Production, 223*, 147–162. https://doi.org/10.1016/j.jclepro.2019.02.269

Gunawan, J., Permatasari, P., & Tilt, C. (2020). Sustainable development goal disclosures: Do they support responsible consumption and production? *Journal of Cleaner Production, 246*, 118989. https://doi.org/10.1016/j.jclepro.2019.118989

Gutberlet, J. (2021). Grassroots waste picker organizations addressing the un sustainable development goals. *World Development, 138*, 105195. https://doi.org/10.1016/j.worlddev.2020.105195

Halkos, G., & Gkampoura, E. C. (2021). Where do we stand on the 17 sustainable development goals? An overview on progress. *Economic Analysis and Policy, 70*, 94–122. https://doi.org/10.1016/j.eap.2021.02.001

Hannan, M. A., Hossain Lipu, M. S., Akhtar, M., Begum, R. A., Al Mamun, M. A., Hussain, A., Mia, M. S., & Basri, H. (2020). Solid waste collection optimization objectives, constraints, modeling approaches, and their challenges toward achieving sustainable development goals. *Journal of Cleaner Production, 277*, 123557. https://doi.org/10.1016/j.jclepro.2020.123557

Haryati, Z., Subramaniam, V., Noor, Z. Z., Loh, S. K., & Aziz, A. A. (2021). Complementing social life cycle assessment to reach sustainable development goals—A case study from the Malaysian oil palm industry. *Chemical Engineering Transactions, 83*, 331–336. https://doi.org/10.3303/CET2183056

Hassan, S. H. et al. (2012). Waste management issues in the northern region of malaysia. *Procedia-Social and Behavioral Sciences, 42*, 175–181. https://doi.org/10.1016/j.sbspro.2012.04.179

Herrera Almanza, A. M., & Corona, B. (2020). Using social life cycle assessment to analyze the contribution of products to the sustainable development goals: A case study in the textile sector. *International Journal of Life Cycle Assessment, 25*(9), 1833–1845. https://doi.org/10.1007/s11367-020-01789-7

Hondo, D., Arthur, L., & Gamaralalage, P. J. D. (2020). Solid waste management in developing Asia: Prioritizing waste separation. *ADBI Policy Brief, 7*, 10. https://www.adb.org/sites/default/files/publication/652121/adbi-pb2020-7.pdf

Huertas-Valdivia, I., Ferrari, A. M., Settembre-Blundo, D., & García-Muiña, F. E. (2020). Social life-cycle assessment: A review by bibliometric analysis. *Sustainability, 12*(15), 1–25. https://doi.org/10.3390/su12156211

Ibáñez-Forés, V. et al. (2019). Assessing the social performance of municipal solid waste management systems in developing countries: Proposal of indicators and a case study. *Ecological Indicators, 98*, 164– 178. https://doi.org/10.1016/j.ecolind.2018.10.031

Iyamu, H. O., Anda, M. and Ho, G. (2020). A review of municipal solid waste management in the BRIC and highincome countries: A thematic framework for low-income countries. *Habitat International, 95*. https://doi.org/10.1016/j.habitatint.2019.102097

Jain, A. (2017). *Summary report waste management* (Vol. 14).

Kurniawan, T. A., Avtar, R., Singh, D., Xue, W., Dzarfan Othman, M. H., Hwang, G. H., Iswanto, I., Albadarin, A. B., & Kern, A. O. (2021). Reforming MSWM in Sukunan (Yogjakarta, Indonesia): A case-study of applying a zero-waste approach based on circular economy paradigm. *Journal of Cleaner Production, 284*, 124775. https://doi.org/10.1016/j.jclepro.2020.124775

Leal Filho, W., Lovren, V. O., Will, M., Salvia, A. L., & Frankenberger, F. (2021). Poverty: A central barrier to the implementation of the UN sustainable development goals. *Environmental Science and Policy, 125*(April), 96–104. https://doi.org/10.1016/j.envsci.2021.08.020

Mathe, S. (2014). Integrating participatory approaches into social life cycle assessment: The SLCA participatory approach. *International Journal of Life Cycle Assessment, 19*(8), 1506–1514. https://doi.org/10.1007/s11367-014-0758-6

Padilla-Rivera, A., Paredes, M. G., & Güereca, L. P. (2019). A systematic review of the sustainability assessment of bioenergy: The case of gaseous biofuels. *Biomass and Bioenergy, 125*, 79–94. https://doi.org/10.1016/j.biombioe.2019.03.014

Page, M. J., McKenzie, J. E., Bossuyt, P. M., Boutron, I., Hoffmann, T. C., Mulrow, C. D., Shamseer, L., Tetzlaff, J. M., Akl, E. A., Brennan, S. E., Chou, R., Glanville, J., Grimshaw, J. M., Hróbjartsson, A., Lalu, M. M., Li, T., Loder, E. W., Mayo-Wilson, E., McDonald, S., & Moher, D. (2021). The PRISMA 2020 statement: An updated guideline for reporting systematic reviews. *Journal of Clinical Epidemiology. The BMJ, 134*, 178–189. https://doi.org/10.1016/j.jclinepi.2021.03.001

Pagunsan, J., & Shimada, K. (2014). *Efficiency evaluation of Philippines waste management sector: A two stage approach (September 2012)* (Vol. 24).

Paul, J. G., Arce-Jaque, J., Ravena, N., & Villamor, S. P. (2012). Integration of the informal sector into municipal solid waste management in the Philippines—What does it need? *Waste Management, 32*(11), 2018–2028. https://doi.org/10.1016/j.wasman.2012.05.026

Permana, A. S., Towolioe, S., Aziz, N. A., & Ho, C. S. (2015). Sustainable solid waste management practices and perceived cleanliness in a low income city. *Habitat International, 49*, 197–205. https://doi.org/10.1016/j.habitatint.2015.05.028

Petti, L., Serreli, M., & Di Cesare, S. (2018). Systematic literature review in social life cycle assessment. *International Journal of Life Cycle Assessment, 23*(3), 422–431. https://doi.org/10.1007/s11367-016-1135-4

Ramos Huarachi, D. A., Piekarski, C. M., Puglieri, F. N., & de Francisco, A. C. (2020). Past and future of social life cycle assessment: Historical evolution and research trends. *Journal of Cleaner Production, 264*, 121506. https://doi.org/10.1016/j.jclepro.2020.121506

Rodić, L., & Wilson, D. C. (2017). Resolving governance issues to achieve priority sustainable development goals related to solid waste management in developing countries. *Sustainability, 9*(3). https://doi.org/10.3390/su9030404

Santos, A. C., Mendes, P., & Ribau Teixeira, M. (2019). Social life cycle analysis as a tool for sustainable management of illegal waste dumping in municipal services. *Journal of Cleaner Production, 210*, 1141–1149. https://doi.org/10.1016/j.jclepro.2018.11.042

Sapuay, G. P. (2016). Resource recovery through RDF: Current trends in solid waste management in the Philippines. *Procedia Environmental Sciences, 35*, 464–473. https://doi.org/10.1016/j.proenv.2016.07.030

Sasaki, S., Araki, T., Tambunan, A. H., & Prasadja, H. (2014). Household income, living and working conditions of dumpsite waste pickers in Bantar Gebang: Toward integrated waste management in Indonesia. *Resources, Conservation and Recycling, 89*, 11–21. https://doi.org/10.1016/j.resconrec.2014.05.006

Scheinberg, Simpson, M, & Gupt, Y. (2010). The economics of the informal sector in solid waste management. *Collaborative Working Group Series, 5*, 36.

Sharma, H. B., Vanapalli, K. R., Samal, B., Cheela, V. R. S., Dubey, B. K., & Bhattacharya, J. (2021). Circular economy approach in solid waste management system to achieve UN-SDGs: Solutions for post-COVID recovery. *Science of the Total Environment, 800*, 149605. https://doi.org/10.1016/j.scitotenv.2021.149605

Tong, Y. D., Huynh, T. D. X., & Khong, T. D. (2021). Understanding the role of informal sector for sustainable development of municipal solid waste management system: A case study in Vietnam. *Waste Management, 124*, 118–127. https://doi.org/10.1016/j.wasman.2021.01.033

Tsai, F. M., Bui, T. D., Tseng, M. L., Lim, M. K., Wu, K. J., & Mashud, A. H. M. (2021). Assessing a hierarchical sustainable solid waste management structure with qualitative information: Policy and regulations drive social impacts and stakeholder participation. *Resources, Conservation and Recycling, 168*, 105285. https://doi.org/10.1016/j.resconrec.2020.105285

Tsai, F. M., Bui, T. D., Tseng, M. L., & Wu, K. J. (2020). A causal municipal solid waste management model for sustainable cities in Vietnam under uncertainty: A comparison. *Resources, Conservation and Recycling, 154*, 104599. https://doi.org/10.1016/j.resconrec.2019.104599

Verma, R. L., Borongan, G., & Memon, M. (2016). Municipal solid waste management in Ho Chi Minh City, Viet Nam, current practices and future recommendation. *Procedia Environmental Sciences, 35*, 127–139. https://doi.org/10.1016/j.proenv.2016.07.059

Weidema, B., Goedkoop, M., Meijer, E., & Harmens, R. (2020). *LCA-based assessment of the sustainable development goals development update and preliminary findings of the project 'Linking the UN Sustainable Development Goals to Life Cycle Impact Frameworks'*. United Nations Environment Programme.

Wibisono, H., Firdausi, F., & Kusuma, M. E. (2020). Municipal solid waste management in small and metropolitan cities in Indonesia: A review of Surabaya and Mojokerto. *IOP Conference Series: Earth and Environmental Science, 447*(1). https://doi.org/10.1088/1755-1315/447/1/012050

Wulf, C., Werker, J., Zapp, P., Schreiber, A., Schlör, H., & Kuckshinrichs, W. (2018). Sustainable development goals as a guideline for indicator selection in life cycle sustainability assessment. *Procedia CIRP, 69*(May), 59–65. https://doi.org/10.1016/j.procir.2017.11.144

Yıldız-Geyhan, E., Altun-Çiftçioğlu, G. A., & Kadırgan, M. A. N. (2017). Social life cycle assessment of different packaging waste collection system. *Resources, Conservation and Recycling, 124*, 1–12. https://doi.org/10.1016/j.resconrec.2017.04.003

Yukalang, N., Clarke, B. and Ross, K. (2017) Barriers to effective municipal solid waste management in a rapidly urbanizing area in Thailand. *International Journal of Environmental Research and Public Health, 14*(9), 9–14. https://doi.org/10.3390/ijerph14091013

Chapter 11
Waste Management as an Opportunity for the Inclusion of Vulnerable Groups

Priyanka Devi, Shipa Rani Dey, Khushbu Sharma, Prasann Kumar, and Joginder Singh

11.1 Introduction

The industrialization process results in the generation of wastes and it is increasing on a regular basis. Its management may be a proactive approach for vulnerable groups, both from an environmental and socio-economic standpoint. Although there is no universally accepted definition of the term "waste material", municipal solid waste is typically considered to include trash from all sectors of the economy, whether commercial, residential, institutional, or industrial. As a result of urbanization, industrialization, and increased population, the amount of urban and municipal solid waste produced is inversely proportional to the rate at which urbanization, industrialization, and population growth are taking place. Due to the rapid increase in the world's population over the past few decades, Kathiravale and Muhd Yunus (2008) observed that there is a rapid rise in the demand for food, housing, and a variety of natural resources. The rapid rate of waste generation today is outpacing both the emission of greenhouse gases (GHG) and the release of toxic substances (Hoornweg et al., 2013). MSWM is an extremely important public service provided by the Ministry of India, but this service is often neglected. Depending on the country, the region, or even the neighbourhood within a city, municipal solid waste (MSW) may be generated in different ways. A study conducted by Pattnaik and Reddy (2010) has shown a large variation in the generation of MSW in Indian cities ranging from 0.3 to 0.6 kg per person per day, with a continuous increase in the amount of waste generated (volume) per person per year. The study also found an annual increase

P. Devi · S. R. Dey · K. Sharma · P. Kumar
Department of Agronomy, School of Agriculture, Lovely Professional University, Phagwara 144411, Punjab, India

J. Singh (✉)
Department of Microbiology, School of Bioengineering and Biosciences, Lovely Professional University, Phagwara 144411, Punjab, India
e-mail: joginder.15005@lpu.co.in

S. A. Bandh and F. A. Malla (eds.), *Waste Management in the Circular Economy*,
https://doi.org/10.1007/978-3-031-42426-7_11

of 1.33% in the MSW generated per person per day. Municipal solid waste (MSW) management is often challenging in developing countries like India because these countries lack the infrastructure to provide their citizens with reliable and adaptable services. As discussed earlier, inadequate collection systems, lack of technical knowledge, and budget constraints are some of the roadblocks in its way to success (Guerrero et al., 2013). In many municipalities, the collection and disposal of municipal solid wastes (MSW) (both primary and secondary) consume a significant amount of municipal budgets, leaving little time or money to be spent on the management of MSWs (Collivignarelli & Sorlini, 2004). There are instances in which a city's per capita GDP in a developing country is not adequate to cover more than half of the cost of the municipal solid waste management facilities. It is important to note that a variety of factors heavily influence the MSWM strategy in addition to political, legal, economic, and organizational ones.

It is important to note that several government agencies, including the Ministry of Environment and Forests (MoEF), the Ministry of Urban Development (MoUD), the Ministry of Agriculture (MoAF), the Ministry of New and Renewable Energy (MNRE), and the Ministry of Non-Conventional Energy Sources (MNES) are involved in the management of municipal solid wastes (MSWM). Furthermore, it is also the responsibility of the informal as well as the formal sectors of society to significantly contribute towards MSWM (Sharholy et al., 2008). A comprehensive strategy, such as integrated solid waste management (ISWM), is crucial to solve this problem, as it is the duty of all sections of society to accomplish this goal (Fig. 11.1).

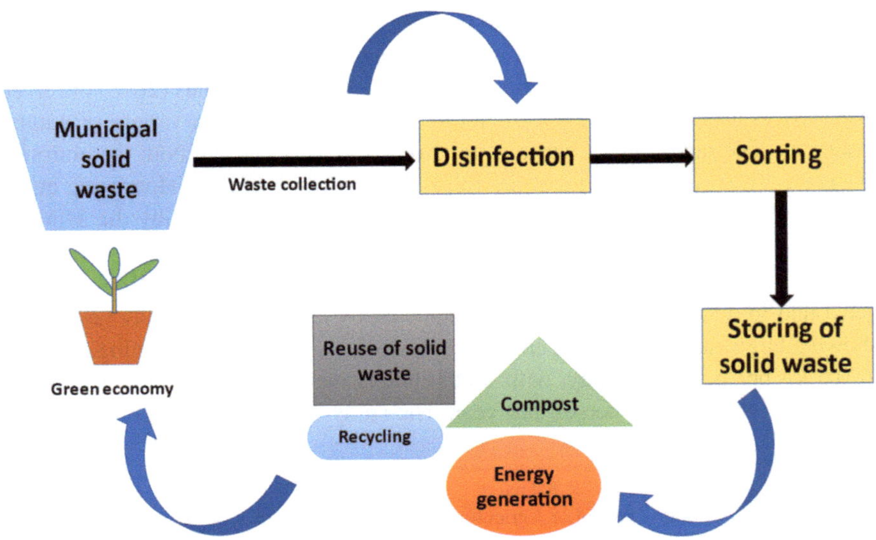

Fig. 11.1 Waste management for the green economy

11.2 World Scenario of Solid Waste

Urbanization and economic development in emerging nations have contributed to a dramatic increase in the generation of MSW in these countries (Devi et al., 2016). According to the World Bank (2012), the amount of waste generated in the urban areas amounts to 1,300,000,000 metric tonnes (1.2 kg per capita/day), with the amount expected to rise to 2,200,000,000 metric tonnes by the year 2025 (Hoornweg & Bhada-Tada, 2012). It is further expected that the MSW production rate in Asia will reach 1.8 million tonnes a day. There is a significant correlation between a country's per capita Gross National Income (GNI) and its MSW generation rate. It is observed that in advanced countries such as the USA, the increase in the per capita gross national product (GNP) has been matched by an increase in the per capita municipal solid waste (MSW). It is therefore, doubtless to mention that the amount of MSW produced by HI economies with high per capita GNI (e.g. Germany, Singapore, Japan, etc.) is much higher than LI economies with low per capita GNI (e.g. Nepal and Bangladesh), which produces tiny amounts of MSW.

11.3 Indian Scenario

India, the second most populous country of the world has undergone significant urbanization and industrialization over the past few decades. Home to more than 1.2 billion people (Census of India, 2011), India represents 17.5% of the global population, making it the most populous country in the world. It is estimated that the urban population of India accounts for approximately 31.16% of the country's total population (Census of India, 2011; Sudhir & Gururaja, 2012). It is believed that the two main reasons for the rapid increase in waste products are the growth in the number of urban residents and the movement of people from rural to urban areas. In parts of India usually open dumps or low-lying areas are used to dispose the municipal solid waste without taking the necessary safety measures. India's megacities pose several environmental challenges due to the handling of municipal solid waste. In India, there are currently 109,598 tonnes of urban MSW generated each day, which equates to a per capita generation of 0.34 kg of MSW on a daily basis. Hoornweg and Bhada-Tada (2012) predict that by 2025, this rate will increase to 376,639 tonnes per day. Based on the findings of a survey conducted by the Central Pollution Control Board (CPCB) (through NEERI), which was conducted by the Central Pollution Control Board (CPCB), 59 cities in India produced approximately 35,401 tonnes of municipal solid waste during 2004–2005. Using this survey as a starting point, the Central Institute of Plastic Engineering and Technology (CIPET) created the survey during 2010–2011 in the same 59 cities and observed that these cities produced approximately 50,592 tonnes of municipal solid waste per day, a quantity much higher than the city's average.

11.4 Integrating Solid Waste Management

Integrated Solid Waste Management (ISWM) is a widely recognized method of managing municipal solid waste (MSW) in a way that is efficient and effective. Using ISWM, it is possible to carry out a comprehensive analysis of the many dimensions of the waste management system. This analysis aims to assess its effectiveness and efficiency through a comprehensive analysis of the many dimensions of the system. The term integrated waste management has been defined by Tchobanoglous et al. (1993) as the selection and implementation of techniques, technologies, and management systems in order to achieve specific waste management objectives and missions. There are several key aspects of ISWM system that works together to maximize the benefit for the least cost, the use of resources and the recycling, reusing, and repurposing of items, to improve safety, health standards, and social acceptance. It comprises three primary components including stakeholders government and non-government sectors, elements (technical aspects of SWM), and issues (policies and effects). In order to achieve success in ISWM, UN-HABITAT has identified three essential elements of the system that must be addressed for it to be successful: health promotion, environmental protection, and the development of natural resources. The information on generated waste as well as its forecasting throughout the entire lifespan of such a project strategy, the physical and chemical properties of waste, the identification of the most appropriate alternative based on their analysis, the establishment of the overall cost of the plan, the assessment of accounting, revenue, and its environmental impact such as greenhouse gas emissions must all be considered before designing an ISWM plan for a city (Hoornweg & Bhada-Tada, 2012). The construction of an ISWM system needs to be designed to maintain a balance between societal, ecological, health, organizational, technological, financial, and legal aspects to offer sustainability to the system (van de Klundert & Anschutz, 2001).

11.5 Approaches Based on Nutritional Composition of Waste

It has been found that economic status, living standards, eating habits, rituals, literacy rate, type of energy source, meteorological and topographical conditions, and geographical location have a significant impact on the composition and characteristics of municipal solid waste (MSW) (Jin et al., 2006). To develop an effective waste management system, it is important to have a reliable data about the generation and composition of the solid waste generated. In high-income countries, organic waste makes up only a very small fraction of the total municipal solid waste compared to the low-income countries where the organic waste makes up 39% of the total MSW generated in Georgia but 62% in Indonesia) (Hoornweg & Bhada-Tada, 2012). Aside from that, the waste that is produced in countries with low and intermediate incomes has a high percentage of moisture and a high density. The makeup of municipal solid waste is different in Europe than it is in Asia. In Europe, MSW includes waste from households and businesses in addition to waste generated by public building areas (Eurostat, 2003). It does not contain the sewage sludge and feces that are considered to be part of the Asian standard. On the other hand, in Asia, in addition to waste generated from human settlements and industries, it sometimes gets contaminated with hospital waste due to negligence and unscientific waste management practices, leading to more hazardous potential than in European cities. This is a problem because hospitals generate a lot of waste (Singh et al., 2011). It appears that the composition of municipal solid wallow-income income countries is characterized by a relatively high percentage of organic matter (40–85%), a relatively high moisture content (40–80%), a relatively high density (250–500 kg/m^3), and a relatively low calorific value (CV) (800–1,100 kcal/kg), whereas the composition of high-income countries demonstrates a relatively low organic matter (20–30%), a relatively low moisture content (5–20%). Table 4 provides the breakdown of the many waste streams that are produced in the various emerging nations of Asia.

The organic stuff makes up the vast majority of the trash. The largest concentration of organic material was found in Indonesia (74%), followed by China (67.8%) and Kathmandu, Nepal (67.8%, 59%). In a similar vein, the table below provides an analysis of the waste generated in a number of different cities across India. Because the urbanization and industry in India have been steadily rising since 1947, the country's cities today produce eight times the amount of waste they did back then (Sharholy et al., 2008). Compared to the western portion of the world, the features of waste in India display a considerable amount of variance regarding their composition and the degree of danger they provide (Sharholy et al., 2008). In every instance, organic waste makes up most of the total trash. The city that reported the most organic waste was Chandigarh, which was followed by Mumbai, which reported the largest amount of organic waste (62 and 57%). In addition, the amount of moisture present was quite high in every instance (except for Ahmedabad), ranging from 41 to 64%. The coefficient of variation is quite low and ranges from 742 to 2,632 kcal/kg, and the carbon-to-nitrogen ratio ranges from 18 to 37.

11.6 Threats from the Generation of Waste

When solid waste is thrown out without any proper treatment, a number of problems like damage to the soil, water, air as people's health and the environment occurs.

11.6.1 Environmental Impacts

There are several types of wastes that people throw away, and each of these wastes needs to be disposed of in a right manner. When choosing the location for a landfill in India, not much of attention is paid to the potential impact the landfill's location may have on the surrounding environment. The improper management and disposal of waste can lead to air, soil, and water pollution.

11.6.2 Effects on Health

People are being put at risk due to the rising volume of waste, the shifting nature of waste, the inadequate management of waste, and the negative perception of solid waste in the community. The methods used for managing waste have direct and indirect effects on the health of the general public (Giusti, 2009). Apathy on the part of waste producers in emerging nations has further worsened this problem. As a result of both people searching for recyclable materials and scavenging animals, there is a large amount of garbage thrown around on the streets A lot of diseases spread by mosquitoes and other insects which generate on sewer lines blocked by wastes, making it easier for mosquitoes to breed and spread diseases like malaria, lymphatic filariasis, and other diseases that have a negative impact on human health (Castro et al., 2010). It has been found that rag pickers have a higher risk of contracting toxocariasis than anyone else, according to research conducted by Alvarado-Esquivel (2013).

Furthermore, in addition to groundwater contamination and air pollution, there are also health concerns resulting from open burning and the illegal movement of biomedical and other hazardous waste to dumping sites, which further contribute to health problems (Kathiravale & Yunus, 2008). The link between being sick and hanging around dumps has been widely accepted. People who live near a landfill or an incinerator are more likely to suffer from congenital disabilities, malignant tumours, and respiratory diseases (Johnson, 1997, 1999; Ray et al., 2009). However, according to the World Health Organization (WHO, 2000, 2007), there is insufficient evidence that trash landfills and incinerators are causing cancer, reproductive effects, and mortality. Several studies conducted by Ray et al. (2009) have suggested that workers at the Okhla dump site, a site for illegal dumping in Delhi, have impaired lung function, which can be evidenced by the fact that 62% of them have impaired lung function compared to 27% of controls with similar age, gender, and socio-economic status. Those who work in landfills are more likely to suffer tissue damage and cardiovascular disorders due to the activation of leukocytes and platelets, as well as inflammation in the airways of the workers.

11.6.3 Act as Soil-Forming Factor

The amount of MSW being deposited on land has been increasing in recent years due to the urbanization and industrialization of our planet, and therefore the soil quality (both biotic and abiotic), as well as its yield, are being negatively affected as a result of the increased MSW deposition land. In city areas of underdeveloped countries, as a consequence of the growth of the population, there are plenty of cases of this kind of thing happening. A recent report from Rawat et al. (2009) detected heavy metal concentrations (Mn, Zn, Cu, Cd, Ni, Pb, and Cr) in soil and road dust samples collected from Kanpur city.

11.6.4 Reuse of Water

Water is essential for survival, economic growth, food safety, and long-term progress. On the one hand, the world is confronting freshwater scarcity. On the other hand, whatever the remaining groundwater resources are available, they are facing significant stress in quality owing to improper urbanization and industrialization. Inadequate upkeep of the distribution system is another source of water pollution. Researcher Nagarajan et al. (2012) compared various physicochemical parameters of groundwater quality in Erode City, Tamil Nadu, India to BIS and WHO standards, finding concentrations of constituents such as total dissolved solids (TDS), total hardness (TH), total alkalinity (TA), sodium (Na^+), magnesium (Mg^{2+}), chloride (Cl), fluoride (F), and nitrate (NO_3) to be excessive. Moreover, the leachate, a consequence of organic waste decomposition, percolates through the soil and finally pollutes underlying and nearby aquifers at landfill sites (Mor et al., 2006). As a result of this leachate, the electrical conductivity (EC), total dissolved solid (TDS), chloride (Cl), sulphate, etc., of the groundwater is changed (Nagarajan et al., 2012). There is also a need to evaluate both the resource availability and the possibility of subsurface water pollution prior to the design of a landfill site. Groundwater quality studies conducted by Vasanthi et al. (2008) in the areas surrounding the Perungudi dumping yard in Chennai also revealed elevated pollutant concentrations, including TDS, EC, TH, Cl, chemical oxygen demand (COD), NO_3^-, and SO_4^{-2}. Groundwater tests from the Delhi region around the Gazipur landfill site and its neighbouring areas showed somewhat high concentrations of Cl, NO_3, NH^{4+}, phenol, Fe, Zn, and COD (Mor et al., 2006).

11.7 Production of Landfill Gas from Municipal Waste

Due to the high density of municipal solid waste (MSW), landfill gas is produced during the anaerobic decomposition of the wastes due to a high level of biodegradable organic matter present in the wastes. Gases from landfills are primarily composed of CH_4 and CO_2, as well as trace amounts of volatile organic compounds and other trace gases (Hegde et al., 2003). Both CH_4 and CO_2 are greenhouse gases, but CH_4's global warming potential is 25 times stronger and CO_2's atmospheric residence duration is just 123 years. Jha et al. (2008) measured GHG emissions from two landfill sites in Chennai and reported that emission flux ranged from 1.0 to 23.5 mg CH_4 m^2/h, 6 to 460 g N_2O m^2/h and 39 to 906 mg CO_2 m^2/h at Kodungaiyur and

0.9 to 433 mg CH_4 m^2/h, 2.7 to 1,200 g N_2O m^2/h and 12.3 to 964.4 mg CO_2 m^2/h at Perungudi. Comparatively, Kumar et al. evaluated the methane emission inventory at the Okhla dumping site in Delhi and found that over the course of eight years, from 1994 to 2001, the site produced a total of 102.006 Gigagrams (Gg) of gas, wherein the total amount of waste (excluding inert) deposited in the landfill up until 2001 was used as the basis (i.e. 3,311.867 Gg). As of 2007, Indian landfills were responsible for approximately 604.5 Gg of CH4 emissions. However, there is substantial uncertainty concerning the quantity of landfill gas emissions due to a lack of available data. Currently, there is a lack of data on landfill gas emissions, and as a result, there is substantial uncertainty regarding the number of landfill gases released.

11.8 Incineration of Waste to Generate Power

Incineration is a method for managing waste through the use of heat. In incineration, the burning of raw or unprocessed waste takes place under regulated circumstances at 850°C in the presence of air. The by-products are carbon dioxide, sulphur dioxide, carbon monoxide, particulate matter, dioxins, furans, water vapour, ash, heat, and non-combustible material. Incinerator bottom ash (IBA) refers to the ash generated during the combustion process. Although incineration results in the greatest possible decrease in waste volume, it is given low priority in an integrated waste management strategy due to ecological considerations. This process is extremely exothermic, creating heat that might be used to generate steam and power. Low moisture content (50%) and high heating value (>5 MJ/kg) are both desirable in solid waste if high efficiency is to be achieved. (Vergara & Tchobanoglous, 2012). Because developing countries produce so much organic waste with high moisture content and low CV, it is more common in developed countries. In India, the first incinerator plant was established in Timarpur, Delhi, in 1987, but was shut down barely a few months after installation. Because waste generated in India is unsuitable for incineration due to high moisture content (around 50%) and low CV (3,350–4,200 kJ/kg) (Sharholy et al., 2008). However, in many cities, tiny incinerator plants are operational for burning biomedical waste. The installed capacity of incineration plants in India is about 83 MW (Jain & Sharma, 2011). Incinerating municipal solid waste (MSW) is unsuccessful in India because of the high organic matter (40–60%), high moisture content (40–60%), and low caloric value (CV; 800–1,100 kcal/kg) of MSW. Another plant was established at BARC (near Mumbai) for burning primarily institutional waste. However, there is a persistent worry that waste incinerators may have adverse effects on human health. According to research, liver cancer risk is higher for Britons who reside near municipal incinerators.

11.9 The Breakdown into Heat and Gas

It is possible to extract energy from trash by burning it in oxygen-depleted conditions. Several endothermic reactions characterize both the process of pyrolysis and the process of gasification. In pyrolysis, organic substances are decomposed by heating them to a temperature range of 400 to 1,000°C in an oxygen-less environment. The process of partially burning organic material at temperatures between 1,000 and 1,400°C with a reduced amount of oxygen or air is called gasification. Syngas (a type of gas), acetic acid, acetone, and methanol are the by-products of the two processes, as well as char (composed of carbon combined with inert material), which are the by-products of the other process. As a result of the process of synthesis, carbon monoxide, methane, and hydrogen dominate the final product, while carbon dioxide, nitrogen, and hydrocarbons have smaller amounts. When syngas is cleaned, it can be burned as fuel for internal combustion engine generators or turbines to generate electricity, and char can be used as a fuel or as a soil conditioner, as has been done in the past by indigenous people in the Amazon basin. It is worth noting that there are two types of gasifiers available in India, each with its unique design. There are only a few plants like it in the world, and this plant is the first of its kind; it has been installed by Navreet Energy Research and Information (NERI) in Nohar, Hanungarh, and Rajasthan with the primary purpose of burning agricultural wastes. This machine processes waste at 50–150 kg/h, corresponding to an efficiency of 70–80%. About 25% of the fuel gas produced during the gasification process may be recycled back into the system to assist in the process of further gasification. At the same time, the remainder of it is used to generate electricity. The second system is operational at the TERI, a think tank for energy (CPCB, 2004; Sharholy et al., 2008). Approximately 70 tonnes of garbage have been processed so far through the plasma arc gasification plant in Pune, India, which was set up in 2010. The plant only generated 1.7 MW instead of the planned 2.4 MW. Despite its limitations, this technique could still be a viable option for creating energy from garbage, despite its performance limitations.

11.10 Composting

Composting is a process of aerobic biological decomposition of organic material in a controlled, regulated environment. It is regulated by several environmental factors, such as temperature, humidity, and pH. Native microbes in the composting process transform organic matter into a stable product called compost. Gardening, farming, and landscaping can benefit from the compost that is produced in the process as it can be used as a soil conditioner in gardening, farming, and landscaping (Singh et al., 2011). As mentioned above, there are a number of variables that can have a significant influence on the quality of compost made with municipal solid waste, including the kind and quantity of waste used in the composting process, the composting design, the length of time that the compost is allowed to mature, and the composting techniques used. According to Hashemimajd et al. (2004), composting has been categorized as a "batch" process since it involves a sequential involvement of many microorganisms during its decomposition process. When the temperature rises to 45 to 65°C, the thermophilic phase begins to appear, and sanitization is initiated. The mesophilic phase, also known as the maturation phase, is characterized by a gradual decay of the remaining organic chemicals as it progresses through the organism. Approximately 60% of municipal solid waste (MSW) in developing countries is organic, has a high moisture content, and has traditionally been used as a tool to improve the soil quality in these countries by using compost. The current mindset has changed as a result of the contamination of MSW with hazardous wastes that have a detrimental effect on the environment (Jain & Sharma, 2011). By converting carbon to nitrogen, phosphorus, and potassium in the soil as a result of proper soil segregation methods, a good choice can be made that helps recycle nutrients back into the soil, including carbon, nitrogen, and phosphorus. Moreover, it also reduces the amount of waste that ends up in landfills, thereby extending the life of landfills (Singh et al., 2011) and, as a consequence, assists in reducing the amount of landfill required for the transportation of waste to landfills, as well as lowering the amount of fuel consumed during that process. It can be argued that composting is a sustainable and environmentally friendly method for dealing with waste since it is environmentally friendly (Fig. 11.2).

The composting of MSW is one of the most promising and cost-effective options for disposing of MSW in the near future. In the early 1960s, the Government of

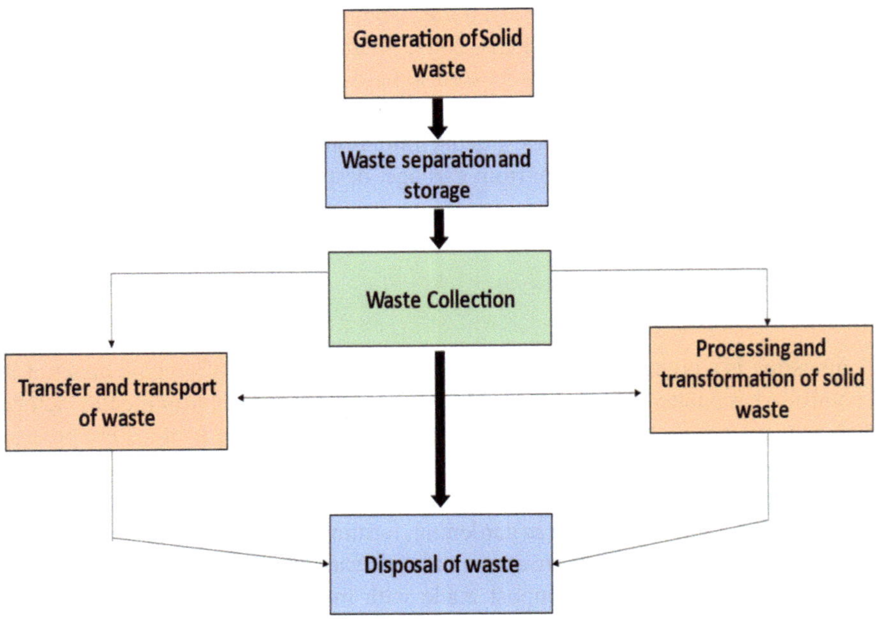

Fig. 11.2 Schematic diagram of the solid waste management system

India supported the idea when it was first floated, but it was stifled during the country's fourth five-year development plan (1969–1974). In 1974, the government of India (GOI) reintroduced municipal solid waste (MSW) composting as a result of a period of decline, focusing on cities with a population of over 0.3 million people. Composting municipal solid waste (MSW) occurs both on a large scale and in decentralized settings in India as a significant amount of MSW is composted. The Excel Industries Ltd. company in Mumbai has recently constructed India's first industrial composting facility on a large scale capable of processing 500 tonnes of municipal solid waste each day.

Furthermore, since the beginning of the year, a facility in Vijayawada has been operating at 150 tonnes per day. In the present day, composting represents about 9% of the total treatment of municipal solid waste worldwide (Kansal, 2002). Every single day, about 700 tonnes of municipal solid waste (MSW) are composted by the Kolkata Municipal Corporation (KMC) and M/S Eastern Organic Fertilizers (India) Private Limited. It has been estimated that there is a market price of 3.50 INR per kilogram for MSW compost (Chattopadhyay et al., 2009).

11.11 Vermicomposting

Vermicomposting is a bio-oxidative, environmentally beneficial, and biotechnological process that converts the organic wastes, such as food and household wastes into a useful bio-product known as vermicompost. In order to accomplish this, earthworms and microbes work together in a collaborative manner. Furthermore, the biochemical degradation of organic matter is carried out by the microbial biomass within the gut of the earthworm. The earthworms increase the surface area available to microorganisms to improve the enzymatic activity, that is, directly and indirectly, responsible for changing the physical and chemical characteristics of the organic waste (Fornes et al., 2012; Malley et al., 2006). There is also some evidence to suggest that earthworm faeces provide bacteria with a nourishing organic substrate for growth and activity in their immediate environment. The earthworm mortality step can be avoided in vermicomposting by performing a pre-composting step as part of the vermicomposting process. Increasing levels of humic acid, aromaticity, and acid phosphatase activity, along with a decrease in the C/N ratio, volatile solids (VS), aliphatic, lignocellulose, protein, and carbohydrate contents of vermicompost, are all indicators of the maturity and mineralization of vermicompost. After the production of vermicast, it contains high levels of nutrients such as carbon, nitrogen, phosphorus, and potassium, alongside some antimicrobial and enzymatic properties that protect plants against disease after (Bhattacharya et al., 2012; Yasir et al., 2009). Several studies (Suthar, 2008) that have confirmed the importance of earthworm faeces, mucus, body fluids, and enzymes in raising the nitrogen content of vermicompost. The epigeic earthworm species such as *Eisenia fetida, Eudrilus eugeniae, Perionyx excavatus*, and *Perionyx sansibaricus* are commonly used in vermicomposting (Oyedele et al., 2006), though *Eudrilus eugeniae* and *Perionyx excavates* are the most effective earthworm species for stabilizing organic material. They can be found in both tropical and subtropical areas. Although both composting and vermicomposting are considered beneficial for the health and fertility of the soil, vermicomposting is preferred because it produces a wider variety of beneficial bacteria and other creatures (Vivas et al., 2009). As per a 2003 report, in India Hyderabad, Bangalore, Mumbai, and Faridabad implemented vermicomposting systems as part of their waste management programmes (Jha et al., 2003).

11.12 Landfilling

Landfills are simply empty plots of land where waste is disposed of. It is an essential component of any MSW management strategy. After the exhaustion of other waste management strategies, municipal landfills become the only place in which a city can dispose of its municipal solid waste. In most underdeveloped countries, open dumping is used to dispose of waste, which is the most widespread, logical, and cost-effective way to do so. It has been estimated that over half (51%) of all management practices in Asia are characterized by open dumping (Hoornweg & Bhada-Tada, 2012). In spite of the fact that open dumping has fallen out of favour in recent years due to environmental concerns and the availability of other engineered techniques such as sanitary landfills, pyrolysis, and incinerators, these are much less common in developing countries due to economic and technical obstacles as well as the nature of the waste they need to deal with. There is a widespread problem of unregulated open dumping in many Indian cities which pollutes the environment and threatens human health and safety. It is well established that greenhouse gases such as CO_2 and methane found in landfill gas play a significant role in global warming and the degradation of the environment (CPHEEO, MoUD, GoI, & Manual on MSWM, 2000).

Further to this, it causes an unhygienic environment, which has a harmful impact on the health of the people living in it. To ensure that landfills function safely and efficiently an MSWM system must be carefully designed, implemented, and monitored.

11.13 Free-Form Waste Dumps and Landfills

It is the most common non-scientific, non-engineered MSWM method used in third-world countries. Open dumping is, as the name implies, the practice of dumping trash as haphazardly and indiscriminately as possible in low-lying areas in such a way as to pose a threat to the health of people and the environment, as well as being detrimental to the area's visual appeal. An increasing amount of municipal waste is being dumped untreated onto lands that have not been treated in urban India (Narayana, 2009), causing a huge problem. It is to be noted that when landfills do not adhere to regulations and are poorly managed, there is an accumulation of waste that makes them more likely to burn openly. As a result, they release poisonous gases that are not only harmful to the environment but also harmful to human health. It is undeniable that rats, fleas, and mosquitoes all find landfills to be a perfect breeding ground for the spread of diseases, particularly malaria, cholera, and dengue fever. Moreover, it is also known to produce leachate, which, due to its taint, may also pose health risks to nearby residents due to groundwater contamination (Mor et al., 2006). According to Joseph et al., active landfill sites in India make up approximately 70–90% of the total number of active waste disposal sites in the country.

11.14 Landfills that Are Only Partially Regulated or Operated Landfills

There are no effective measures to deal with the disposal of leachates and the unregulated release of landfill gases from open landfills, similar to the situation with open landfills. Whenever landfills are being managed, topsoil is routinely added to these landfills to prevent toxic waste from being released into the environment.

11.15 Refuse Dumps that Adhere to Sanitation Standards

These dumps result from careful planning and design on the part of experts who are well-versed in this field. The amount of landfill gases and leachates that are produced by decomposing organic waste can be minimized in a sanitary landfill. While landfills are given the lowest priority in the ISWM system, they are nevertheless necessary since they are where all of the waste that cannot be recycled or composted ends up. This happens a lot more frequently in the developed world.

11.16 Wasteland Application

It is not uncommon for untreated organic wastes, such as solid waste and sewage sludge, to be disposed on farmland without being treated. A significant improvement to the nutritional profile of agricultural and horticultural soils can be achieved by adding sewage sludge as part of the soil amendment process. The biodegradable component of MSW, in particular, is often quite high in organic matter, nitrogen, phosphorus, potassium, micronutrients, and macronutrients, all of which can contribute to reviving the soil's fertility if used in the right way. It is common for hazardous and biological wastes to be mixed up with municipal solid wastes in the majority of cases. Due to this fact, long-term application of fertilizers on fields can cause heavy

Fig. 11.3 Waste management model

metal deposition in the soil being subsequently absorbed by the living organisms. It could further adversely affect the crop yields, livestock health, and human health (Wang et al., 2003) (Fig. 11.3).

11.17 Waste as a Resource

Due to urbanization, we generate a lot of garbage, also known as municipal solid waste, which could be viewed as an opportunity rather than a burden due to the (Kathiravale & Yunus, 2008) benefits we can extract from it.

11.17.1 Energy Source

Energy plays a crucial role in the development of every nation and cannot be overstated in terms of its importance. Keeping up with the rising energy demand is one of the most challenging tasks that developing countries have to deal with in order to keep up with the rising economic growth. Since there is a growing concern over their environmental impacts, renewable energy sources are becoming more popular as an alternative energy source to conventional energy sources that cannot be used indefinitely. In recent years, scientists all over the world have become increasingly interested in the possibility of generating electricity from municipal solid waste, through a variety of methods. It is significant to note that the use of municipal solid waste (MSW) as a source of electricity not only helps to meet the rising energy demand but also reduces the amount of trash that would otherwise be disposed of in landfills by drastically reducing the amount of waste that ends up there. This practice not only assists in compliance with pollution control regulations but also improves the quality of the waste before it is ultimately disposed of, thereby contributing to pollution control and a cleaner environment. It is being recognized that the process of converting MSW into ethanol has been gaining scientific and technical support in recent years due to the large percentage of lignocelluloses in MSW (Kalogo et al., 2007). Several businesses are also looking into more efficient ways of recycling trash into useful products such as polylactic acid, an industrial chemical, and "drop-in" fuels like butanol, which can be used instead of more traditional fuels like gasoline diesel, and even jet fuel. Unfortunately, this method hasn't been used in underdeveloped countries because it lacks the necessary infrastructure for waste segregation. Constructing these plants would have been prohibitively expensive. Currently, the energy generation potential of India is at approximately 1,460 megawatts (MW) from MSW. Contrastingly, 226 megawatts (MW) can be generated from sewage, according to the Ministry of New and Renewable Energy (MNRE). The MNRE report, however, indicates that only 24 MW of this capacity has been capitalized in accordance with the report. There is a possibility that waste-to-energy plants can displace 86.16% of India's total installed thermal power capacity each year, or about a million tonnes of coal.

11.17.2 Resources for Reuse

Taking second place in the hierarchy of effective waste management, reprocessing plays a crucial role in a solid waste management system. It is a practice of reusing resources that would otherwise be thrown away by repurposing them into something new in order to reduce the amount of waste produced. In addition to this, it provides raw or crude materials that can be used by businesses that manufacture goods, which reduces the need for other natural resources and, in turn, reduces the number of emissions and the amount of energy needed to extract and process raw materials. In many third-world countries, informal recycling is a way for people to make a living. As far as recycling is concerned, rag pickers in India are the ones who are responsible for the majority of it. It has been suggested that rag pickers, who rummage through trash heaps in search of recyclables, are crucial for the smooth working of the municipal solid waste (MSW) system (Agarwal et al., 2005).

11.17.3 Improving Soil Quality and Restoring Plant Nutrients

It is relatively well known that among the organic components of municipal solid waste (MSW), there is a great deal of potential for being used as a soil additive in arid or desert regions to revitalize them. It can solve not only the problem of waste disposal but also the problem of soil fertility control. According to Kizilkaya et al. (2012), vermicompost, when applied to soil as an amendment, has a more significant effect on plant production than non-vermicomposted organic waste disposal, in addition to having a direct impact on soil fertility control. It has a greater impact on plant growth and development when added as an amendment to soil significantly improves the NPK, aeration, porosity, structure, moisture, and nutrient-holding capacity when compared to their non-composted counterparts (Hashemimajd et al., 2004). It has a higher content of humic acid, responsible for stimulating the proton pumps in the plasma membrane of root cells, which in turn is responsible for promoting root growth (Aguiar et al., 2012), which can significantly increase yields of both ornamental plants and crops (Aguiar et al., 2012).

11.17.4 Employment Generation Source

Due to growing population and workforce, developing nations like India are experiencing a significant increase in the number of people looking for jobs. It is estimated that many people in India (e.g. rag pickers, scrap dealers, sweepers, etc.) rely on the recycling and sanitation assistance services provided by the government to survive. There is no doubt that informal recyclers often encounter hazardous working conditions due to the conditions in which they operate. Still, it is more crucial to recognize that waste can provide a means of subsistence and income for most people. Recycling of municipal solid waste (MSW) is considered by the Environmental Protection Agency (EPA) to be one of the "most environmentally sound" methods of managing trash, second only to reducing waste generation in the construction industry through increased resource efficiency. India being one of the world's most environmentally conscious countries employs a number of people in this sector, e.g. as per Datta (1997), more than 85,000 people are involved in this kind of activity in Delhi. Similarly, Agarwal et al. (2005) found that in Delhi, nearly 89,600 recyclists, many of whom come from the lowest income groups of society, rely on the recycling sector for a living. It is estimated that the Municipal Corporation of Delhi employs between 150,000 and 250,000 ragpickers. A recent estimate shows that more than 20,000 women are employed as paper pickers in the Ahmadabad metropolitan area.

11.17.5 Planning for Municipal Solid Waste Management

In order to create or enhance a system for managing trash, many aspects have to be taken into account, but one of the most crucial aspects is planning. For a solid waste management system to be successful, it is important to note that careful planning must be undertaken to ensure its success. There can be differences in the approach required from one city to the other, or even from one country to the other, depending on the situation. According to the USA, Environmental Protection Agency, an optimal waste management plan requires careful coordination between several aspects (social, institutional, environmental, economic, and technical) to achieve the best possible outcomes. To create a successful SWM strategy for a city, it is crucial that the following details are kept in mind before the strategy is created in order to make it as effective as possible:

- The city's demographics and economy.
- The data on solid waste creation.
- The data on the composition of MSW and its characteristics, including moisture content and density.
- The data on the identification and evaluation of the best possible choices.
- An assessment of the planned infrastructure and services' initial capital expenditures and ongoing operating expenses (Hoornweg & Bhada-Tada, 2012).
- Social acceptance;
- Collaboration among different levels of government as well as between neighbouring regions and states or provinces.

The use of geographic information system (GIS) data to optimize MSW routes, select landfill sites, and determine MSW plans is becoming increasingly critical due to the lack of funding for these services, which has a negative impact on SWM ambitions. There seems to be a consensus that when it comes to municipal solid waste management (MSWM), the majority of funding in poor nations is invested in the collection and transportation of waste, which leaves very little money for other services to be provided. It remains a challenge, especially in cities in developing nations, to be able to provide quality sustainable urban management services while also ensuring the financial viability of the system at the same time, which is a challenging task (Lohri et al., 2014). In order to ensure the long-term viability of the SWM system in Bahir Dar, Ethiopia, Lohri et al. (2014) offered four solutions: increasing the value chain through the sale of organic waste recycling products; diversifying revenue streams and financing mechanisms (polluter-pays-, cross-subsidy, and business-principles); reducing costs and increasing cost-effectiveness; and linking the fees for solid waste collection and water supply. Yadav and Garg (2009), reporting on empirical studies and real data acquired from MCC, described the financial difficulty the Mysore City Corporation (MCC) is experiencing in managing MSW. According to his research, MCC spent Rs. 365.7 million on capital projects in 2005 despite receiving only Rs. 320.1 million in capital receipts (state government and financial institutions' grants).

The following suggestions were made by him to be taken in order to improve the financial status of MCC (applicable to most of the local governing bodies in Indian cities):

- improvements in property tax collection efficiency (Madon et al., 2004);
- implementation of the Polluter-pays principle (Karagiannidis et al., 2008);
- installation of a "pay as you throw" system (Karagiannidis et al., 2008);
- relatively improved tax collection, property tax reform, and optimal use of assets/land to produce revenue (through impact fees on new development, parking fees, etc.). (Yadav & Garg, 2009).

According to Rathi (2006), the Municipal Corporation of Greater Mumbai is able to cover the costs associated with the management of municipal solid waste (MSW) thanks to the engagement of community-based organizations (CBOs) and public–private partnerships (PPPs) (MCGM). It was determined that the per-ton cost of waste management is Rs. 1,518 (US$35) when CBOs are involved, Rs. 1,797 (US$41) when PPPs are involved, and Rs. 1,908 (US$44) when MCGM is the sole waste manager.

11.18 Problems and Concerns with MSWM

Some of the issues with MSWM are discussed here.

11.18.1 Classifying the Origin of a Problem

When waste management in underdeveloped countries is carried out in an unscientific and unplanned manner, source segregation becomes the biggest obstacle to achieving a sustainable waste management system. Many Indians do not bother to sort their garbage when they live in a city, relying largely on professional waste men and ragpickers (informal sector) to do the job for them. The waste collectors and rag pickers of impoverished nations prioritize sorting the materials that have a high economic value to the recycling industry to maximize their profits. Because of this, the efficiency of the segregation process is quite low. It has been observed that Indian cities produce a variety of trash, some of which is not suitable for incineration, while other types of waste, such as municipal solid waste (MSW), are sometimes contaminated with biomedical waste and, as such, not suitable for composting.

11.18.2 Problems Due to Technical Aspects

Municipal councils and government bodies may be hampered in their solid waste management efforts by a lack of technical knowledge and planning prowess in their staff. Although it exists in some very large and very large cities, it is not being used to its full potential in terms of informing policy. The failure of a city's solid waste management (SWM) strategy is generally attributable to the chosen technology without considering the city's socio-economic state, climate, waste kind, etc. Therefore, prior to selecting the waste management technology, it is important to keep in mind the waste's characteristics and the operating environment at the local level. Certain treatment techniques, such as burning, may not be viable depending on the physical qualities of the trash, such as the waste's composition, moisture content, and density. Sanitary landfill designs, waste-to-energy conversion technologies, etc., all suffer from a lack of funding. Inadequate research and development, as well as a lack of common standards for the definition, measurement, categorization, and data creation, all have a negative impact on solid waste management systems (Guerrero et al., 2013).

11.18.3 Issues with Financial Aspects

In the introduction of a solid waste management system, financing plays a crucial role in the success of the programme. In developing nations, however, there seems to be a problem where urban local governments, or ULBs, are challenged to raise sufficient funds from local taxes to meet their citizens' needs. There is a lack of private enterprise investment in this area due to residents' reluctance to pay for the services, as very little profit is made in this area. It is difficult for local governments to meet the needs of society at the same time as maintaining financial stability even though they are trying to meet societal demands simultaneously. Therefore, it is vital to encourage the participation of the private sector (PPP) if we are to overcome the limitations imposed by a lack of financial resources and expert knowledge in the country.

11.18.4 Concerning Society

MSWM faces adversity in the form of haphazardly expanding, low-income residential neighbourhoods. Inadequate infrastructural services, such as roads, sewers, and sanitary facilities, are indicative of this. Drains as they currently exist are frequently clogged with waste and plastic bags, creating an ideal environment for the proliferation of disease-causing vectors and other dangerous bacteria. As was previously said, the waste management sector is a significant source of income for many people in the informal economic sector in developing countries. They are economically disadvantaged and social outcasts who toil in dangerous, unhealthy environments. Because of this, it is crucial to enhance their working conditions, earnings, and social standing. A creative strategy is required to include this unstructured or informal industry in the ISWM plan. The key social challenges are the low social standing and salaries of waste employees, a lack of interest among the public and stakeholders, and the absence of awareness initiatives for disseminating public knowledge about ISWM.

11.18.5 Programmes Related to Awareness and Education

The lack of solid waste management education programmes at secondary and tertiary levels has led to waste being improperly managed, and those currently employed in the field do not have the necessary experience and exposure to deal with the issue of solid waste management.

11.18.6 Policy and Legislation

The lack of cooperation between environmental and industrial regulators often results in waste management problems. Legal impediments associated with waste management can be attributed to a variety of factors, including a lack of enabling laws, regulations, standards, and policies, a lack of enforcement of existing laws, political intervention, and weak punishment structures, all of which contribute to the existence of legal impediments.

11.19 Conclusions

Managing municipal solid waste is one of the most difficult tasks for any country. For countries like India and China to join the ranks of the industrialized world, they must continue their rapid rate of urbanization and industrialization. It would lead to increased waste all across the world and disorganized urban development. Therefore, the need for healthier and proper disposal of solid waste is warranted, and emerging countries will face increasing challenges in the near future in dealing with this massive amount of MSW. In developing countries, open dumping is commonplace, endangering local ecosystems and putting people's health at risk. Due to its high moisture content and density, municipal solid waste (MSW) from poor countries is unsuitable for energy conversion. Land application of waste is being considered as a viable option because of its high nutrient value and high organic carbon content. Its long-term use is discouraged due to the presence of heavy metals and other hazardous substances. Due to its high nutritional content and lack of pathogens, composting or vermicomposting of MSW could be a good solution for MSW in developing nations. If not handled correctly, this waste can have serious consequences for people's health, the environment, and the value of their homes. However, its capacity to provide energy and jobs also presents us with an opportunity. In addition, it could be used as a raw material in the production of items and in composting and vermicomposting. Therefore, the management of municipal solid waste (MSW) not only reduces negative consequences but also has the potential to contribute to the supply of energy and the creation of new jobs. It's recommended that the potential for using the by-products of one business as a raw material in another be explored. For MSWM to be successful, careful preparation is required. Population, weather, socioeconomics, solid waste composition, funding, environmental pollution, health impacts, etc. should be considered before a SWM plan is developed. In addition to this methodical approach, other measures such as designing an integrated solid waste management (ISWM) system, source segregation, a 4Rs (reduce, reuse, recycle, and recover) strategy for waste minimization, the use of a geographic information system (GIS) in MSW management, decentralized composting at the micro level, compliance with MSW rules, a user-charge system based on income, a ban on open burning, a separate treatment system for biomedical waste, awareness Therefore, the management of municipal solid waste (MSW) not only reduces negative consequences but also has the potential to contribute to ecological and economic needs.

Acknowledgements With great appreciation, the authors acknowledge the Department of Agronomy at Lovely Professional University for their constant support and encouragement throughout the research process.

Author Contributions In addition to contributing to the outline, the author is responsible for leading the draft and editing of the manuscript. It was Priyanka Devi, Shipa Rani Dey, Khushbu Sharma, Prasann Kumar, and Joginder Singh who conducted an extensive literature search and contributed to the writing of sections and the construction of figures and tables. In addition, they contributed to the construction of figures and tables. In addition to providing professional advice, Prasann Kumar, and Joginder Singh helped revise the final version of the document and participated in its revision. In the end, all authors read and approved the final version of the manuscript.

Conflicts of Interest None.

References

Agarwal, A., Singhmar, A., Kulshrestha, M., & Mittal, A. K. (2005). Municipal solid waste recycling and associated markets in Delhi, India. *Resources, Conservation and Recycling, 44*(1), 73–90. https://doi.org/10.1016/j.resconrec.2004.09.007

Aguiar, N. O., et al. (2012). Bioactivity of humic acids isolated from vermincomposts at different maturation stages. *Plant and Soil, 362*(1–2), 161–174.

Alvarado-Esquivel, C. (2013). Toxocariasis in waste pickers: A case control seroprevalence study. *PLoS ONE, 8*(1), e54897. https://doi.org/10.1371/journal.pone.0054897

Bhattacharya, S. S., Iftikar, W., Sahariah, B., & Chattopadhyay, G. N. (2012). Vermicomposting converts fly ash to enrich soil fertility and sustain crop growth in red and lateritic soils. *Resources, Conservation and Recycling, 65*, 100–106. https://doi.org/10.1016/j.resconrec.2012.05.008

Castro, M. C., Kanamori, S., Kannady, K., Mkude, S., Killeen, G. F., & Fillinger, U. (2010). The importance of drains for the larval development of lymphatic filariasis and malaria vectors in Dar es Salaam, United Republic of Tanzania. *PLOS Neglected Tropical Diseases, 4*(5), e693. https://doi.org/10.1371/journal.pntd.0000693

Census of India. (2011). Retrieved from http://censusindia.gov.in. Ministry of Home Affairs.

Central Pollution Control Board (CPCB). (2004). *Management of municipal solid waste*. Ministry of Environment and Forests (MoEF).

Central Public Health and Environmental Engineering Organization (CPHEEO). (2000). *Manual on municipal solid waste management*. Ministry of Urban Development, Government of India.

Chattopadhyay, S., Dutta, A., & Ray, S. (2009). Municipal solid waste management in Kolkata, India—A review. *Waste Management, 29*(4), 1449–1458. https://doi.org/10.1016/j.wasman.2008.08.030

Collivignarelli, C., & Sorlini, S. (2004). AOPs with ozone and UV radiation in drinking water: Contaminants removal and effects on disinfection by-products formation. *Water Science and Technology, 49*(4), 51–56. https://doi.org/10.2166/wst.2004.0218

Datta, M. (1997). *Waste disposal in engineered landfills*. India: Narosa Publishing House.

Devi, K., Swamy, A. V. V., & Nailofer, S. (2016). Municipal solid waste management in India: An overview. *Asia Pacific Journal of Research, 1*, 118–126.

Eurostat. (2003). *Waste generated and treated in Europe*.

Fornes, F., Mendoza-Hernández, D., García-de-la-Fuente, R., Abad, M., & Belda, R. M. (2012). Composting versus vermicomposting: A comparative study of organic matter evolution through straight and combined processes. *Bioresource Technology, 118*, 296–305. https://doi.org/10.1016/j.biortech.2012.05.028

Giusti, L. (2009). A review of waste management practices and their impact on human health. *Waste Management, 29*(8), 2227–2239. https://doi.org/10.1016/j.wasman.2009.03.028

Guerrero, L. A., Maas, G., & Hogland, W. (2013). Solid waste management challenges for cities in developing countries. *Waste Management, 33*(1), 220–232. https://doi.org/10.1016/j.wasman.2012.09.008

Hashemimajd, K., Kalbasi, M., Golchin, A., & Shariatmadari, H. (2004). Comparison of vermicompost and composts as potting media for growth of tomatoes. *Journal of Plant Nutrition, 27*(6), 1107–1123. https://doi.org/10.1081/PLN-120037538

Hegde, U., Chang, T. C., & Yang, S. S. (2003). Methane and carbon dioxide emissions from Shan-Chu-Ku landfill site in Northern Taiwan. *Chemosphere, 52*(8), 1275–1285. https://doi.org/10.1016/S0045-6535(03)00352-7

Hoornweg, D., & Bhada-Tada, P. (2012) What a waste: A global review of solid waste management. In *Urban development series knowledge papers (No. 15). Washington, D.C.: World Bank Group.*

Hoornweg, D., Bhada-Tata, P., & Kennedy, C. (2013). Environment: Waste production must peak this century. *Nature, 502*(7473), 615–617. https://doi.org/10.1038/502615a

Jain, S., & Sharma, M. P. (2011). Power generation from MSW of Haridwar city: A feasibility study. *Renewable and Sustainable Energy Reviews, 15*(1), 69–90. https://doi.org/10.1016/j.rser.2010.09.007

Jha, M. K., Sondhi, O. A. K., & Pansare, M. (2003). Solid waste management—A case study. *Indian Journal of Environmental Protection, 23*(10), 1153–1160.

Jha, A. K., Sharma, C., Singh, N., Ramesh, R., Purvaja, R., & Gupta, P. K. (2008). Greenhouse gas emissions from municipal solid waste management in Indian mega-cities: A case study of Chennai landfill sites. *Chemosphere, 71*(4), 750–758. https://doi.org/10.1016/j.chemosphere.2007.10.024

Jin, J., Wang, Z., & Ran, S. (2006). Solid waste management in Macao: Practices and Challenges. *Waste Management, 26*(9), 1045–1051. https://doi.org/10.1016/j.wasman.2005.08.006

Johnson, B. L. (1997). Hazardous waste: Human health effects. *Toxicology and Industrial Health, 13*(2–3), 121–143. https://doi.org/10.1177/074823379701300203

Johnson, B. L. (1999). A review of the effects of hazardous waste on reproductive health. *American Journal of Obstetrics and Gynecology, 181*(1), S12–S16. https://doi.org/10.1016/S0002-9378(99)70466-7

Kalogo, Y., Habibi, S., MacLean, H. L., & Joshi, S. V. (2007). Environmental implications of municipal solid waste-derived ethanol. *Environmental Science and Technology, 41*(1), 35–41. https://doi.org/10.1021/es061117b

Kansal, A. (2002). Solid waste management strategies for India. *Indian Journal of Environmental Protection, 22*(4), 444–448.

Karagiannidis, A., Xirogiannopoulou, A., & Tchobanoglous, G. (2008). Full cost accounting as a tool for the financial assessment of pay-as-you-throw schemes: A case study for the Panorama municipality, Greece. *Waste Management, 28*(12), 2801–2808. https://doi.org/10.1016/j.wasman.2008.03.031

Kathiravale, S., & Yunus, M. N. M. (2008). Waste to wealth. *Asia Europe Journal, 6*(2), 359–371. https://doi.org/10.1007/s10308-008-0179-x

Kizilkaya, R., Turkay, F. S. H., Turkmen, C., & Durmus, M. (2012). Vermicompost effects on wheat yield and nutrient contents in soil and plant. *Archives of Agronomy and Soil Science, 58*(sup1), S175–S179.

Lohri, C. R., Camenzind, E. J., & Zurbrügg, C. (2014). Financial sustainability in municipal solid waste management—Costs and revenues in Bahir Dar, Ethiopia. *Waste Management, 34*(2), 542–552. https://doi.org/10.1016/j.wasman.2013.10.014

Madon, S., Sahay, S., & Sahay, J. (2004). Implementing property tax reforms in Bangalore: An actor-network perspective. *Information and Organization, 14*(4), 269–295. https://doi.org/10.1016/j.infoandorg.2004.07.002

Malley, C., Nair, J., & Ho, G. (2006). Impact of heavy metals on enzymatic activity of substrate and on composting worms *Eisenia fetida. Bioresource Technology, 97*(13), 1498–1502. https://doi.org/10.1016/j.biortech.2005.06.012

Mor, S., Ravindra, K., De Visscher, A., Dahiya, R. P., & Chandra, A. (2006). Municipal solid waste characterization and its assessment for potential methane generation: A case study. *Science of the Total Environment, 371*(1–3), 1–10. https://doi.org/10.1016/j.scitotenv.2006.04.014

Nagarajan, R., Thirumalaisamy, S., & Lakshumanan, E. (2012). Impact of leachate on groundwater pollution due to non-engineered municipal solid waste landfill sites of Erode city, Tamil Nadu, India. *Iranian Journal of Environmental Health Science and Engineering, 9*(1), 35. https://doi.org/10.1186/1735-2746-9-35

Narayana, T. (2009). Municipal solid waste management in India: From waste disposal to recovery of resources? *Waste Management, 29*(3), 1163–1166. https://doi.org/10.1016/j.wasman.2008.06.038

Oyedele, D. J., Schjonning, P., & Amussan, A. A. (2006). Physicochemical properties of earthworm casts and uningested parent soil from selected sites in southwestern Nigeria. *Ecological Engineering, 28*(2), 106–113.

Pattnaik, S., & Reddy, M. V. (2010). Assessment of municipal solid waste management in Puducherry (Pondicherry), India. *Resources, Conservation and Recycling, 54*(8), 512–520. https://doi.org/10.1016/j.resconrec.2009.10.008

Rathi, S. (2006). Alternative approaches for better municipal solid waste management in Mumbai, India. *Waste Management, 26*(10), 1192–1200. https://doi.org/10.1016/j.wasman.2005.09.006

Rawat, M., Ramanathan, A. L., & Subramanian, V. (2009). Quantification and distribution of heavy metals from small-scale industrial areas of Kanpur city, India. *Journal of Hazardous Materials, 172*(2–3), 1145–1149. https://doi.org/10.1016/j.jhazmat.2009.07.115

Ray, M. R., Roychoudhury, S., Mukherjee, S., Siddique, S., Banerjee, M., Akolkar, A. B., Sengupta, B., & Lahiri, T. (2009). Airway inflammation and upregulation of β_2 Mac-1 integrin expression on circulating leukocytes of female ragpickers in India. *Journal of Occupational Health, 51*(3), 232–238. https://doi.org/10.1539/joh.L8116

Sharholy, M., Ahmad, K., Mahmood, G., & Trivedi, R. C. (2008). Municipal solid waste management in Indian cities—A review. *Waste Management, 28*(2), 459–467. https://doi.org/10.1016/j.wasman.2007.02.008

Singh, R. P., Singh, P., Araujo, A. S. F., Ibrahim, M. H., & Sulaiman, O. (2011). Management of urban solid waste: Vermicomposting a sustainable option. *Resources, Conservation and Recycling, 55*(7), 719–729. https://doi.org/10.1016/j.resconrec.2011.02.005

Sudhir, H. S., & Gururaja, K. V. (2012). Population crunch in India: Is it urban or still rural? *Current Science, 103*(1), 37–40.

Suthar, S. (2008). Bioremediation of aerobically treated distillery sludge mixed with cow dung by using epigeic earthworm *Eisenia fetida. The Environmentalist, 28*(2), 76–84. https://doi.org/10.1007/s10669-007-9031-x

Tchobanoglous, G., Theisen, H., & Vigil, S. A. (1993). *Integrated solid waste management, engineering principles and management issues.* McGraw-Hill.

van de Klundert, A., & Anschutz, J. (2001) *Integrated sustainable waste management—The concept. Tools for decision-makers—Experiences from the urban waste expertise programme (1995–2001).* WASTE, Gouda, The Netherlands.

Vasanthi, P., Kaliappan, S., & Srinivasaraghavan, R. (2008). Impact of poor solid waste management on ground water. *Environmental Monitoring and Assessment, 143*(1–3), 227–238. https://doi.org/10.1007/s10661-007-9971-0

Vergara, S. E., & Tchobanoglous, G. (2012). Municipal solid waste and the environment: A global perspective. *Annual Review of Environment and Resources, 37*(1), 277–309. https://doi.org/10.1146/annurev-environ-050511-122532

Vivas, A., Moreno, B., Garcia-Rodriguez, S., & Benitez, E. (2009). Assessing the impact of composting and vermicomposting on bacterial community size and structure, and microbial functional diversity of an olive-mill waste. *Bioresource Technology, 100*(3), 1319–1326. https://doi.org/10.1016/j.biortech.2008.08.014

Wang, Q. R., Cui, Y. S., Liu, X. M., Dong, Y. T., & Christie, P. (2003). Soil contamination and plant uptake of heavy metals at polluted sites in China. *Journal of Environmental Science and Health—Part A Toxic/Hazardous Substances and Environmental Engineering, A38*(5), 823–838. Waste Integrated Sustainable Waste Management. https://doi.org/10.1081/ESE-120018594

World Health Organization. (2000). *Methods for assessing risk to health from exposure to hazards released from waste landfills.* Report from a WHO meeting, Lódz, Poland. World Health Organization (WHO) European Centre for Environment and Health.

World Health Organization. (2007). *Population health and waste management: Scientific data and policy options.* Report of a WHO workshop, Rome, Italy.

Yadav, A., & Garg, V. K. (2009). Feasibility of nutrient recovery from industrial sludge by vermin composting technology. *Journal of Hazardous Materials, 168*(1), 262–268. https://doi.org/10.1016/j.jhazmat.2009.02.035

Yasir, M., Aslam, Z., Kim, S. W., Lee, S. W., Jeon, C. O., & Chung, Y. R. (2009). Bacterial community composition and chitinase gene diversity of vermicompost with antifungal activity. *Bioresource Technology, 100*(19), 4396–4403. https://doi.org/10.1016/j.biortech.2009.04.015

Chapter 12
High Potential Organic Feedstocks for Production of Renewable Solid Briquettes—A Comprehensive Review

Gokul Raghavendra Srinivasan, Aditya Mahajan, Rajiv Seth, and Rakesh Mahajan

12.1 Introduction

Depletion of resources and unstable capital background for coal has put the entire power generating units at stake and has raised concerns over the large large-scale power generation and the fate of small-scale units depending on this solid fossil fuel. Besides this, emissions from these non-renewable fuels have resulted in increased concentrations of greenhouse gases and have also been found to be a primary source for increasing global warming (Chen, 2021). Considering these setbacks, scientists and researchers have focused on developing renewable solid biofuel from organic biomasses, with significant energy density and zero sulphur emissions (Yuliansyah et al., 2010). One such solid biofuel is briquette, which is made up of any dry organic biomass compacted in high-density product under the application of heavy pressure (Chen et al., 2009).

In general, briquettes are the resultant compacted products produced using briquetting technique and are regarded as an ideal customisable solid biofuel. Also, these briquettes are seen as viable replacements or successful alternatives for existing charcoals (Dinesha et al., 2019). In fact, Rhén et al. (2007) have pointed out the similarities between briquettes and petroleum products, in terms of their opportunities like automation and optimization, citing their higher combustion efficiencies and lower combustion residues (Rhén et al., 2007). Technically, these briquettes are solid biomass agglomerated or densified under the influence of internal forces exerted from the biomass itself and the external forces, which signifies the pressure applied for their compacting. Often, biomass briquettes are developed in different geometries and have been used in numerous industries which include food and industrial

G. R. Srinivasan (✉) · A. Mahajan · R. Seth · R. Mahajan
Research and Development Department, Steamax Envirocare Pvt. Ltd., Delhi, India
e-mail: gokusrinivasan@gmail.com

S. A. Bandh and F. A. Malla (eds.), *Waste Management in the Circular Economy*,
https://doi.org/10.1007/978-3-031-42426-7_12

fodder industries, mining and chemical industries, besides the biomass industries (Ugwu & Agbo, 2013). Commonly known for their application in special boilers or co-firing with unconventional energy carriers (Kihedu, 2015; Kubica et al., 2016), these briquettes are easy to handle and need very least attention among other known biofuels. These renewable briquettes work effectively in both grate and fluidised bed furnaces and showcase a higher degree of combustion (Barneto et al., 2010). In spite of their lower efficiency than coal, these densified solids are widely preferred owing to their simplicity, less cost and need for less maintenance, which enables it for remote applications like cooking, heating and even electricity power generation (Demirbas, 2004; Dziedzic et al., 2018). Accounting for these, this briquette technology can address the problems and challenges related to bio-residue management and is seen as effective "waste-to-energy method".

In common practice, briquetting of biomass increases its volumetric density and calorific value, simultaneously reduces its moisture content (MC) (Sriram et al., 2014) and overcomes the limitations associated with the use of biomass with low bulk density (Ujjinappa & Sreepathi, 2018). In fact, densifying biomasses improvises their overall physical and combustion behaviour, thereby allowing different types of lignocellulosic materials to be used as fuel (Wang et al., 2017). Additionally, it enhances their handling characteristics, reduces their transportation costs, and improvises their suitability for wide variety of applications. Besides, briquetting also induces longer burning time to the biomass, in view of its increased volumetric density (Olorunnisola, 2007).

Looking into the science behind it, briquettes are produced as a result of solid bridge developed between two macroscopic particles of the biomass upon the application of high loads. In general, briquettes undergo densification in three different stages. During the loose stage, biomass particles are re-organized, and are squeezed out for air and moisture content from their porous structures. Moving on to the transition stage, these particles break down further and start filling up the empty voids available between the biomass particles. Finally in the compaction stage, these biomass particles undergo plastic deformation and establish contact with each other through meshing, thereby becoming more compact. This deformation occurs perpendicularly to the principal stress and is limited up to the pressure beyond which these briquettes cannot be compacted (Zhang et al., 2014). Adding to this, MC in their dough induces van der Waals forces between their particles, which help in their agglomeration (Patil et al., 2021). In fact, binding between these particles is enhanced and accomplished by using briquette binders, whose effectivity depends on the type of binder used, which in turn is decided based on the nature of biomass; and the concentration of the binder used. In general, any briquette binder requires the following properties to be regarded as an ideal binding agent (i) owing to its renewability, both feedstock and binder should be organic and environmental-friendly, (ii) must have good ability to induce strong bonding between the biomass particles, (iii) should not affect the heat release and combustibility of the briquette, and (iv) must be economically feasible (Zhang et al., 2018; Zhao et al., 2001). Binders can be classified into organic, inorganic and compound binders based on their material composition. The most commonly preferred binders include starch, protein, fibre, fat/oil, lignin, cattle dung, press mud, molasses, and pulp and paper (Patil et al., 2021). Interestingly, any biomass with high lignin content needs very less to no binders during their briquetting process, as they release these lignin under very high loads, which then acts as binding agent and helps in forming a solid bridge between their biomass particles (Kaliyan & Morey, 2009).

Apparently, these biomasses and binders give rise to a briquette upon its densification; however, the quality of these briquettes can be judged by understanding their fuel characteristics. Hence, it is always necessary to assess these characteristics before commercialising or bringing them into real-time applications. Accordingly, the characteristics deciding the quality of briquettes have been studied in the following section.

12.2 Thermo-Physicochemical Properties of Briquettes

As mentioned earlier, briquetting is regarded as an effective technique for processing biomass feedstocks with medium to high MC into uniform-sized solid biofuels, which can be transported and handled easily. More often, any compacted biomass with good volumetric and net energy density along with significant durability is deemed as a good quality briquette. In turn, these qualities are evaluated based on their briquetting conditions, apart from the physical and chemical properties of the feedstocks (Okot et al., 2018). The chemical properties evaluated for these biomasses include their proximate composition such as moisture content (MC), volatile matter (VM), fixed carbon (FC), ash content (AC) and their lignocellulosic (LC) composition which includes cellulose, hemicellulose and lignin content. On the other hand, the physical properties include their calorific value (gross and net)/heating value (higher/lower), particle size and diameter, and bulk and particle density. In case of briquettes, the chemical properties evaluated include their proximate composition such as MC, VM, FC and AC, and their ultimate composition that includes carbon (C), hydrogen (H), oxygen (O), nitrogen (N) and sulphur (S) content (Sukarta et al., 2018). While, the evaluated physical properties include their calorific value (gross and net)/heating value (higher/lower), compressed and relaxed density, bulk density, compressive strength, tumbling and shattering resistance, and durability, which decides their physical integrity. Further, biomass binders also contribute significantly to these qualities and are also evaluated based on their physical and chemical characteristics. Although these briquettes are assessed from various fuel characteristics discussed above, only certain characteristics like CV, VM, MC and AC, density, compressive strength and durability predominantly decide their quality and usage (Dziedzic et al., 2018).

To begin with, all biomasses comprise lignocellulosic components like cellulose, hemicellulose and lignin, and these organic compounds are widely distributed in the form of hydrocarbons. Following this, the average composition of these lignocellulosic compounds in plant biomass range between 30 and 50% for cellulose, 20 and 30% for hemicellulose and 15 and 30% for lignin (Achinas & Euverink, 2016). On average, typical plant biomass reports its volatile content and calorific content in a range of 65–80% (Rybak, 2006) and 17.31–18.84 MJ/kg, respectively, with its AC ranging between 4 and 5% (2% in case of solid biofuels), in their analytical and working state (Dziedzic et al., 2018). In case of moisture-free and ash-free bases, this biomass records 80% of VM and 20% of FC and in contrast to a bituminous grade coal that states its VM and FC as 20–30%, and 70–80%, respectively (Maciejewska et al., 2006). Comparatively, coal samples exhibit superior calorific value and higher AC, owing to their higher concentration of non-combustible and inorganic content. On the contrary, biomass samples exhibited rapid ignitability and faster burning rate citing their higher porosity, which ensures the infiltration of oxygen and outflow of combustion products (Akuma & Charles, 2017). Worth mentioning, lignin and cellulose content in biomass play a crucial role, by acting as natural binder during their high-load briquetting process (Gangil, 2014).

Moisture content (MC) in briquettes quantifies the amount of water available in it (Yuliah et al., 2017). As per ASTM standards, MC for any briquette must be maintained below 8% (Waluyo & Pratiwi, 2018); however, commercially used briquettes can report their MC up to 30%, following which they can be classified as low moisture (12%) and high moisture (29%) briquettes (Mkini & Bakari, 2015). The briquettes with low MC exhibit improvised rate of combustion and enhanced rate of ignition, in addition to its prolonged storage life (Suprianto et al., 2017). On the contrary, briquette with high MC displays relatively low calorific value and reduced combustion temperature, and prolonged residence time, uneventfully resulting in partial combustion along with increased quantity of flue gas (Maciejewska et al., 2006). Further, the higher MC increases the density of briquettes and favours mould growth, thereby shortening their storage life (Ishii & Furuichi, 2014; Waweru & Chirchir, 2017). Also, finely ground biomass with reduced particle size tends to absorb and retain more water content into them and hence requires additional drying and more conditioning (Adapa et al., 2009). About drying these briquettes, sun drying is seen as the most simple and cost-effective technique, with an efficiency of reducing the MC from 70.56 to 8% in a time span of 2 weeks (Sawadogo et al., 2018). However, rate of drying depends on the weather conditions and is adversely affected during rainy and winter seasons; this can be overcome by using mechanical dryers, which are proven to be effective irrespective of the ambient and weather conditions (Ifa et al., 2020). In short, the MC of the biomass must be limited to 10–15%, so as to allow their briquettes to undergo complete combustion (Adapa et al., 2009; Maciejewska et al., 2006).

Next up, volatile matter (VM) of any briquette illustrates the volume of organic matter with boiling temperature lesser than or equivalent to 250°C available in it and in turn predominantly contributes to their thermal behaviour. In general, biomass exhibiting high VM has less concentration of FC, and vice versa, thereby stating that both biomass and their briquettes torrefied at high temperatures (>250°C) will yield high FC and calorific content. The high volatility in these solid briquettes enhances their rate of ignitability but causes them to burn with smokey flame citing the presence of combustible gases in the form of methane and other volatile hydrocarbons (Thabuot et al., 2015). As mentioned earlier, VM in both biomass and their briquettes can reach up to 80% (Dziedzic et al., 2018) and is also contributed by the binders used during compaction. Relating to this, the usage of starch as binders at higher dosage can lead to increased concentration of VM in the resultant briquettes (Yuliah et al., 2017).

About their ash content (AC), these residues comprise of metal oxides formed as a result of combustion of non-evaporable minerals in the briquettes and are used as indicator for deciding the quality of briquettes. Explaining this, briquettes with high AC tend to report low calorific content and produce less amount of heat. On average, briquettes developed from plant biomass produce their average AC in between 1 and 20% and are entirely dependent on the mineral composition of both biomass and binders. Starch-based binders help reducing the AC; however, higher concentration of both starch-based and mineral-based binders increases it owing to their inorganic content like silica, Fe, MgO and Fe_2O_3 (Yuliah et al., 2017). Besides, these ashes are highly hygroscopic and absorb MC, which solidifies it upon cooling. In addition, these ashes exhibit their fusion temperature ranging between 1200 and 1300°C; and briquettes with high calorific value, cellulose and inorganic content produce high AC and result in higher ash fusion temperature.

In relevance to their thermal behaviour, heating value (HV) of briquettes defines the maximum amount of heat liberated per unit mass of the biofuel during its combustion. In other words, it demonstrates the maximum amount of energy stored in the briquettes and is determined using a bomb calorimeter. In general, the HV ranges between 14.23 and 23.01 MJ/kg for the commercially available biomass briquettes and in turn is decided by the calorific value of the biomass itself. In common practice, HV of these briquettes can be improved by mixing two biomasses together prior to compaction, with either one reporting higher calorific value than the other or formulating their blending composition. It can also be improvised by following certain pre-treatment techniques on biomass during preparation stages like carbonization/ torrefaction and pulverization (Akogun et al., 2022; Yuliah et al., 2017). Again, the binders also contribute to the overall calorific value (CV) of briquettes; hence, identifying the most suitable binder is a highly selective process during the early stages of briquette production. Though small dosages of binders contribute positively to the CVs, high dosage of binders, especially water-activated binders have undesirable effects on these calorific content, as a part of heat is used for evaporating the water content in the briquettes (Yuliah et al., 2017). In addition, high calorific content of these briquettes increases their burning temperature (Haryati et al., 2018).

Besides CV, briquettes are also evaluated for their ignitability and ignition time, with former defining how fast the briquettes ignite, while the latter specifies the rate at which the briquettes burn or combust. Being influenced by the organic matter in the biomass used in briquetting, ignitability increases with increasing organic matter in the biomass, whereas ignition time (also known as burning time) decreases for the same. Together, both CV and burning characteristics decide the water boiling time of the briquette, where lesser time signifies higher efficiency. Explaining this, water boiling test defines the volume of fuel consumed and time taken in raising a known quantity of water up to its boiling point (Akuma & Charles, 2017).

Into their physical characteristics, density of briquettes relates to the amount of biomass that occupies for a unit volume and is classified as bulk density, relaxed density and particle density. Bulk density refers to the value measured immediately after compaction, while relaxed density refers to the value measured after a time period (Falemara et al., 2018), whereas particle density refers to the mass of biomass particles alone excluding the pore space, for an unit volume. Moreover, density has significant influence on their burning characteristics, with high-density briquettes reporting longer burning time and high HRR. Besides, it is also influenced by the particle grain size of biomass used and their MC. Explaining this, briquettes developed from biomass with smaller particle size accommodates larger volume, thereby increasing their density. On the other hand, MC had a negative effect on density as it causes voids and empty pockets post-evaporation, thereby reducing its density and also making the briquettes brittle. Also, high dosage of binders leads to the filling of pores in the biomass which unfortunately increases the overall density of briquettes (Yuliah et al., 2017). In fact, length of briquettes and pellets is also deeply influenced by the bulk density of both biomass and their briquettes (Dziedzic et al., 2018). Put another way, grain size distribution is yet another property of briquettes that explain the particle size of biomass used in its production. Larger grain size accommodates less biomass, thereby reducing its overall volumetric and net energy density; however, use of small particle-sized grains overcomes this and produces high-quality briquettes.

Likewise, durability of briquettes defines its ability to withstand any external force or pressure and is very high for any briquettes having fine particle size, high compaction pressure and high particle density. In addition, high concentration of lignin and cellulosic content in biomass increases the structural integrity of the briquettes and enhances their durability. Besides, other parameters like binder type, dosage and composition, glass transition temperature and compressibility were also found to be contributing in deciding the durability of these briquettes (Karunanithy et al., 2012). On average, the durability of an ideal briquette would range between 85 and 95% and is seen as the most optimal value for any commercially used briquettes.

Besides durability, compressive strength estimates the magnitude of force required to deform or crush the briquettes partially or completely. These strengths generally depend on particle size, briquette shape, binder dosage, curing temperature and AC produced during combustion. Based on the literature, it is seen that briquettes exhibiting high binder dosage, poor curing temperature, larger particle size and briquettes with high AC perform poorly under external force. On the contrary, other briquetting parameters like compaction pressure, retention time and MC have no significant effect on the compressive strength of these briquettes (Taulbee et al., 2009). In addition, biomass with high lignin and cellulosic content tends to impart high structural strength to these densified briquettes, thus increasing their compressive strength.

Fairly evident that the fuel characteristics of any briquette depend on the chemical and physical properties of the biomass used. Table 12.1 summarises the fuel characteristics of both biomass and their briquettes, ASTM standards for their evaluation, and the empirical formula used for calculation.

Table 12.1 Fuel characteristics of both biomass and their briquettes, ASTM standards for their evaluation, and the empirical formula used for calculation (Cunliffe & Williams, 1998; Du et al., 2016)

Fuel property	ASTM standards	Formula
Biomass feedstocks		
Moisture content (MC)	ASTM D3173	$MC(\text{in}\,\%) = \frac{\text{Initial weight} - \text{Dry weight}}{\text{Initial weight}} * 100$
Ash content (AC)	ASTM D3174	$AC(\text{in}\,\%) = \frac{\text{Weight of Ash}}{\text{Dry weight}} * 100$
Fixed carbon (FC)	ASTM D3172	$FC\,(\text{in}\,\%) = 100\% - MC(\%) - VM(\%) - AC(\%)$
Volatile matter (VM)	ASTM D3175	$VM(\text{in}\,\%) = \frac{\text{Dry weight} - \text{weight @ } 300\,^{\circ}\text{C}}{\text{Initial weight}} * 100$
Ultimate composition (CHNSO)	ASTM D3176	–
Calorific value	ASTM D 3286–77	Using bomb calorimeter
Particle size	ASTM E 1037–84	–
Cellulose	ASTM D 1103-55T	$\text{Cellulose(in\%)} = \text{Acid detergent fibre(ADF)} - \text{Lignin(1)}$
Hemicellulose	ASTM D1104-56	$\text{Hemicellulose(in\%)} =$ $\text{Neutral detergent fibre(NDF)} - \text{Acid detergent fibre(ADF)}$
Lignin	ASTM D 1106–56	$\text{Lignin(in\%)} = \text{Acid detergent lignin(ADL)}$
Densified briquettes		
Moisture content (MC)	ASTM D3173	$MC(\text{in}\,\%) = \frac{\text{Initial weight} - \text{Dry weight}}{\text{Initial weight}} * 100$
Ash content (AC)	ASTM D3174	$AC(\text{in}\,\%) = \frac{\text{Weight of Ash}}{\text{Dry weight}} * 100$
Calorific value (CV)	ASTM D 3286–77	Using bomb calorimeter
Density	ASTM D 2395–83	$\text{Density} = \frac{\text{Weight of Briquettes}}{\text{Volume of Briquettes}} * 100$
Compressive strength	ASTM D2166-85	$\sigma\left(\text{in}\,\frac{KN}{m^2}\right) = \frac{\text{Load(in KN)} * \text{Strain}}{\text{Area of briquette (in } m^2)}$
Durability/ Shattering Index	ASTM D440-86	$\frac{\text{Weight before shatter} - \text{Weight of shatter}}{\text{Weight before shatter}}$

12.3 Biomasses for Briquetting

Approximately 60 exajoules of total energy demand were supplied from biomass in 2015, and biomass resources contributed about 9–13% of the total global energy supply (Energy, 2016; Wang et al., 2017). In other words, any fuel developed from the biomass displays its potential in contributing to this energy supply; however, the amount of energy supplied depends on the fuel type and its characteristics. Accounting for their simplicity, energy equivalence, and flexibility towards accommodating any biomass as raw material, these briquettes have proven to contribute significantly to the energy supply, especially to meet the demand of domestic and industrial heating applications that require the replacement of fossil coals. In general, any biomass with good lignocellulosic content can be deemed suitable for briquetting; however, the suitability of these biomasses can be decided based on its non-edibility, availability, geographical distribution and even their suitability for the process. Besides using these biomasses in their raw form, they are also pre-treated through carbonization or torrefaction (low-temperature pyrolysis), which helps in altering the chemical characteristics of the biomasses without altering their calorific content (Basu, 2018). These alterations include depolymerisation of long chain polysaccharide from lignin and hemi-cellulose and also reduce their oxygen-to-carbon (O/C) ratio, which significantly improves their hygroscopic nature and energy content (Basu, 2018; Pimchuai et al., 2010). The most commonly used biomasses, identified as high potential feedstocks, are discussed briefly in the following section in alphabetical order.

12.3.1 Almond Shell

Moisture content (%)	3.3
Ash content (%)	0.6–3.29
Fixed carbon content (%)	15.8–20.7
Volatile matter content (%)	76.0–80.3
Calorific value (MJ/kg)	18.2–19.8
Cellulosic content (%)	38.47–50.7
Hemi-cellulosic content (%)	28.82–31.1
Lignin content (%)	20.4–29.54

Almond (*Prunus amygdalus*) is an Iranian native tree belonging to genus prunus and amygdalus subgenus, whose fruit being a drupe has an outer hull and a hard shell enclosing its seed (Ladizinsky, 1999). However, almond seed is edible, its hard shell is always discarded as wastes. Since almond holds a remarkable spot in multi-cuisine markets, the amount of shells generated as wastes also increase proportionally. Though these wastes are returned back to the soil to maintain its potassium balance, they can be used as solid biofuel for burning owing to their significant biomass content. Supporting this, proximate and LC composition of these shells exhibited higher FC and VM besides their lower MC and AC in addition to their identical lignocellulosic contents. Besides, study on thermal decomposition of these almond shells based on the data characterised from thermogravimetry (TGA), derivate thermogravimetry (DTG) and differential thermal analysis (DTA), reported that these shells report evaporation and volatilization of lighter molecules between 282 and 398°C, devolatilisation involving degradation of biomass, and oxidation between 397 and 522°C; with major weight loss of ~ 50% (9% per min.) noticed during the second stage (Allouch et al., 2014).

Following this, Allouch et al. (2014) developed briquettes from almond shell-based biochar by mixing with 20% green clay, 5% wheat flour, and water, and compacting them manually. Results explained that these briquettes displayed increased holding time and better combustion characteristics, claiming strong devolatilisation and reduction in particle size enhanced their fuel properties (Allouch et al., 2014). Meanwhile, Orhevba and Olatunji (2021) used these shells as additives for developing groundnut shells-based compound briquettes reporting their highest CV as 29.99 MJ/kg; upon mixing them in the following concentration: groundnut shell–52%, almond shell–10%, rice husk–10%, cassava starch–20%, clay (filler)–5%, and water–3%. Here, optimum parameters are as follows: compaction pressure as 250 MPa, dwell time as 300 s, carbonising and drying temperature as 650 and 160°C, respectively (Orhevba & Olatunji, 2021).

12.3.2 Barley Straw

Moisture content (%)	5.7–20.2
Ash content (%)	2.18–9.87
Fixed carbon content (%)	13.29–24.8
Volatile matter content (%)	65.2–82.41
Calorific value (MJ/kg)	16.42–17.65
Cellulosic content (%)	31.1–35.5
Hemi-cellulosic content (%)	20.36–29
Lignin content (%)	10.1–17.13

Barley (*Hordeum vulgare*), a cereal grain-based grass, grown in temperate climates is most commonly used as animal fodder, besides being used as fermentable biomass feedstock for distilleries. As a good source for dietary fibre and vitamins, barley is also consumed globally and is regarded as the fourth majorly harvested crop (Zohary & Hopf, 2000). Though these cereals are edible, other parts of the plants are treated as wastes and are disposed during stubble burning. The barley straws are highly effective for algal growth control in water eco-systems, besides exhibiting promising features in supplying energy. Looking into their proximate and LC composition, these straws exhibit good HV, which can be used for heating purposes instead of burning it for disposal. In fact, the net energy content of these straws can be enhanced by compacting these wastes and converting them into solid biofuels.

Accordingly, Adapa et al. (2009) developed briquettes from ground barley straws by utilising their protein and lignin as natural binders and compacted them using a compaction apparatus operated at 63.2 MPa. Preliminary findings on these ground biomass measured their geometric mean particle diameter, bulk density and particle density as 0.384 ± 0.003 mm, 261 ± 2 kg/m^3 and 1484 ± 3 kg/m^3, respectively. Post-briquetting, the mean compact density, specific energy and total specific energy were calculated as 978 ± 14 kg/m^3, 5.42 ± 0.33 MJ/t and 5.81 ± 0.36 MJ/t, respectively (Adapa et al., 2009). Again, briquettes from raw barley straws with its MC and bulk density as 9 wt.% (W.B.), and 36–67 kg/m^3, reported its density (immediately after compaction) as 755.52 kg/m^3 and durability as 95.62%. Besides, this study also concluded that physical properties like density, MC and durability of these briquettes were also dependent on operating parameters like die temperature, compression pressure and physical properties of the feedstock used (MC and particle size) (Tumuluru et al., 2015).

12.3.3 Canola Straw

Moisture content (%)	7.64
Ash content (%)	2.1–6.47
Fixed carbon content (%)	NA
Volatile matter content (%)	NA
Calorific value (MJ/kg)	NA
Cellulosic content (%)	42.39
Hemi-cellulosic content (%)	16.41
Lignin content (%)	14.15

Canola, a cultivar of rapeseed (*Brassica napus*), belongs to Brassicaceae family, usually occurring as flowering plant and has been cultivated exclusively for its uric acid-rich oil seeds. Among canola and rapeseed-based seeds, the former reports low concentration of uric acid and is widely consumed by both humans and animals, citing its high protein and oil content (Tan et al., 2011). Apart from seeds, these plants don't hold any use and are commonly disposed using stubble burning. However, results from proximate and LC composition showcased these canola straws as potential biomass with significant CV, which can be utilised for energy purposes.

Likewise, Adapa et al. (2009) developed briquettes from ground canola straws by utilising their protein and lignin as natural binders and compacted them using a compaction apparatus operated at 94.7 MPa. Preliminary findings on these ground biomass measured their geometric mean particle diameter, bulk density and particle density as 0.391 ± 0.017 mm, 273 ± 11 kg/m^3 and 1551 ± 47 kg/m^3, respectively. Post-briquetting, the mean compact density, specific energy and total specific energy were calculated as 980 ± 17 kg/m^3, 6.91 ± 0.25 MJ/t and 7.26 ± 0.3 MJ/t, respectively (Adapa et al., 2009). Again, briquettes from raw canola straws with its MC and bulk density as 9 wt.% (W.B.), and 48–58 kg/m^3, reported its density (immediately after compaction) as 976.61 kg/m^3 and durability as 99.7%. Besides, this study also concluded that physical properties like density, MC and durability of these briquettes were also dependent on operating parameters like die temperature, compression pressure and physical properties of the feedstock used (MC and particle size) (Tumuluru et al., 2015).

12.3.4 Cashew Nut Shell (CNS) Wastes

Moisture content (%)	6.47
Ash content (%)	1.05
Fixed carbon content (%)	20.48
Volatile matter content (%)	72.0
Calorific value (MJ/kg)	20.18–20.46
Cellulosic content (%)	9.6 ± 1.3
Hemi-cellulosic content (%)	28.3 ± 0.2
Lignin content (%)	28.8 ± 2.3

Cashew tree (*Anacardium occidentale*), a tropical evergreen tree belonging to Anacardiaceae family, under genus Anacardium and is widely grown for its cashew seed, a snack nut, and occasionally for its cashew apple, an accessory fruit (Morton, 2007). Here, the edible kernels are separated from its shell by means of roasting, leaving behind a large amount of waste cashew nut shells (CNS), which are then processed into press cake and nut shell liquid (CNSL). Though the proximate and LC composition suggests the suitability of these shells for converting into solid biofuel, it cannot be used for heating directly owing to the higher concentration of anacardic acid and smoke from its CNSL causing carcinogenic effects. In fact, 20% of CNSL must be removed from these shells prior processing them into fuel, in order to reduce the toxicity of the raw material and production of tars (Sawadogo et al., 2018). Eventually, this can be overcome by carbonising these wastes CNS through pyrolysis at 350°C, which yields about 41% of CNS biochar and later on can be used for briquetting (Ifa et al., 2020).

With this, Sawadogo et al. (2018) used 55% of torrefied CNS char (0.5 mm), 10% of cassava starch, and 35% of water to develop briquettes using mechanical screw press, and presented their CV, density, compressive strength and impact resistance index as 25.7 MJ/kg, 0.91 kg/m^3, 382.89 and 61.10 kPa, respectively, with water boiling tests reporting similar performance to that of wood charcoal (Sawadogo et al., 2018). Following this, Ifa et al. (2020) used torrefied CNS biochar for developing briquettes by adding tapioca flour as binder and was compacted using simple hydraulic press at 29.4 MPa (300 kgf/cm^2) for 5 min, followed by oven drying for 4 to 6 h at 50°C. Evaluation of fuel properties of these briquettes accounted for their CV as 29.49 MJ/kg, MC as 5.3%, AC as 4.96%, VM as 17.16% and C as 72.62% and were in accordance with the bio-briquettes standard (SNI 016,235–2000, Japanese, English and ISO 17225) (Ifa et al., 2020). Meanwhile, briquettes developed optimum blend (by weight) of 65% CNS char (carbonised at 300°C), 25% areca nut shells and 10% cassava flour, reported their fuel characteristics as follows, maximum hardness value: 141 HB (62.7 N), CV: 18–21 MJ/kg, density: 2–3 g/cm^3, flame temperature: 570°C, AC: 2.4–5.8%, FC: 17.23–20.62%, VM: 70–75% and MC below 10% (Chungcharoen & Srisang, 2020).

12.3.5 Cassava Stalks

Moisture content (%)	8.0–23.46
Ash content (%)	4.0–5.48
Fixed carbon content (%)	13.23–27
Volatile matter content (%)	45.72–69.1
Calorific value (MJ/kg)	12.4–25.83
Cellulosic content (%)	33.7
Hemi-cellulosic content (%)	31.61
Lignin content (%)	27.04

Cassava (*Manihot esculenta*), a native south American woody shrub plant belonging to spurge family, whose starchy tuberous root is globally consumed in its boiled form besides being used as raw material for tapioca (cassava starch) on account of its high carbohydrate content. Being the third-largest source of food carbohydrates, cassava is identified as the major staple food for many developing and famine-driven countries owing to their extreme drought tolerance (Fauquet & Fargette, 1990). Since only cassava roots are consumed, its stalks are often treated as wastes and used as fuel for heating. However, these stalks require more time for their ignition and perform poorly, thus requiring certain pre-treatment techniques for enhancing its fuel properties. Though these stems serve as source for nutrients to soil proximate and LC composition suggested the use of these stalks as fuel in form of briquettes. Especially, after Wilaipon, 2008, assessed the net energy availability from these wastes across the northern province of Thailand as 289 TJ/year, accounting for their availability and CV as 18 k tonne/year and 16.39 MJ/kg, respectively (Wilaipon, 2008).

Thus, Ikelle et al. (2020) blended 40% of torrefied cassava stalk with 60% of coal dust, along with starch and $Ca(OH)_2$ as binder and desulphurising agent to produce composite briquettes using manual briquetting machine operated at 276.36 N/ 31.67 N/m^2. Post-compaction, the sun-dried briquettes presented the following fuel properties: MC–3.36%, AC–25.13%, VM–30.15%, FC–41.36%, CV–25.9 MJ/kg, water boiling test–2.98 min, burning time–22.79 g/min, ignition time–36.68 s and compressive strength–10.78 N/mm^2. On the other hand, raw cassava stalk briquettes had their densities ranging between 0.40 and 0.77 g/cm^3 (Wilaipon, 2008). Besides this, Akogun et al. (2020) also proposed cassava peels as potential feedstock by producing composite briquettes by adding saw dust as additive. These briquettes reported their FC as 26.42%, VM as 59.9%, AC as 5.42%, MC as 8.26%, CV as 15.56 MJ/kg, compressive strength, water resistance and density ranging between 0.55–0.8 N/mm^2, 86.5–89.3% and 0.9–1.0 g/cm^3 of density (Akogun et al., 2020). On the other hand, Anggraeni et al. (2021) used torrefied cassava peels as feedstock for developing briquettes with significant fuel qualities (Anggraeni et al., 2021).

12.3.6 Coconut Shell

Moisture content (%)	6.34–9.4
Ash content (%)	0.655–0.66
Fixed carbon content (%)	21.94–22.1
Volatile matter content (%)	73.96–77.7
Calorific value (MJ/kg)	17.31–21.41
Cellulosic content (%)	44.12 ± 0.2
Hemi-cellulosic content (%)	30.21 ± 0.11
Lignin content (%)	21.24 ± 0.31

Coconut (*Cocos nucifera*) belongs to Arecaceae family, under the genus cocos, and its palm is regarded botanically as drupe, citing its inner flesh enveloped with hard shell, which in turn is enclosed by fibrous husks. The coconut palm trees need regular rainfall, moderate temperature, high humidity and abundant sunlight for its enhanced growth (Dziedzic et al., 2018). These trees serve many useful purposes with the coconut palm holding both cultural and religious significance (Nayar, 2016). In specific, its edible flesh is used in both medicinal and cooking while discarding its outer shell as wastes. Occasionally, these shells are used as fuel in traditional stoves in rural parts of the globe; however, its potential can be improvised by following certain pre-treatment techniques. Most commonly followed technique involves carbonising the shells prior to briquetting and has been seen to be highly effective. Numerous studies have used the coconut shells for producing either raw or composite briquettes citing their significant proximate and LC composition (Dziedzic et al., 2018; Rodiah et al., 2021).

To begin with, Dziedzic et al. (2018) compacted ground coconut shell along with their lignin and protein content as natural binder into briquettes using a briquetting press and reported its bulk density and mechanical durability in between 312.5 and 341.5 kg/m^3, and 92.1 and 96.8% for their particle size and compaction pressure maintained between 8 and 12 mm, and 37–47 MPa, respectively (Dziedzic et al., 2018). Meanwhile, the proximate composition of coconut shell char briquettes was measured as follows: FC–67.02%, VM–22.58%, AC–5.91%, MC–4.49% and CV–28.38 MJ/kg. On the other hand, proximate analysis on torrified coconut shell–rice husk char briquettes, reported as VM–48.99%, FC–44.05%, MC–3.55%, AC–3.41% and CV–26.18 MJ/Kg; and their average CO and NOx emission as 416 and 2 PPM, respectively (Rodiah et al., 2021). Similar results were noted for compound briquettes developed from coconut shell and rice husk char briquettes, developed under hydraulic compression tool using tapioca starch as binder (Yuliah et al., 2017). It is highly evident that coconut shell can be either taken as primary feedstock or as an additive for enhancing the overall fuel properties of resultant briquettes. Besides shells, fibrous husk in their fruit also serves similar purpose by proving significant amount of energy during its combustion.

12.3.7 *Spent Coffee Ground (SCG) Wastes*

Moisture content (%)	1.3–6.64
Ash content (%)	0.66–1.78
Fixed carbon content (%)	16.53–19.83
Volatile matter content (%)	72.15–81.98
Calorific value (MJ/kg)	21.22–25.4
Cellulosic content (%)	11.6–13.2
Hemi-cellulosic content (%)	37.2–41.0
Lignin content (%)	22.2–25.6

Coffee (*Coffea arabica*), a tropical and southern African and tropical Asian shrub is falling under the family Rubiaceae and genus coffee and is most commonly known for yielding caffeine-rich coffee beans, which are predominantly used as numerous edible and non-edible products. On average, one tonne of coffee beans produces about 650 kg of these spend coffee ground wastes and almost 7–9 million tonnes of SCG were produced against 166.63 million 60 kg bags of fresh coffee powder, during the year 2021 (García-García et al., 2015; Girotto et al., 2018; Kim et al., 2022; Santos et al., 2017). Eventually, these wastes are widely celebrated for its high energy content and have received less attention in processing them into fuel source. From their proximate and LC composition, higher HV and elemental composition (C: 54.33–57.29%, H: 6.59–7.52%, O: 33.18–35.66%, N: 2.01–3.97%, on D.B.), it is fairly evident that these wastes can be suggested as promising feature to meet the energy demand.

In relevance to that, Espuelas et al. (2020) used these raw spent coffee ground (SCG) wastes for producing briquettes with the help of lab scale briquetting press operated at room temperature and 12 MPa compaction pressure, using xanthan and guar gum as binders. Interestingly, this study produced results related to fuel properties of both raw SCG and gum added SCG samples (in Table 12.2) and showcased significant variation, thus concluding that addition of binders improvised the overall quality of these briquettes (Espuelas et al., 2020). Moving further, Fehse et al. (2021) developed briquettes from raw SCG, binder added SCG, solvent-treated SCG and SCG pyrolyzed at 850°C using a hydraulic stamp press operated at 60°C with a compression rate of 10 mm/s. Highest briquette densities (1097–1114 kg/m^3) were reported for the following combinations: raw SCG with 5% MC and 5% cellulose, SCG biochar with 10% tapioca starch in addition to solvent-treated SCGs (Fehse et al., 2021). Another study by Potip and Wongwuttanasatian (2018), focused on developing briquettes from 90% SCG and 10% crude glycerol with the help of hydraulic briquetting machine operated at 10 MPa and yielded briquettes with density and CV as 872.12 kg/m^3 and 21.55 MJ/kg, respectively. These briquettes required 850% of theoretical air for effective combustion and resulted in highest temperature and combustion rate of 533.4°C and 0.20 g/s, respectively, along with emission levels in the following concentrations: CO–1,262.3 mg/m^3, NOx–38.1 mg/m^3, HC–270.3 mg/m^3, O_2–20.6% and CO_{2max}–19.0% (Potip & Wongwuttanasatian, 2018).

Table 12.2 Proximate and elemental composition of raw and gum-binded briquette samples (Espuelas et al., 2020)

Analysis	Raw SCG	SCG + 5% of Xanthan	SCG + 10% of Xanthan	SCG + 5% of Guar	SCG + 10% of Guar
Proximate analysis (wt.%)					
MC	2.46	3.84	4.14	3.93	3.96
AC	0.66	0.81	0.97	0.57	0.52
FC	18	19.01	19.05	19.07	19.2
VM	78.88	76.35	75.85	76.43	76.33
CV	25.4	24.5	23.5	24.4	24.32
Ultimate analysis (wt.%)					
C	57.29	57.23	57.2	55.79	54.5
O	33.18	33.25	33.24	35.04	36.46
H	7.52	7.78	7.84	7.48	7.37
N	2.01	1.74	1.72	1.69	1.67

12.3.8 Cotton Waste Residues

Moisture content (%)	8.22–13.34
Ash content (%)	2.93–14.74
Fixed carbon content (%)	11.31–20.53
Volatile matter content (%)	61.88–75.14
Calorific value (MJ/kg)	16.01
Cellulosic content (%)	34.92–66.2
Hemi-cellulosic content (%)	15.35–18.4
Lignin content (%)	15.4–26.81

Cotton (*Gossypium species*), a shrub plant native to tropical and subtropical regions, is belonging to family Malvaceae under the genus Gossypium and is cultivated exclusively for its soft, fluffy staple fibre, rich in cellulose, with traces of water, waxes, pectins and fats (Khadi et al., 2010). In general, these staple fibres are spun into yarn for making cotton fabrics and hold a significant place in the textile markets and industries. With the growing demand for these fabrics, the number of plants grown also increases proportionally, thereby accommodating up to 2.5% of arable lands, which in turn produces proportionate quantity of both cotton stubble and its residues. Owing to their VM and CV, these wastes can be used for energy applications in form of briquettes.

Following this, Song et al. (2020) used hydrothermally treated cotton stalk for developing high-quality briquettes using briquetting press operated at 75°C and 80 MPa. Here, the cotton stalks were treated via hydrothermal treatment for 30 min at 200–260°C and used their feedstock bound lignin and protein content as natural binders. Ultimate analysis on these cotton stalks reported their chemical composition (D.B.) as: carbon–49.56%, hydrogen–3.2%, oxygen–45.94%, nitrogen–0.86%, sulphur–0.44% and their ash composition as: SiO_2–11.7%, Al_2O_3–1.8%, Fe_2O_3–1.1%, CaO–30.8%, MgO–13.9%, TiO_2–1.4%, SO_3–6.3%, K_2O–18.9%, Na_2O–10%, P_2O_5–4.1%. Meanwhile, their briquettes with length and diameter measuring 20 $\pm$ 1 and 33 $\pm$ 1 mm, respectively, reported their fuel properties as follows: MC–6.9 to 8.4%, AC–2.1 to 3.1%, VM–69.9 to 78.2%, FC–11.8 to 19.6%, HV–18.41 to 20.23 MJ/kg, ignition temperature–267.8 to 269.41°C, and burnout temperature–457.8 to 469.88°C. Lastly, this study concluded that the hydrothermal pretreatment significantly improvised the physical characteristics of both biomass and its briquettes, by softening their lignin content (Gilbert et al., 2009; Song et al., 2020).

12.3.9 Cow Dung Wastes

Moisture content (%)	4.57
Ash content (%)	21.42
Fixed carbon content (%)	13.34
Volatile matter content (%)	60.67
Calorific value (MJ/kg)	15.43
Cellulosic content (%)	27.2
Hemi-cellulosic content (%)	17.2
Lignin content (%)	20.6

Cow dung or cow pats are the faecal wastes of bovine animal species, which includes domesticated cattle (cows and buffalos), rarely yaks. For ages, these wastes have been used in numerous applications such as biogas production, fermentation medium, compost, fuel and carbon material (Bhattacharjya & Yu, 2014; Vijayaraghavan et al., 2015; Wan et al., 2018). In general, these wastes consist of undigested plant residues passed through their guts and are usually enriched with minerals besides their high biomass content. Even though these mineral-rich wastes have high potential for natural manure and fertilisers, they are used as fuel for supplying heat in rural and remote geographies for a very long time. Supporting this, proximate and LC composition displayed higher FC and VM content for this biomass, along with its CV as 15.43 MJ/kg which suggested it as high potential raw feedstock for developing solid biofuel with promising features.

In response to that, Song et al. (2019) developed briquettes from cow dung wastes with the help of cold-press briquetting technique and added anthracite as additive (1:3.5), composite (1:0.7:0.3) of potassium nitrate (KNO_3), manganese dioxide (MnO_2) and citric acid monohydrate ($C_6H_8O_7 \cdot H_2O$) as combustion promoter, mixture (2:1) of calmogastrin ($Al(OH)_3$) and ammonium molybdate ($(NH)_4Mo_7O_{24} \cdot 4H_2O$) as smoking suppressor, sodium humate and red clay acted as binder, and acidified calcium oxide as desulphuriser. Upon maintaining the dosage of binder, combustion promoter, smoke suppressor and desulphuriser as 17.0, 6.0, 4.5 and 6.5%, respectively, the fuel characteristics were measured as follows: VM–17.1%, AC–28.6%, CV–19.8 MJ/kg, comprehensive combustion characteristic index (SN)–$2.11 \times 10^{-7}\%^2\ min^{-2}{}^{\circ}C^{-3}$, stable burning characteristic index (DW)–$4.77 \times 10^{-}5\%\ min^{-1}{}^{\circ}C^{-2}$, burnout index ($C_b$)–$50.95\%\ min^{-1}$, specific optical density (DS)–3.58 (Song et al., 2019). Besides being used as raw feedstock for briquetting, these wastes can also be used as binders while compaction, and consequently, Patil et al. (2021) used cow and buffalo dung as binders while compacting sugarcane bagasse into briquettes (Patil et al., 2021).

12.3.10 Durian Peel

Moisture content (%)	0.01–15.30
Ash content (%)	3.94
Fixed carbon content (%)	18.18
Volatile matter content (%)	77.87
Calorific value (MJ/kg)	26.25
Cellulosic content (%)	60.45
Hemi-cellulosic content (%)	13.09
Lignin content (%)	15.45

Durian (*Durio zibethinus*), an edible fruit belonging to genus Durio, grown in trees that are native to south Asian countries and are predominantly found in Thailand and Malaysia. Looking into its anatomy, this round to oblong-shaped fruits consists of sweet fragrant flesh enclosed in a thorn covered rind. These fleshes hold a special recipe in Southeast Asian cuisines and are consumed at different stages of its ripeness (Morton, 1987). In view of increasing consumption of this fruit, the volume of its peel also increases considerably. Owing to its strong odour, these waste peels are left unattended and rarely used as remediating medium for various dyes and pollutants. In most cases, these durian shell or peel wastes are used for fuelling furnaces or else simply burned away; nonetheless, both practices lead to air pollution. However, valorising these wastes into fuel helps reducing its exploitation and can help meeting the energy needs (Nuriana et al., 2014). Agreeing to this, these wastes exhibit higher HV and increased concentration of FC and cellulose content.

Considering these, Nuriana et al. (2014), developed briquettes from waste durian peel carbonised at 450°C for 1.5 h, and 10% of cooked starch water as binder with the help of a simple briquetting press. Evaluation on the fuel properties of these briquettes reported their FC as 77.87%, MC as 0.01%, VM as 3.94% of and AC as 18.18%, density as 990 kg/m^3, CV as 26.25 MJ/kg, and compressive strength as 15.10 N/cm^2, for the grain size maintained at 100 mesh or 150 microns (Nuriana et al., 2014). Moving on, Haryati et al. (2018), used durian peel wastes torrefied at 350°C for 30 min and tapioca glue as binder for developing briquettes using manual cylinder tube pressing technique. Here, the developed briquettes are having diameter as 63.5 mm and length as 100 mm exhibited higher CV upto 25.76 MJ/kg, along with a combustion rate of 0.0398 g/s and was explained by their high FC and cellulose content (Haryati et al., 2018).

12.3.11 Groundnut Shell/Husks

Moisture content (%)	9.2–11.12
Ash content (%)	2.89–6.0
Fixed carbon content (%)	19.3–29.0
Volatile matter content (%)	54.7–73.02
Calorific value (MJ/kg)	18.46–19.8
Cellulosic content (%)	44.8
Hemi-cellulosic content (%)	5.6
Lignin content (%)	36.1

Groundnut or peanut (*Arachis hypogaea*), belonging to family Fabaceae (or Leguminosae) under the genus Arachis, is grown as legume crop for its edible seeds. Interestingly, the root nodules of the groundnut plant serve as host for symbiotic nitrogen-fixing bacteria, which reduces the plants nitrogen necessity remarkably and increases the fertility of the soil (Tekulu et al., 2020). Looking into their consumption, these nuts are consumed globally across various cuisines under different recipes; however, they are predominantly regarded as culinary nut, just like walnuts and almonds. Besides, these nuts are treated as oil crops and used for extracting groundnut oil, which are rich in mono- and polyunsaturated C18 fatty acids. In common, only these nuts are edible, while their shells/husk are discarded as wastes; yet, they hold good HV (18 MJ/kg) which can be utilised as fuel for heating (Okello et al., 2013). In spite of their proximate and LC composition suggesting higher HV of 18 MJ/kg, a part of this energy is lost due to its reduced volumetric density (around 258.8 kg/m^3) and light weightiness. However, it can be overcome by compacting these wastes into briquettes.

Accordingly, Ikelle et al. (2020) developed compound briquettes from ground groundnut husk and coal dust with the help of a manual hydraulic briquetting machine, using starch and $Ca(OH)_2$ as binder and desulphurising agent. In post-compaction, these briquettes were evaluated for its fuel properties and were measured as follows: MC: 2.43–6.44%, compressive strength: 7.72–10.85 N/mm^2, AC: 24.18–29.15%, CV: 21.71–25.02 MJ/kg, FC: 16.77–53.22%, ignition time: 22.23–45.20 s, water boiling test: 1.50–4.99 min and burning rate: 16.10–28.32 g/min. Eventually, all reported properties were in good agreement with thermal properties of the bio-briquettes (Ikelle et al., 2020). Again, Akuma and Charles (2017) developed compound briquettes from carbonised groundnut shells (40%) and coal dust (60%) using clay/rice and $Ca(OH)_2$ as binder and desulphurising agent and reported their fuel properties as follows: AC–32.5%, FC–51.50%, MC–7%, density–0.71 g/cm^3, VM–9.0%, porosity index–48.12%, higher CV–12.10 MJ/kg, lower CV–8.04 MJ/Kg, water boiling test–20 min, ignition time–13 min and burning time–56.14 min (Akuma & Charles, 2017). Other work includes development of compound briquettes from carbonised groundnut shells and bagasse with cassava and wheat starch as binders and reported their highest HVs between 21 and 23 MJ/kg and flame temperature as 890°C (Lubwama & Yiga, 2017). It is highly evident that these groundnut shells can be recommended as potential feedstock for developing briquettes, citing their higher MC and VM.

12.3.12 Hazel Nut Shell

Moisture content (%)	8.7
Ash content (%)	1.3–1.4
Fixed carbon content (%)	27.6–28.3
Volatile matter content (%)	62.4–69.3
Calorific value (MJ/kg)	17.36–19.5
Cellulosic content (%)	25.9–42.6
Hemi-cellulosic content (%)	28.7–29.9
Lignin content (%)	42.5–44.4

Hazel nut trees (*Corylus avellana*), belonging to genus Corylus, have their origin in European and West Asian countries and yield spherical and oval shaped edible nuts (Martins et al., 2014). With Turkey and Italy being their prime producers, these fruits comprise of a seed kernel enveloped inside a hard shell and are used in desserts and bakery products, along with chocolates owing to their rich protein, monounsaturated fat, multivitamins and nutrient contents. On the other hand, these hard shells, lacking no proper applications, are discarded as wastes; however, they can be used as fuel for heating applications based on their favourable proximate and LC composition. In specific, these nut shell wastes can be converted into briquettes for prolonged burning time and enhanced CVs.

Responding to this, Demirbas (1999), produced torrified hazelnut shells briquettes using a briquetting press operated at 800 MPa and 400 K, adding 18 wt.% of pyrolysis oil or tar as binding agent. Here, the pyrolyzed hazel nut shells reported their proximate composition as FC–77.11%; VM–20.61%; AC–2.28%; HHV–28.71 MJ/kg. Post-compacting, these briquettes reported their density and compressive strength as 0.7–0.8 g/cm^3 and 40–45 MPa, respectively, wherein raw hazel briquettes reported 0.75–0.8 g/cm^3 and 35–40 MPa, respectively. Worth mentioning, these briquettes exhibited reduced ignitability owing to their reduced porosity as a result of high-level compaction of these shell particles (Demirbas, 1999).

12.3.13 Kenari Shell

Moisture content (%)	4.3–8.19
Ash content (%)	1.08–12.85
Fixed carbon content (%)	55.9–69.68
Volatile matter content (%)	21.05–27
Calorific value (MJ/kg)	18.36
Cellulosic content (%)	39.24
Hemi-cellulosic content (%)	9.25
Lignin content (%)	38.0

Kenari (*Canarium ovatum*), a fruit bearing tropical tree native to Philippines, belonging to family Burseraceae under the genus Canarium, is used as edible nuts (Pham & Dumandan, 2015). Into its anatomy, these fruits have fibrous fleshy pulp enclosing a hard shell (endocarp), which in turn envelopes the edible seed wrapped in a fibrous seed coat. Since only seed is taken for consumption, the shell enveloping it is discarded and thrown as wastes, which has been identified as high potential feedstock for solid biofuel based on their proximate and LC composition. Besides, torrified char of these Kenari shells was found to be highly calorific than its raw counterpart and their proximate composition was calculated as follows: MC–1.92%, AC–3.83%, FC–66.46%, VM–27.79%, CV–28.41 MJ/kg (Widodo et al., 2021).

Based on this, Papuangan and Jabid (2019) used the carbonised Kenari shell char for developing briquettes, using starch, cocoa pulp and sago flour as binders with the help of a briquetting press. Post-evaluation, the fuel properties of the developed briquettes were reported as MC: 3.03–4.3%, AC: 8.55–12.85%, FC: 55.9–67.18%, VM: 20.5–27% and time taken to reach boiling point for 100 ml of water: 26–27 min. Also, this study commented on similarities in physical and chemical characteristics between Kenari shell and coconut shell (Papuangan & Jabid, 2019). In another study, Widodo et al. (2021) mixed 250 g of carbonised Kenari shell char with coal dust in varying proportions along with 12.5 g of tapioca starch as binder and 50 ml of water, to develop composite briquettes using a simple push tool. Assessment on these briquettes reported their calorific content as follows: 23.18 MJ/kg (250 g of plain charcoal dust), 26.59 MJ/kg (250 g of Kenari shell char), 25.19 MJ/kg (62.5 g of canary shell char and 62.5 g of coal dust), and 25.51 MJ/kg (93.75 g of canary shell char and 31.25 g of coal dust) (Widodo et al., 2021). Meanwhile, mixing carbonised canary shell and coal of particle size 80 mesh, in equivalent ratio with tapioca flour as binder yielded briquettes with a compressive strength of 10.110 kg/cm^2, CV of 25.04 MJ/kg, and burning time as 1 h 56 min (Widodo et al., 2019).

12.3.14 *Maize Residues (Cob and Straw)*

	Maize cob	Maize straw
Moisture content (%)	8.73	5.36–8.22
Ash content (%)	1.1–3.2	6.78–14.74
Fixed carbon content (%)	11.5–20.7	13.44–15.16
Volatile matter content (%)	76.1–87.4	61.88–75.95
Calorific value (MJ/kg)	17.11–17.99	17.31–18.2
Cellulosic content (%)	39.74–52.49	35.0–51.53
Hemi-cellulosic content (%)	32.32–37.38	21.0–30.88
Lignin content (%)	14.74–15.19	11.0–19.0

Maize (*Zea mays*), a cereal grain native to South Mexico belonging to genus Zea under family Poaceae (Benz, 2001), is a staple food consumed by both humans and animals in form of direct maize, corn-based oil and ethanol, maize-based products like starch and syrup, and even as animal feed. In specific, sweet corn, rich in sugar content, is cultivated for human consumption and is edible while field corns are used as raw feedstock in oil companies and alcoholic beverages-based distilleries. Besides that, it is regarded as the primary raw material for producing variety of liquid biofuels (especially, bio-ethanol and biodiesel). On average, the global maize production was estimated as 1,162 million thousand tonnes during year 2020 and is expected to increase in forthcoming decades. Despite this volume, only maize kernels are found edible and are predominantly used; whereas, other parts of this plants which includes cob, stalk or straw are either discarded as wastes or burned as stubbles. Eventually, these wastes store significant energy content, and accordingly, both proximate and LC composition of these wastes showed higher VM and cellulosic content, along with higher CV (~17 MJ/kg).

Moreover, these cobs tend to exhibit high MC between 30.3 and 73.9% (Umogbai & Iorter, 2013), due to its freshness, along with high volatile content (76%) and CV (18.9 M/kg), besides their low ash content (3.2%) (Du et al., 2015; Shah et al., 2012). Looking into their chemical composition (on D.B.), these cobs showed their carbon, oxygen, hydrogen and nitrogen content as 46.9 $\pm$ 0.01%, 42.2 $\pm$ 0.33%, 8.1 $\pm$ 0.39% and 2.8 $\pm$ 0.06% respectively. As a result, Okot et al. (2018) compacted these cobs into briquettes under a hydraulic bench press operated at 200–250 MPa and 80°C, using their lignin as natural binders. In post-evaluation, these briquettes with relatively low MC (7–8%) exhibited their density, impact resistance and mechanical strength ranging in between 516 and 1058.2 kg/m^3, 17.7 and 99.8%, and 10 and 40 MPa, respectively (Okot et al., 2018). Moving on, these cobs were added with bean straws in an optimal ratio of 75:25 w/w, followed by compaction in hydraulic bench press using their lignin as natural binders. For a compaction pressure and temperature of 200 MPa and 80°C, the density, impact resistance and compressive strength were measured as 1154.2 kg/m^3, 99.4%, 83.6 MPa, respectively; for

150 MPa and 50°C, it was measured as 1052.9 kg/m^3, 99.5%, 69 MPa. Besides, these briquettes recorded its proximate composition as 5.4, 76.9, 17.7% of AC, VM and FC; and its CV as 17 MJ/kg (Okot et al., 2019).

Besides, maize straws were also noted as high potential biomass waste feedstock accounting for their ultimate composition estimated as carbon–41.84 ± 0.11%, hydrogen–5.41 ± 0.06%, oxygen–28.22 ± 0.15%, nitrogen–1.30 ± 0.08%, sulphur–0.27 ± 0.02%. Acknowledging this, Wang et al. (2017) used these maize straws pyrolyzed at 550°C, for producing briquettes with the help of laboratory compaction apparatus compacted at 15 MPa. Following this, these briquettes presented their FC as 53.37%, AC as 25.82%, VM as 20.81%, carbon as 57.75%, oxygen as 10.79%, hydrogen as 2.53%, nitrogen as 1.28%, sulphur as 0.25%, HHV as 21.05 MJ/kg, density as 525 kg/m^3, volumetric energy density as 11.05 GJ/m^3, mass yield ratio as 33.32%, energy yield ratio as 41.46% and energy-mass co-benefit (EMCI) index as 8.14 (Wang et al., 2017).

12.3.15 Oat Straws

Moisture content (%)	1.5
Ash content (%)	6.3–8.0
Fixed carbon content (%)	16.1
Volatile matter content (%)	76.1
Calorific value (MJ/kg)	15.52
Cellulosic content (%)	31.0–37.6
Hemi-cellulosic content (%)	23.34–38.0
Lignin content (%)	12.85–19.0

Oat (*Avena sativa*) plants, native to Middle East and European countries, are grown for their cereal grains and belong to family Poaceae under genus Avena (Zhou et al., 1999). In general, oats are consumed as oatmeal and rolled oats by both humans and livestock, while their straws and roots are usually discarded as wastes. In general, these straws are used in the treatment of hard waters to soften it; besides being used as bedding for cattle and horse owing to their high absorbance and softness. Though these straws are used for domestic purposes, data related to their proximate and LC composition have suggested these straws as a promising biomass with significant energy content.

Following this, Adapa et al. (2009) used these straws for developing briquettes by using their protein and lignin as natural binders and compacted them using a compaction apparatus operated at 94.7 MPa. Preliminary findings on these ground biomass measured their geometric mean particle diameter, bulk density and particle density as 0.347 ± 0.003 mm, 268 ± ± 4 kg/m^3 and 1523 ± 15 kg/m^3, respectively. Post-briquetting, the mean compact density, specific energy and total specific energy were calculated as 991 ± 63 kg/m^3, 7.09 ± 0.43 MJ/t and 7.62 ± 0.49 MJ/t, respectively (Adapa et al., 2009). Again, briquettes from raw oat straws with its MC and bulk density as 9 wt.% (W.B.), and 40–58 kg/m^3, reported its density (immediately after compaction) as 716.65 kg/m^3 and durability as 99.23% upon maintaining the die temperature and compaction pressure between 118.96–123.99oC, and 8.27–12.16 MPa, respectively (Tumuluru et al., 2015). Besides, this study also concluded that physical properties like density, MC and durability of these briquettes were also dependent on operating parameters like die temperature, compression pressure and physical properties of the feedstock used (MC and particle size) (Tumuluru et al., 2015).

12.3.16 Olive Residues (Pomace and Husk)

Moisture content (%)	10.0
Ash content (%)	3.6–9.49
Fixed carbon content (%)	15.30–26.1
Volatile matter content (%)	61.86–79.10
Calorific value (MJ/kg)	15.77–21.8
Cellulosic content (%)	23.08–32.7
Hemi-cellulosic content (%)	16.2–23.6
Lignin content (%)	30.6–55.29

Olive (*Olea europaea*) shrub is a tree native to Mediterranean Basin, belonging to family Oleaceae under genus olea and is cultivated for its oil-bearing fruits. Spread across Australian, American and African continent, these fruits are mainly sourced for their oil, besides table olives with former being used in cooking, cosmetics and pharmaceutical industries (Fernández et al., 1997). Besides, this oil has its unique medical benefits, which further increases its market value and also its demand. Accordingly, this demand boosted up the production rate of olive oil, leaving behind a large volume of husk and pomace as wastes (upto 70% of fruit weight). From the proximate and LC composition, it was evident that olive pomace exhibits higher CV and VM, along with high lignin content which made these wastes as ideal feedstock for producing high-quality briquettes.

With this, Khlifi et al., (2020a, 2020b) developed high-quality briquettes from these olive mill solid wastes with the help of hydraulic briquetting press operated at 150 MPa and 35°C, respectively, using 15% corn starch as binder. Results suggested their fuel characteristics as follows: VM–64.65%, FC–18.39%, MC–10.44%, AC–6.72%, HHV–16.92 MJ/kg, unit density–2950 kg/m^3, bulk density–1200 kg/m^3, and compressive strength–4581 KN. Besides, these briquettes were made up of 43.43% of carbon, 6.87% of hydrogen, 48.5% of oxygen, 0.78% of sulphur and 0.42% of nitrogen; while plain olive mill solid waste briquettes comprised about 44.99% of carbon, 7.17% of hydrogen, 46.49% of oxygen, 0.89% of sulphur and 0.46% of nitrogen (Khlifi et al., 2020a, 2020b). Another study used waste olive press cakes for producing briquettes using a simple hydraulic press for the following operating parameters: pressure–35 MPa, MC of feedstock–30 to 35% and dwell time–5 s and reported their relaxed density ranging between 1100 and 1300 kg/m^3, twice the density of loose press cakes (Al-Widyan et al., 2002). Meanwhile, olive residue briquettes bound using paraffin waxes reported their breaking force, durability and density as 40.18–349.39 N, 70.98–460.66 N/mm and 844–970 kg/m^3, respectively (Fennir et al., 2014). In conclusion, any olive biomass with particles size less than 100 μm can be used for producing briquettes for domestic and industrial purposes (Khlifi et al., 2020a, 2020b).

12.3.17 Orange Peels

Moisture content (%)	8.62
Ash content (%)	3.6
Fixed carbon content (%)	20.6
Volatile matter content (%)	75.8
Calorific value (MJ/kg)	15.22
Cellulosic content (%)	11.93
Hemi-cellulosic content %)	14.46
Lignin content (%)	2.17

Orange (*Citrus sinensis*), a tree yielding citrus fruit belonging to family Rutaceae under the genus citrus, is native to Southern China, Northeast India and Myanmar region. These fruits are mainly cultivated for its juice-filled vesicles, often found in segments that are delimited by membrane and are consumed for its rich C-vitamins, besides for its anti-oxidants and anti-inflammatory nature (Morton, 1987; Velasco & Licciardello, 2014; Xu et al., 2013). Besides, these vesicles are attached and enveloped inside the fruit's rind, which are discarded as wastes, in most cases except being used as an active ingredient in natural cosmetics. To be precise, the annual production of orange peel was around 79 million tonnes (during 2019), with Brazil leading by 22% followed by China and India (Mohsin et al., 2021). Interestingly, proximate and LC composition of this waste peel suggest high VM and CV, thereby making it ideal for making briquettes.

As a result, Aliyu et al. (2020) developed composite briquettes from the pulverised powders of orange rind/peels and maize cobs in varying mixing ratios (peel: cob) (20:80, 80:20 and 50:50), along with 80 g of pasty starch as binder and compacted using a manually operated hydraulic jack briquetting machine. Post-compacting, the fuel properties of these sun-dried briquettes are list out in the following order (20:80,

80:20 and 50:50): MC–4.64, 4.19, 5.28%; VM–4.05, 0.76, 4.77%; AC–2.39, 4.64, 3.92%; FC–88.93, 90.42, 86.05%; CV–31.89, 31.3, 31.14 MJ/kg; ignition time–3.58, 5.35, 4.56 min; and water boiling test (for 350 ml using 90 g fuel)–8.2, 14.28, 12.51 min. Ultimately, this study encourages for the addition of any suitable biomass as additive upon choosing orange peels as biomass for briquetting, to achieve solid biofuel with superior fuel characteristics (Aliyu et al., 2020).

12.3.18 Palm Wastes (Empty Fruit Bunch, Fibres and Kernel Shells)

	Empty fruit bunch	Kernel shell	Palm fibre
Moisture content (%)	2.44	4.7–7.96	2.33
Ash content (%)	5.26	1.1–11.75	4.28
Fixed carbon content (%)	18.67	13.7–19.1	21.26
Volatile matter content (%)	73.63	62.82–79.2	72.13
Calorific value (MJ/kg)	17.85–18.84	16.14–23.61	14.51
Cellulosic content (%)	37.26	20.7–35.64	21.8 ± 1.2
Hemi-cellulosic content (%)	14.62	21.39–47.7	36.3 ± 0.3
Lignin content (%)	31.68	42.97–53.4	36 ± 0.7

Palms (*Elaeis guineensis*), the grown Arecaceae in form of climbers, shrubs, tree-like and stemless plants, are cultivated in almost all known habitats (from rain forests to deserts), especially for their fruit bunches. With the exception of using these fruits for extracting oil, other parts of this tree are discarded as wastes citing their poor usage and market value in their raw form; and these wastes account up to 140 million tonnes per year (Uemura et al., 2013). However, processed wastes are used for various purposes like constructions and developing fibrous materials; yet, these wastes are underutilised if considered as raw materials for producing energy. Accordingly, proximate and LC composition suggests that wastes from palm tree-like their empty fruit bunch, kernel shell and waste fibres are ideal raw material for developing briquettes citing their high CV and VM content, besides their significant lignin content.

Following this, Maitah et al. (2016) used these palm oil empty fruit bunches (EFB) for developing briquettes using a briquette machine, with their lignin content as natural binder; and estimated their fuel characteristics as follows: Dry mass content–93.70%, net CV–17.61 MJ/kg, AC–5.9%, nitrogen–1.7%, carbon–47.10%, hydrogen–6.20%, oxygen–40.54%, gross CV–18.96 kJ/kg, density between 0.83 and 0.92 g/cm^3 (Maitah et al., 2016). Next up, Nyakuma et al. (2015) developed briquettes from empty fruit bunch of oil palm trees and torrified at 250, 275 and 300°C in a muffle furnace for 60 min. Post-torrefaction, both oxygen-to-carbon and hydrogen-to-carbon ratio reduced from 1.07 to 0.48 and 0.13 to 0.08, respectively while nitrogen and sulphur content increased from 0.54 to 1.07, and 0.2 to 0.43, respectively. In addition, HV of these briquettes improved significantly from 17.57 to 26.24 MJ/kg; whereas, their mass and energy yields reduced from 79.70 to 43.03%, and 89.44 to 64.27%, respectively due to the drying, partial devolatisation and the breakdown of hemicellulose (Basu, 2018; Nyakuma et al., 2015). In addition, Yuhazri et al. (2012) used palm fibres along with waste papers as additive and binder, to develop compound briquettes, whose mean CV ranged between 13.01 and 14.93 MJ/kg (Yuhazri et al., 2012). Meanwhile, Thabuot et al. (2015) used 20 wt.% of palm fibre as additive for developing bamboo saw dust briquettes having density between 260–416 kg/m^3, CV as 21.26 MJ/kg; and rubber wood residue briquettes having the slowest burning rate of 2.01 g/min (Thabuot et al., 2015).

Besides their fibre and empty bunches, palm kernel shells were also used in briquetting citing their ultimate composition as carbon–45.19 ± 0.78%, oxygen–48.49 ± 0.64%, hydrogen–5.95 ± 0.36%, nitrogen–0.33 ± 0.01%, sulphur–0.04 ± 0.01% (Faizal et al., 2018); and their specific heat, specific gravity, bulk density, thermal conductivity and phase change measured as 1.983 ± 0.01 kJ/kg-K, 1.26 ± 0.07, 560 ± 17.4, 0.68 ± 0.05 W/mk and 101.4°C, respectively (Ikumapayi & Akinlabi, 2018). Accordingly, Mohammed and Olugbade (2015), developed composite briquettes from these shells along with rice bran in manually operated briquette machine, using cassava starch as binder. These briquettes recorded their MC, VM, fixed carbon, AC, HV, compressive strength and density as 18.97, 64.54, 21.30, 14.16%, 14.25 MJ/kg and 1.08 kN/m^2, respectively; and their elemental composition as 45.67, 5.80, 0.05, 1.78 and 46.70% for carbon, hydrogen, sulphur, nitrogen and oxygen, respectively (Mohammed & Olugbade, 2015). Even, carbonised palm kernel shell char has also been used for developing briquettes, which yielded the following fuel characteristics, density–1.65 g/cm^3; CV–23.6 MJ/kg, MC–6.67%, burning rate–3.2 g/min, specific fuel consumption–0.7 kg (Bazargan et al., 2014; Faizal et al., 2018; Ugwu & Agbo, 2011). Inferring from these results, it was fairly evident that these palm wastes can be valorised into good energy content biofuel, suitable for both domestic and industrial applications (Chiew & Shimada, 2013).

12.3.19 Pongamia and Tamarind Shells

	Pongamia shell	Tamarind shell
Moisture content (%)	9.97–12	8.44
Ash content (%)	4.09–5.7	9.56
Fixed carbon content (%)	11.71–18.95	13.44
Volatile matter content (%)	66.99–71.21	68.56

(continued)

(continued)

	Pongamia shell	Tamarind shell
Calorific value (MJ/kg)	16.81–17.65	16.3
Cellulosic content (%)	51.73	18.55
Hemi-cellulosic content (%)	–	47.6
Lignin content (%)	21.71	4.04

Tamarind (*Tamarindus indica*) is a leguminous-hardwood tree native to tropical Africa belonging to family Fabaceae, under genus Tamarindus. These trees bear edible fruits usually brown in colour, with appearance resembling like pods, and contains sweet and tangy pulp, used globally across different cuisines as an active ingredient. Besides this, these pulps are widely used in metal polishing works and also in traditional medicines. Beyond this, wood from this hardwood tree is widely used for various wood and furniture works, whereas its seed oil is used for various medicinal purposes. Hence, these trees are cultivated globally, predominantly in tropical and subtropical zones (Rao & Mathew, 2012). Considering their global consumption, these fruits tend to produce a large of volume of wastes, in form of their peels, which can be used as feedstock for developing biofuel, upon considering their proximate and LC composition.

Likewise, Pongamia (*Millettia pinnata*) is also a leguminous tree belonging to family Fabaceae under genus Millettia and is native to tropical zones on the earth, especially to Asia, Australia and Pacific islands. Often termed as Indian beech and Pongame oil tree, these trees are highly tolerant to drought, intense heat and sunlight and help retarding the surface water evaporation due to their dense shades, besides promoting nitrogen fixation. Though drought resistance, these trees can also survive in freshwater floods for a longer time, thus enabling their existences even in swamp forests. In spite of all these benefits, these trees are widely preferred for its oil-rich seeds, which normally exists inside a shell. On global scale, Pongamia oil biodiesel has a very good value in renewable biofuel markets and is seen as the most commonly preferred and cheap alternative to existing diesel. With increasing energy demand, a large volume of seeds are being used for oil production, leaving behind a proportionate volume of shells as wastes (Yadav et al., 2011). With good CV and cellulosic content, these wastes have also been seen as a potential feedstock for developing solid biofuels.

Following this, Ujjinappa and Sreepathi (2018), used these Pongamia shell (PS) and tamarind shell (TS) along with Pongamia cake (PC) as binders for developing composite briquettes using a universal testing machine, by varying their mixing ratios as follows: sample 1 (60:40:00), sample 2 (60:30:10), sample 3 (60:20:20) and sample 4 (60:10:30). For an ideal compaction pressure of 200 MPa, sample 1 briquettes, found highly durable, reported their proximate and elemental composition as follows: MC–5.03%, VM–76.86%, FC–15.4%, AC–2.71%, CV–16.74 MJ/kg, carbon–65.79%, hydrogen–7.06%, nitrogen–0.56%, oxygen–18.67%, sulphur–0.18%. Meanwhile, their compressed and relaxed density, and compressive strength were in-between 1026 and 1108 kg/m^3, 947 and 1023 kg/m^3, 6.26 and 20.18 N/mm^2, respectively, with a shattering index of 96.42%. However, this study discouraged the use of Pongamia cake as binders, as it created an adverse effect on the properties of these briquettes (Ujjinappa & Sreepathi, 2018). On the other hand, briquettes developed from tamarind shell as an additive with onion peel and cassava starch as binder, blended in different concentrations, reported their fuel properties in the following range–MC: 4.01 and 8.42%, AC: 3.83 and 11.14%, VM: 64.28 and 80.12%, FC: 9.36 and 13.87%, CV: 18.24 and 21.05 MJ/kg, carbon: 39.79 and 56.27%, hydrogen: 4.8 and 5.60%, and oxygen: 35.87 and 45.94% (Velusamy et al., 2022).

12.3.20 Rice Wastes (Straw and Husks)

	Rice husk	Rice straw
Moisture content (%)	4.07	10.3–26.63
Ash content (%)	17.0–21.24	7.56–18.67
Fixed carbon content (%)	7.06–16.22	11.23–15.86
Volatile matter content (%)	61.81–71.47	64.07–79.71
Calorific value (MJ/kg)	13.38–13.52	6.38–18.7
Cellulosic content (%)	33.43–35.0	32.1–38.02
Hemi-cellulosic content (%)	20.99–25.0	18.3–24.0
Lignin content (%)	18.25–20.0	18.0–21.6

Rice (*Oryza sativa or Oryza glaberrima*) are monocotic seed-yielding grass species belonging to family Poaceae under the genus Oryza and are grown for its cereal grain annually, especially for consumption as staple food. In fact, rice is regarded as the third-largest cultivated agricultural commodity, after sugarcane and maize and is consumed by almost over half of the world's human population, especially in Asia and Africa for its nutrition and caloric intake. In spite of its nativity to Asia and certain parts of Africa, this crop can be grown in almost every known terrain to humans (Smith, 1995) and is estimated to produce 4% of global greenhouse gas emissions during various stages of its production and consumption during 2010. Looking into their edibility, only the rice cereal is consumed as food, followed by its bran being used as raw material for extracting rice bran oil, while other parts are often discarded as wastes. Especially, these wastes include rice husk and straw, which are disposed by means of stubble burning and, however, exhibit good potential as fuel for energy applications. Accordingly, proximate and LC composition suggested significant CV and VM content, thereby making these wastes as ideal feedstocks for making briquettes.

Supporting this, Oladeji and Enweremadu (2012), densified these husks into briquettes with the help of a prototype briquetting machine, by adding starch as binder. These square-shaped briquettes with sides and thickness as

75 and 8 mm, reported their fuel characteristics as follows: briquette weight–0.025 kg, compaction pressure–2.1 MPa, carbon–42.1%, hydrogen–5.8%, oxygen–51.67%, sulphur–0.05%, AC–18.6%, nitrogen–0.38%, VM–67.98%, FC–13.4%, MC–12.67%, compressive strength–1.07 KN/m^2, HV–13.389 MJ/kg, initial density–138 kg/m^3, maximum density–524 kg/m^3, relaxed density–240 kg/m^3, density ratio–0.45, compaction ratio–3.80, relaxation ratio–2.22, after glow time–354 s, flame propagation rate–0.10 cm/s (Oladeji & Enweremadu, 2012). Besides, both rice husk and bran were compacted together into briquettes using a manual briquetting press operated at 4.2 MPa, with cassava wastewater and okra stem gum taken as binders and reported their HV in between 16.01 and 16.45 MJ/kg, and highest density as 471.3 kg/m^3 (bran concentration–10%) (Yank et al., 2016).

In addition, numerous biomasses like banana resides, coconut shells and sugarcane leaves have been identified as potential additive for these rice husks and have produced briquettes with fuel properties having good agreement with permissible ranges allowed for commercial briquettes (Jittabut, 2015; Nazari et al., 2019; Rodiah et al., 2021). Following this, Rhofita et al. (2018) used ground rice straw to produce briquettes using a manual hydraulic press briquetting machine operated at 10–40 MPa and 175°C. Upon evaluating their fuel characteristics, the highest density was measured as 1178 kg/m^3 (corresponding compressive strength 2.05 MPa), for the compaction pressure of 40 MPa while the highest compressive strength was calculated as 2.45 MPa (corresponding density: 888 kg/m^3) for the compaction pressure of 10 MPa. Meanwhile, these straws presented their bulk density between 106.89 and 112.87 kg/m^3 (Rhofita et al., 2018). Summing up this, both discussed results and conclusions drawn suggest these rice wastes, especially their husk/bran and straw, as biomass with good potential for valorising into solid biofuels that can be used for combustion-oriented applications.

12.3.21 Sorghum (Guinea Corn) Straw

Moisture content (%)	7.72–9.08
Ash content (%)	7.13
Fixed carbon content (%)	19.77
Volatile matter content (%)	65.38
Calorific value (MJ/kg)	16.05–20.58
Cellulosic content (%)	35.4
Hemi-cellulosic content (%)	19.4
Lignin content (%)	10.35

Sorghum (*Sorghum vulgare*), a genus comprising of 25 different flowering plants under the family poaceae and subfamily Panicoideae, is either cultivated for its cereals, consumed by humans or for its fodder/pastures, to be consumed by animals.

Most commonly known species include sorghum bicolour, an African native plant grown globally for its grains and fodder and are predominantly cultivated in warm climates or in pasture lands (Bhattacharya et al., 2011). The above discussions briefly describe its unsuitability for any edible purposes, especially their stalks or straw portions, thereby signifying their potential as raw material for briquettes. In relevance to that proximate and LC composition suggests higher concentration of VM and lignin content, and CV for these wastes, thus making them idle for producing briquettes. In addition, this straw biomass weighs their volume weight between 800 and 870 kg/m^3, compressive force between 40 and 60 N/mm and compaction pressure between 31 and 35 MPa (Plíštil et al., 2005).

With this, Wang et al. (2013) compacted these sorghum straws into briquettes with the help of hydraulic operated universal testing machine operated at 50 MPa and 90°C, using their lignin content as natural binders. As a result, these briquettes presented their confined and relaxed density as 2.25 and 0.78 g/cm^3, respectively, whereas briquettes compacted at 10 MPa reported 1.4 and 0.58 g/cm^3, respectively. Moreover, MC of dough as 20% and compaction pressure of 50 MPa helped in softening and plasticising the lignin content available in them (Wang et al., 2013). Likewise, Bamgboye and Bolufawi (2009) developed wide varieties of briquettes from ground sorghum straws for varying particle sizes, binder concentration and compaction pressure. Preliminary findings showed the MC (D.B.) and bulk density of ground straw particles as 9.08% and 46.03 kg/m^3 and for developed briquettes as 7.15% and 208.15 kg/m^3. Meanwhile, briquettes with particle size 4.7 and 0.6 mm exhibited their densities in range of 789 and 1372 kg/m^3, and 235 and 435 kg/m^3, respectively, for the binder concentration and compaction pressure maintained as 40% and 10.5 MPa. Worth mentioning, this study used hydraulic press for compaction and starch mucilage as binders (Bamgboye & Bolufawi, 2009).

12.3.22 Soybean Straw

Moisture content (%)	6.77–8.45
Ash content (%)	2.8–5.03
Fixed carbon content (%)	14.59–16.95
Volatile matter content (%)	71.8–73.61
Calorific value (MJ/kg)	16.9–18.23
Cellulosic content (%)	44.2
Hemi-cellulosic content (%)	5.9
Lignin content (%)	19.2

Soybean (*Glycine max*), an native east Asian legume plant belonging to family Fabaceae under genus Glycine, that is cultivated for its edible bean which serves many purposes in food industries. These beans are used for preparing soy milk from its beans, tofu from its skins, soy sauce and bean pastes; besides being used as textured vegetable proteins (TVP), an active ingredient in many meat and dairy substitutes, and as animal fodder (Riaz, 2005). Besides, edible beans, these plants don't hold any values, thereby considering its stalks and leaves as wastes, thus disposing them as stubbles. However, these straws hold certain calorific content which can be converted into useful energy, meant for heating or its related applications. Explaining this, Kiš et al. (2009), illustrated about the energy content available in these soybean straws, with data and variables corresponding to Croatia; and were as follows: available mass (@ 15% MC)–2928.884 kg/ha, energy value–15.22 MJ/kg, total energy value–44593.84 MJ/ha, total energy value @ $\eta = 80\%$–35,675 MJ/ha, natural gas equivalent–1002.432 m^3/ha, liquid light fuel equivalent–847.638 kg/ha, Mazut equivalent–870.47 kg/ha (Kiš et al., 2009). Besides their significant calorific content, these straws also exhibited good amount of lignocellulosic content, which makes them as suitable feedstock for producing briquettes.

Supporting this, Makarynska and Turpurova (2020) densified these straws with bulk mass, MC, particle size, milled particle size and density as 55 kg/m^3, 11.5%, 22 mm, 5–6 mm and 750 kg/m^3, respectively, into briquettes with the help of briquetting press using its lignocellulosic compounds as natural binders. Post-briquetting, the fuel characteristics of these compacted briquettes were evaluated as follows: Length–200 mm, diameter–50 mm, mass fraction of moisture (dried at 40 and 105°C)–1.96 and 6.58%, AC–6.96%, sulphur content–0.246%, total carbon content–47.05%, VM–80.20% and their HV ranging between 14.85 and 17.75 MJ/kg. Lastly, this study concluded that briquetting of these soybean straws helped in increasing their CV per unit volume by 10 times than compared to their raw form (Makarynska & Turpurova, 2020).

12.3.23 Sugarcane Wastes (Bagasse and Leaves)

	Sugarcane bagasse	Sugarcane leaves
Moisture content (%)	6.83–9.51	6.61
Ash content (%)	0.9–1.94	6.48–10.6
Fixed carbon content (%)	12.44–13.57	14.9–16.1
Volatile matter content (%)	74.98–86.25	74.5–77.4
Calorific value (MJ/kg)	15.69–18.53	14.73
Cellulosic content (%)	42.5	12.57
Hemi-cellulosic content (%)	33.7	14.01
Lignin content (%)	23.0	5.71

Sugarcane (*Saccharum officinarum*) is a tall, perennial grass belonging to family Poaceae under the genus Saccharum and is cultivated for producing sugar from its fibrous stalks, rich in sucrose. These grass plants are native to warm temperate and tropical regions of India, Southeast Asia and New Guinea (Papini-Terzi et al., 2009). In general, these sucrose are converted into sugar leaving behind a large volume of bagasse as biomass; accordingly, each tonne of sugarcane yields 740 kg of juice (135 kg of sucrose and 605 kg of water) and 260 kg of wet bagasse (130 kg of dry bagasse). Apart from producing sugar, these juices are also used for fermenting into ethanol, especially in Brazil, thereby making them also as an energy crop. Owing to its global consumption and extended applications, volume of sugar/ethanol produced from these canes and their bagasse wastes are always proportional with latter being used commonly as raw material for organic disposable products, animal fodder and even as feedstock for briquettes. In fact, proximate and LC composition have shown higher VM and cellulosic content with good CV for these bagasse; and supporting this, these bagasse showcased high gross CV of 18.72 MJ/kg, energy density of 1.27 GJ/m^3 and bulk density of 95 kg/m^3 (Costa et al., 2019).

Supporting this, Brunerová et al. (2020) compacted these sugarcane bagasses with the help of high-pressure briquetting press, using their lignin content as natural binders. This study estimated the average waste ratio as 35.45 for bagasse to whole stem, and as 8.18 for leaves to whole stem; and reported the following fuel characteristics for these bagasses: carbon–44.35 to 48.16%, hydrogen–5.52 to 5.99%, nitrogen–0.35 to 0.38%, oxygen–41.87 to 45.46%, gross CV–18.35 MJ/kg, and net CV–17.06 MJ/kg. Following this, fuel properties of the developed briquettes were represented as legth–54.08 $\pm$ 2.05 mm, diameter–52.33 $\pm$ 0.28 mm, mass–118.8 $\pm$ 3.98 g, MC–7.54 $\pm$ 0.51%, bulk density–1022.44 $\pm$ 15.59 kg/m^3, compressive strength–150.82 $\pm$ 15.86 N/mm, mechanical durability–99.29 $\pm$ 0.59% (Brunerová et al., 2020). Meanwhile, alkali treated sugarcane bagasse briquettes had MC as 3 $\pm$ 0.5%, AC as 7.3 $\pm$ 1.7%, VM as 80.7 $\pm$ 0.1%, FC as 9 $\pm$ 0.25%, compressive strength as 37.77 $\pm$ 0.68 N/cm^2, density as 0.81 $\pm$ 0.05 g/cm^3, CV as 34.59 $\pm$ 0.57 MJ/kg, ignition time as 5.33 $\pm$ 1.84 min., combustibility test as 2 min, after glow time as 391 $\pm$ 10.1 s, upon using 20% starch binder (Oyibo et al., 2020).

Besides bagasse, even their leaves can also be used as feedstock for developing briquettes, especially due to their high volatile content and lignocellulosic content based on their proximate and LC composition. Subsequently, these sugarcane leaves were compacted into a mould using a simple press operated at different loads, by adding cow dung, buffalo dung and press mud as binders. Here, cow dung binder required the highest load of 22 KN, followed by press mud and buffalo dung binder requiring 12 and 11 KN, respectively. Post-compaction and drying, these briquettes were evaluated for their fuel characteristics and are tabulated in Table 12.3, compared among each other (Patil et al., 2021).

In conclusion, these sugarcane plant wastes are seen as naturally occurring high potential biomass available in abundance owing to their numerous applications and these studies have proposed that these wastes can be valorised into useful biofuels to satisfy the energy demand.

Table 12.3 Fuel characteristics of sugarcane leaf briquettes for different binders (Patil et al., 2021)

	DSL/Cow dung	DSL/Buffalo dung	DSL/Press mud
Proximate analysis			
MC (%)	25.61	33.89	6.52
AC (%)	10.99	9.86	18.88
FC (%)	2.93	7.96	6.87
VM (%)	60.47	48.29	67.73
Gross CV, MJ/kg	16.26	16.23	15.26
Net CV, MJ/kg	15.36	13.47	13.97
Energy density ratio (EDR)	0.93	0.85	0.90
Physical properties			
Bulk density (kg/m^3)	198.1	216.8	191.9
Relaxed density (kg/m^3)	169.47	174.95	171.31
Degree of densification	0.033	0.132	0.002
Compression ratio	1.033	1.131	1.002
Split tensile strength (kN/m^2)	7.164	5.59	6.98
Tumbling resistance (%)	87.84	84.13	86.66
Shatter resistance (%)	12.75	Disintegrated after seven drops	Disintegrated after six drops

*DSL–Dry sugarcane leaves

12.3.24 Switch Grass Wastes

Moisture content (%)	2.65–5.71
Ash content (%)	2.54–7.6
Fixed carbon content (%)	13.81
Volatile matter content (%)	81.2
Calorific value (MJ/kg)	15.99–19.06
Cellulosic content (%)	27.8–44.3
Hemi-cellulosic content (%)	20.30–32.10
Lignin content (%)	7.4–22.5

Switchgrass (*Panicum virgatum*), a perennial bunchgrass grown in warm temperatures belonging to family under genus panicum, is found widely across its native North American continent. In specific, these grasses are found commonly in North

American prairie eco-systems and are primarily used for in soil conservation, environmental phytoremediation, carbon dioxide bio-sequestration and as raw material for fibres. Besides, these grass are used in supplying heat for generating power, producing bio-alcohols (ethanol and butanol) and were accounted for by its VM and CV. Owing to their non-edibility, and superior proximate and LC composition, these wastes are seen as potential feedstock for producing compacted solid biofuels meant for combustion-based applications. In addition, these waste grasses reported their geometrical mean diameter (GMD) as 0.736 mm, bulk density between 115–182 kg/m^3, and their glass transition temperature as 82.5°C (Karunanithy et al., 2012).

Responding to this, Karunanithy et al. (2012) compacted these switch grass into briquettes with the help of simple horizontal briquetting press, using their lignin and protein content as natural binders. Post-compaction, these briquettes recorded their glucose, xylose, and lignin concentration, AC and extractives as 36, 19, 24.8, 3.7 and 16.5%, respectively; with their bulk density and MC found around 946–1173 kg/m^3 and 6–8% (W.B.), respectively (Karunanithy et al., 2012). Moving on, around 80% of switch grass was mixed with 20% of corn stover to develop compound briquettes with the help of a uniaxial, piston-cylinder densification apparatus operated at 150 MPa and 100°C, using corn starch as binding agent. Results included the densities noted immediately and one week after compaction as 1135.5 ± 21.9 and 1096.5 ± 20.8 kg/m^3, MC (W.B.) as 7.9 ± 0.2% and durability after 7 days as 77.8 ± 2.1%, for the particle size of 0.56 ± 0.29 mm and grind MC of 10.4 ± 0.1%. Meanwhile, plain switch grass briquettes without any binders, recorded their densities noted immediately and one week after compaction as 1099.7 ± 12.5 and 1053.5 ± 61.7 kg/m^3, MC (W.B.) as 6.1 ± 0.4% and durability after 7 days as 67.3 ± 1.5%, for the particle size of 0.56 ± 0.29 mm and grind MC of 9.8 ± 0.2% (Kaliyan & Morey, 2009; Kaliyan et al., 2009). In short, it can claimed that briquettes, and even pellets developed from switch grass biomass, hold significant fuel qualities that can be utilised for supplying energy.

12.3.25 Water Hyacinth Wastes

Moisture content (%)	5.7–6.35
Ash content (%)	13.21–38.11
Fixed carbon content (%)	6.28–16.02
Volatile matter content (%)	49.9–64.38
Calorific value (MJ/kg)	13.39–14.4
Cellulosic content (%)	24.0
Hemi-cellulosic content (%)	30.0
Lignin content (%)	16.0

Water hyacinth (*Pontederia crassipes*; formerly *Eichhornia crassipes*) is an aquatic plant belonging to family Pontederiaceae under genus Eichhornia; with amazon basins as its nativity; and is regarded as the most chaotic invasive plant species in the world. Having their habitat ranging from desert regions to rain forests, these aquatic plants have highly buyout leaves connected to its long, spongy and bulbous stalks with their flowers at the top of their stalks and free floating roots at the bottom (Penfound & Earle, 1948). Being regarded as the highly invasive plant, they are often removed from the water bodies to restore its oxygen levels and cause various challenges in disposing it. However, looking into their proximate and LC composition, it can be concluded they have sufficient calorific and lignin content fairly enough for producing high-quality briquettes.

In view of that, Rezania et al. (2016) used 25% water hyacinth (WH) and 75% empty fruit bunch (EFB) fibres, along with cassava starch to develop composite briquettes with the help of a briquetting press. Post evaluation, these briquettes reported their proximate composition as follows: MC–9.3%, FC–15.97%, AC–3.73%, VM–70.87%, and CV–17.17 MJ/Kg while their emission concentrations were measured as O_2–20.60%, CO–25.67 ± 1.45 ppm, CO_2–0.23 ± 0.33 ppm, NO–13 ± 2 ppm, NO_2–0.33 ± 0.33 ppm, SO_2–28.67 ± 6.33 ppm, claiming that increase in WH concentration resulted in decreased O_2 and CO level and increased CO_2 and NO, NO_2 and SO_2 levels (Rezania et al., 2016). Furthermore, Carnaje et al. (2018) mixed 20% of water hyacinth char, carbonised at 425°C with 80% of molasses (binder), to develop briquettes using a briquette moulding machine operated at 8.27 bar. These briquettes reported their proximate composition as 16.8 ± 0.8%, 48 ± 2.2%, 18.6 ± 0.2% and 16.5 ± 2.3% of FC, VM, MC and AC, respectively; together with their fuel characteristics calculated as follows: CV–13.45 ± 1.6 MJ/kg, bulk density–0.85 ± 0.02 g/cm^3, compressive strength–7.4 ± 0.4 kg/cm^2, burning rate–0.007 ± 0.002 g/s, ignition time–198 ± 8 s. Overall, the results have shown positive significance of considering these water hyacinths as potential source for alternative fuels, which also helps in reducing the environmental problems caused by these invasive weeds in water bodies (Carnaje et al., 2018).

12.3.26 Wheat Straws

Moisture content (%)	8.38
Ash content (%)	9.49
Fixed carbon content (%)	8.11
Volatile matter content (%)	74.02
Calorific value (MJ/kg)	17.22
Cellulosic content (%)	28.8–34.2

(continued)

(continued)

Hemi-cellulosic content (%)	23.68–39.10
Lignin content (%)	13.88–18.6

Wheat (*Triticum aestivum*) is a seed-yielding grass plant belonging to family under the genus triticum and is regarded as the highly consumed staple food globally. In specific, these cereals-based grains are primarily used in numerous food industries for producing wide varieties of food recipes and are a well-proclaimed source of carbohydrates, multiple nutrients and dietary fibres. Besides, wheat grains hold vital role in processed food industries owing to their unique viscoelastic and adhesive properties exhibited by their gluten, thereby increasing its global consumption besides their market value (Shewry, 2009). In view of increased consumption, the rate of cultivation of these plants and their wastes generation have also been increasing proportionally, making the disposal of latter very challenging. In fact, major portion of these wastes are constituted by their stalks, which often are disposed during stubble burning; however, this practice increases air pollution. Considering their potential as energy feedstock based on their proximate and LC composition, this raw material can be used for producing briquettes for controlled combustion instead of open burning.

Adapa et al. (2009) developed briquettes from ground wheat straws by utilising their protein and lignin as natural binders and compacted them using a compaction apparatus operated at 63.2 MPa. Preliminary findings on these ground biomass measured their geometric mean particle diameter, bulk density and particle density as 0.398 ± 0.006 mm, 269 ± 9 kg/m^3 and 1585 ± 46 kg/m^3, respectively. Post-briquetting, the mean compact density, specific energy and total specific energy were calculated as 929 ± 30 kg/m^3, 5.28 ± 0.98 MJ/t and 5.53 ± 0.98 MJ/t, respectively (Adapa et al., 2009). Again, briquettes from raw wheat straws with its MC and bulk density as 9 wt.% (W.B.), and 37–58 kg/m^3, reported its density (immediately after compaction) as 795.38 kg/m^3 and durability as 99.2%. Besides, this study also concluded that physical properties like density, MC and durability of these briquettes were also dependent on operating parameters like die temperature, compression pressure and physical properties of the feedstock used (MC and particle size) (Tumuluru et al., 2015). In another study, Smith et al. (1977) developed briquettes from wheat straw pre-heated and heated during compaction for 20 min, by means of a briquetting press operated at 56.3 MN/m^2; and concluded that the presence of MC, until its threshold value helped in developing briquettes with high durability, whereas briquettes with low MC yielded very low durability (Smith et al., 1977).

12.3.27 *Wood Saw Dust*

Moisture content (%)	3.07
Ash content (%)	2.8–3.38
Fixed carbon content (%)	12.68–15.0
Volatile matter content (%)	80.87–82.2
Calorific value (MJ/kg)	18.14–18.88
Cellulosic content (%)	41.58
Hemi-cellulosic content (%)	32.81
Lignin content (%)	33.56

Sawdust (or wood dust) is the by-product produced during various stages and different processes of woodworking operations like sawing, sanding, milling, planing and routing and predominantly consists of wood chips and shavings. In most cases, these dusts tend to be organic biomass, yet, their proximate and LC composition varies with the plant or tree from which the woods are derived from. Besides industrial sources, these wood dust are produced by certain birds and animals like woodpecker and carpenter ants; however, they are least considered as their source. Most importantly, these wood dust acts as primary source for fire hazard in most cases and their particulate nature contributes to air pollution in form of PM emission, thereby creating an occupational dust exposure. Even though these particulate dusts are harmful to humans, they can be processed into value-added products, especially as briquettes owing to their higher CV.

Accordingly, Sánchez et al. (2014) developed sawdust briquettes by following the drying and compaction and produced briquettes with fuel properties as follows: CV: 19.8 MJ/kg, MC: 10%, bulk density: 894 kg/m^3, AC: 1.3%, FC: 15.29% and VM: 83.41%. Here, these briquettes exhibited positive results than compared to their raw materials especially in terms of their emission level, thereby promoting a healthier environment for the both consumer and environment. Comparatively, these briquettes have similar energy potential of a sugarcane bagasse but with higher bulk density, thus recommending these solid fuels for domestic use in low-income sectors (Sánchez et al., 2014).

12.4 Conclusions

Thus, the comprehensive study on understanding the potential of organic biomass feedstocks for developing high-quality renewable solid briquettes has been carried out successfully and has been presented as a detailed review. Following are the major conclusions drawn from this all-inclusive study and are as follows:

(i) Any biomass with significant amount of fixed carbon (7.56–55.59%) and volatile matter (27–81.2%) can be used as raw feedstock for producing high-quality briquettes and tend to exhibit similar fuel characteristics as that of fossil coal.
(ii) And, biomass compacted into briquettes under high pressure and temperature, report superior fuel characteristics like significant energy density and good mechanical properties. Yet, these pressure and temperature are dependent on the physical entities like particle grain size, moisture content of mixture dough, binder material and the briquette geometry.
(iii) Besides pressure and temperature, binders contribute significantly during the biomass briquetting and help in establishing adhesion between the biomass particles. Amidst different binder materials, lignin content from the biomass is seen as the most effective binder and helps introducing high mechanical strength to these briquettes, besides contributing to their calorific value. And, biomass with medium to high lignin content can be used for briquetting without any external binders; however, this may require high compaction pressure and temperature. Moreover, dosage and concentration of these external binders are decided based on the nature of biomass used and their applications.
(iv) Predominantly, any briquette is accounted as high-quality fuel depending on its calorific value, which in turn is contributed by the biomass used in its compaction. In specific, they are contributed predominately by its lignocellulosic compounds, especially by its cellulose and lignin content. Since lignin in biomass contributes to the major share to this calorific value, use of untreated biomass is highly recommended for briquetting as any pre-treated biomass yields reduced lignin content. However, high concentration of lignin might lead to increased ash content, which in turn reduces this calorific value; hence, biomass with good calorific content and adequate lignin content is highly recommended.
(v) From their fuel properties, average density was measured in between (700–1100 kg/m^3), compressive strength calculated in between (6–20 N/mm^2) and ash content between (1–20%), which were in good agreement with permissible standards for commercial solid fuels. With these superior fuel characteristics, these briquettes and also pellets can be simply seen as a viable fuel source for industrial and commercial applications, and as an effective alternative for solid fossil coals.

Fairly evident, any briquette with good fuel quality is attributed by its superior fuel characteristics, which are entirely contributed by its raw biomass feedstock. Thus, having a good understanding on these technical aspects enables one to choose the most efficient biomass for producing briquettes focused exclusively for a particular application along with desired results. Upon bringing this into a common practice, this allows many industrial and energy production sectors to produce necessary amount of energy under controlled emission levels, besides serving as an effective means of disposing abundantly available waste biomass, thereby paving a way for sustainable development.

References

Achinas, S., & Euverink, G. J. W. (2016). Theoretical analysis of biogas potential prediction from agricultural waste. *Resource-Efficient Technologies, 2*(3), 143–147. https://doi.org/10.1016/j.reffit.2016.08.001

Adapa, P., Tabil, L., & Schoenau, G. (2009). Compaction characteristics of barley, canola, oat and wheat straw. *Biosystems Engineering, 104*(3), 335–344. https://doi.org/10.1016/j.biosystemseng.2009.06.022

Akogun, O. A., Waheed, M. A., Ismaila, S. O., & Dairo, O. U. (2020). Co-briquetting characteristics of cassava peel with sawdust at different torrefaction pretreatment conditions. *Energy Sources, Part A: Recovery, Utilization, and Environmental Effects*, pp. 1–19. https://doi.org/10.1080/15567036.2020.1752333

Akogun, O. A., Waheed, M. A., Ismaila, S. O., & Dairo, O. U. (2022). Physical and combustion indices of thermally treated cornhusk and sawdust briquettes for heating applications in Nigeria. *Journal of Natural Fibers, 19*(4), 1201–1216. https://doi.org/10.1080/15440478.2020.1764445

Akuma, O., & Charles, M. (2017). Characteristic analysis of bio-coal briquette (coal and groundnut shell admixtures). *International Journal of Scientific Research in Science and Technology, 2*(3), 30–38.

Aliyu, M., Mohammed, I. S., Usman, M., Dauda, S. M., & Igbetua, J. I. (2020). *Production of composite briquettes (orange peels and corn cobs) and determination of its fuel properties.* CIGR.

Allouch, M., Boukhlifi, F., & Alami, M. (2014). Study of the thermal behavior of almond shells and acorn cups for production of fuel briquettes. *Journal of Energy Technologies and Policy, 4*, 33–39.

Al-Widyan, M. I., Al-Jalil, H. F., Abu-Zreig, M. M., & Abu-Hamdeh, N. H. (2002). Physical durability and stability of olive cake briquettes. *Canadian Biosystems Engineering, 44*, 3–41.

Anggraeni, S., Hofifah, S. N., Nandiyanto, A. B. D., & Bilad, M. R. (2021). Effects of particle size and composition of cassava peels and rice husk on the briquette performance. *Journal of Engineering Science and Technology, 16*(1), 527–542.

Bamgboye, A. I., & Bolufawi, S. J. (2009). Physical characteristics of briquettes from Guinea corn (sorghum bi-color) residue. *Agricultural Engineering International: CIGR Journal.*

Barneto, A. G., Carmona, J. A., Alfonso, J. E. M., & Serrano, R. S. (2010). Simulation of the thermogravimetry analysis of three non-wood pulps. *Bioresource Technology, 101*(9), 3220–3229. https://doi.org/10.1016/j.biortech.2009.12.034

Basu, P. (2018). *Biomass gasification, pyrolysis and torrefaction: Practical design and theory.* Academic Press.

Bazargan, A., Rough, S. L., & McKay, G. (2014). Compaction of palm kernel shell biochars for application as solid fuel. *Biomass and Bioenergy, 70*, 489–497. https://doi.org/10.1016/j.biombioe.2014.08.015

Benz, B. F. (2001). Archaeological evidence of teosinte domestication from Guilá Naquitz, Oaxaca. *Proceedings of the National Academy of Sciences of the United States of America, 98*(4), 2104–2106. https://doi.org/10.1073/pnas.98.4.2104

Bhattacharya, A., Rice, N., Shapter, F. M., Norton, S. L., & Henry, R. J. (2011). Sorghum. In *Wild crop relatives: Genomic and breeding resources* (pp. 397–406). Springer.

Bhattacharjya, D., & Yu, J. S. (2014). Activated carbon made from cow dung as electrode material for electrochemical double layer capacitor. *Journal of Power Sources, 262*, 224–231. https://doi.org/10.1016/j.jpowsour.2014.03.143

Brunerová, A., Roubík, H., Brožek, M., Van Dung, D., Phung, L. D., Hasanudin, U., Iryani, D. A., & Herák, D. (2020). Briquetting of sugarcane bagasse as a proper waste management technology in Vietnam. *Waste Management and Research, 38*(11), 1239–1250. https://doi.org/10.1177/0734242X20938438

Carnaje, N. P., Talagon, R. B., Peralta, J. P., Shah, K., & Paz-Ferreiro, J. (2018). Development and characterisation of charcoal briquettes from water hyacinth (Eichhornia crassipes)-molasses blend. *PLoS ONE, 13*(11), e0207135. https://doi.org/10.1371/journal.pone.0207135

Chen, L., Xing, L., & Han, L. (2009). Renewable energy from agro-residues in China: Solid biofuels and biomass briquetting technology. *Renewable and Sustainable Energy Reviews, 13*(9), 2689–2695. https://doi.org/10.1016/j.rser.2009.06.025

Chen, J. M. (2021). Carbon neutrality: Toward a sustainable future. *Innovation, 2*(3), 100127. https://doi.org/10.1016/j.xinn.2021.100127

Chiew, Y. L., & Shimada, S. (2013). Current state and environmental impact assessment for utilizing oil palm empty fruit bunches for fuel, fiber and fertilizer–A case study of Malaysia. *Biomass and Bioenergy, 51*, 109–124. https://doi.org/10.1016/j.biombioe.2013.01.012

Chungcharoen, T., & Srisang, N. (2020). Preparation and characterization of fuel briquettes made from dual agricultural waste: Cashew nut shells and areca nuts. *Journal of Cleaner Production, 256*, 120434. https://doi.org/10.1016/j.jclepro.2020.120434

Costa, E. V. S., Pereira, M. PdC. F., Silva, C. M. Sd, Pereira, B. L. C., Rocha, M. F. V., & Carneiro, AdC. O. (2019). Torrefied briquettes of sugar cane bagasse and eucalyptus. *Revista Árvore, 43*(1). https://doi.org/10.1590/1806-90882019000100001

Cunliffe, A. M., & Williams, P. T. (1998). Properties of chars and activated carbons derived from the pyrolysis of used tyres. *Environmental Technology, 19*(12), 1177–1190. https://doi.org/10.1080/09593331908616778

Demirbaş, A. (1999). Properties of charcoal derived from hazelnut shell and the production of briquettes using pyrolytic oil. *Energy, 24*(2), 141–150. https://doi.org/10.1016/S0360-5442(98)00077-2

Demirbas, A. (2004). Combustion characteristics of different biomass fuels. *Progress in Energy and Combustion Science, 30*(2), 219–230. https://doi.org/10.1016/j.pecs.2003.10.004

Dinesha, P., Kumar, S., & Rosen, M. A. (2019). Biomass briquettes as an alternative fuel: A comprehensive review. *Energy Technology, 7*(5), 1801011. https://doi.org/10.1002/ente.201801011

Du, C., Wu, J., Ma, D., Liu, Y., Qiu, P., Qiu, R., Liao, S., & Gao, D. (2015). Gasification of corn cob using non-thermal arc plasma. *International Journal of Hydrogen Energy, 40*(37), 12634–12649. https://doi.org/10.1016/j.ijhydene.2015.07.111

Du, C., Li, H., Li, B., Liu, M., & Zhan, H. (2016). Characteristics and properties of cellulose nanofibers prepared by TEMPO oxidation of corn husk. *BioResources, 11*(2), 5276–5284. https://doi.org/10.15376/biores.11.2.5276-5284

Dziedzic, K., Mudryk, K., Hutsol, T., & Dziedzic, B. (2018). *Impact of grinding coconut shell and agglomeration pressure on quality parameters of briquette* (Doctoral Dissertation). Engineering for Rural Development. https://doi.org/10.22616/ERDev2018.17.N461

Energy, B. (2016). *Renewables 2017 global status report*. Renewable Energy Policy Network for the 21st Century. REN21.

Espuelas, S., Marcelino, S., Echeverría, A. M., Del Castillo, J. M., & Seco, A. (2020). Low energy spent coffee grounds briquetting with organic binders for biomass fuel manufacturing. *Fuel, 278*, 118310. https://doi.org/10.1016/j.fuel.2020.118310

Faizal, H. M., Shamsuddin, H. S., Heiree, H. M., Hanaffi, M. F. M. A., Rahman, M. R. A., Rahman, M. Z., & Latiff, Z. A. (2018). Torrefaction of densified mesocarp fibre and palm kernel shell. *Renewable Energy, 122*, 419–428. https://doi.org/10.1016/j.renene.2018.01.118

Falemara, B. C., Joshua, V. I., Aina, O. O., & Nuhu, R. D. (2018). Performance evaluation of the physical and combustion properties of briquettes produced from agro-wastes and wood residues. *Recycling, 3*(3), 37. https://doi.org/10.3390/recycling3030037

Fauquet, C., & Fargette, D. (1990). African cassava mosaic virus: Etiology, epidemiology and control. *Plant Disease, 74*(6), 404–411. https://doi.org/10.1094/PD-74-0404

Fehse, F., Kummich, J., & Schröder, H. W. (2021). Influence of pre-treatment and variation of briquetting parameters on the mechanical refinement of spent coffee grounds. *Biomass and Bioenergy, 152*, 106201. https://doi.org/10.1016/j.biombioe.2021.106201

Fennir, M. A., Raghavan, V. G., Gariépy, Y., & Sotocinal, S. (2014). Strength of pellets and briquettes made from Libyan olive oil solid residues. In *Dalam international conference on advances in agricultural, biological and environmental sciences* (AABES-2014) (pp. 33–36).
Fernández, A. G., Adams, M. R., & Fernandez-Diez, M. J. (1997). *Table olives: Production and processing*. Springer Science+Business Media.
Gangil, S. (2014). Beneficial transitions in thermogravimetric signals and activation energy levels due to briquetting of raw pigeon pea stalk. *Fuel, 128*, 7–13. https://doi.org/10.1016/j.fuel.2014.02.065
García-García, D., Carbonell, A., Samper, M. D., García-Sanoguera, D., & Balart, R. (2015). Green composites based on polypropylene matrix and hydrophobized spend coffee ground (SCG) powder. *Composites Part B Engineering, 78*, 256–265. https://doi.org/10.1016/j.compositesb.2015.03.080
Gilbert, P., Ryu, C., Sharifi, V., & Swithenbank, J. (2009). Effect of process parameters on pelletisation of herbaceous crops. *Fuel, 88*(8), 1491–1497. https://doi.org/10.1016/j.fuel.2009.03.015
Girotto, F., Pivato, A., Cossu, R., Nkeng, G. E., & Lavagnolo, M. C. (2018). The broad spectrum of possibilities for spent coffee grounds valorisation. *Journal of Material Cycles and Waste Management, 20*(1), 695–701. https://doi.org/10.1007/s10163-017-0621-5
Haryati, S., Rahmatullah, Putri, R. W. (2018). Torrefaction of Durian peel and bagasse for bio-briquette as an alternative solid fuel. In *IOP conference series. IOP conference series: Materials science and engineering* (Vol. 334, No. 1, p. 012008). IOP Publishing. https://doi.org/10.1088/1757-899X/334/1/012008
Ifa, L., Yani, S., Nurjannah, N., Darnengsih, D., Rusnaenah, A., Mel, M., Mahfud, M., & Kusuma, H. S. (2020). Techno-economic analysis of bio-briquette from cashew nut shell waste. *Heliyon, 6*(9), e05009. https://doi.org/10.1016/j.heliyon.2020.e05009
Ikelle, I. I., Sunday, N. J., Sunday, N. F., John, J., Okechukwu, O. J., & Elom, N. I. (2020). Thermal analyses of briquette fuels produced from coal dust and groundnut husk. *Acta Chemica Malaysia, 4*(1), 24–27. https://doi.org/10.2478/acmy-2020-0004
Ikumapayi, O. M., & Akinlabi, E. T. (2018). Composition, characteristics and socioeconomic benefits of palm kernel shell exploitation-an overview. *Journal of Environmental Science and Technology, 11*(6), 1–13.
Ishii, K., & Furuichi, T. (2014). Influence of moisture content, particle size and forming temperature on productivity and quality of rice straw pellets. *Waste Management, 34*(12), 2621–2626. https://doi.org/10.1016/j.wasman.2014.08.008
Jittabut, P. (2015). Physical and thermal properties of briquette fuels from rice straw and sugarcane leaves by mixing molasses. *Energy Procedia, 79*, 2–9. https://doi.org/10.1016/j.egypro.2015.11.452
Kaliyan, N., & Morey, R.V. (2009). Strategies to improve durability of switchgrass briquettes. *Transactions of the ASABE*, *52*(6), 1943–1953. https://doi.org/10.13031/2013.29190
Kaliyan, N., Morey, R.V., White, M. D., & Doering, A. (2009). Roll press briquetting and pelleting of corn stover and switchgrass. *Transactions of the ASABE*, *52*(2), 543–555. https://doi.org/10.13031/2013.26812
Kaliyan, N., & Vance Morey, R. V. (2009). Factors affecting strength and durability of densified biomass products. *Biomass and Bioenergy, 33*(3), 337–359. https://doi.org/10.1016/j.biombioe.2008.08.005
Karunanithy, C., Wang, Y., Muthukumarappan, K., & Pugalendhi, S. (2012). Physiochemical characterization of briquettes made from different feedstocks. *Biotechnology Research International, 2012*, 165202. https://doi.org/10.1155/2012/165202
Khadi, B. M., Santhy, V., & Yadav, M. S. (2010). Cotton: An introduction. In *Cotton* (pp. 1–14). Springer. https://doi.org/10.1007/978-3-642-04796-1_1
Khlifi, S., Lajili, M., Belghith, S., Mezlini, S., Tabet, F., & Jeguirim, M. (2020a). Briquettes production from olive mill waste under optimal temperature and pressure conditions: Physico-chemical and mechanical characterizations. *Energies, 13*(5), 1214. https://doi.org/10.3390/en13051214

Khlifi, S., Lajili, M., Tabet, F., Boushaki, T., & Sarh, B. (2020b). Investigation of the combustion characteristics of briquettes prepared from olive mill solid waste blended with and without a natural binder in a fixed bed reactor. *Biomass Conversion and Biorefinery, 10*(2), 535–544. https://doi.org/10.1007/s13399-019-00449-7

Kihedu, J. (2015). Torrefaction and combustion of ligno-cellulosic biomass. *Energy Procedia, 75*, 162–167. https://doi.org/10.1016/j.egypro.2015.07.273

Kim, Y., Park, T., & Hong, D. (2022). Heating and emission characteristics of briquettes developed from spent coffee grounds. *Environmental Engineering Research, 27*(4), 210063. https://doi.org/10.4491/eer.2021.063

Kiš, D., Sučić, B., Guberac, V., Voća, N., Rozman, V., & Šumanovac, L. (2009). Soybean biomass as a renewable energy resource. *Agriculturae Conspectus Scientificus, 74*(3), 201–203.

Kubica, K., Jewiarz, M., Kubica, R., & Szlęk, A. (2016). Straw combustion: Pilot and laboratory studies on a straw-fired grate boiler. *Energy and Fuels, 30*(6), 4405–4410. https://doi.org/10.1021/acs.energyfuels.5b02693

Ladizinsky, G. (1999). On the origin of almond. *Genetic Resources and Crop Evolution, 46*(2), 143–147. https://doi.org/10.1023/A:1008690409554

Lubwama, M., & Yiga, V. A. (2017). Development of groundnut shells and bagasse briquettes as sustainable fuel sources for domestic cooking applications in Uganda. *Renewable Energy, 111*, 532–542. https://doi.org/10.1016/j.renene.2017.04.041

Maciejewska, A. K., Veringa, H., Sanders, J. P. M., & Peteves, S. D. (2006). *Co-firing of biomass with coal: Constraints and role of biomass pretreatment*. Office for Official Publications of the European Communities.

Maitah, M., Prochazka, P., Pachmann, A., Šrédl, K., & Řezbová, H. (2016). Economics of palm oil empty fruit bunches bio briquettes in Indonesia. *International Journal of Energy Economics and Policy, 6*(1), 35–38.

Makarynska, A., & Turpurova, T. (2020). Production of solid biofuels from Grain Processing wastes. *Grain Products and Mixed Fodder's, 20*(1), 48–54.

Martins, S., Simões, F., Matos, J., Silva, A. P., & Carnide, V. (2014). Genetic relationship among wild, landraces and cultivars of hazelnut (*Corylus avellana*) from Portugal revealed through ISSR and AFLP markers. *Plant Systematics and Evolution, 300*(5), 1035–1046. https://doi.org/10.1007/s00606-013-0942-3

Mkini, R. I., & Bakari, Z. (2015). Effect of moisture content on combustion and friability characteristics of biomass waste briquettes made by small scale producers in Tanzania. *International Journal of Engineering Research and Reviews, 3*(1), 66–72.

Mohammed, T. I., & Olugbade, T. O. (2015). Characterisation of briquettes from rice bran and palm kernel shell. *International Journal of Material Science Innovations, 3*(2), 60–67.

Mohsin, A., Hussain, M. H., Zaman, W. Q., Mohsin, M. Z., Zhang, J., Liu, Z., Tian, X., Salim-Ur-Rehman, Khan, I. M., Niazi, S., Zhuang, Y., & Guo, M. (2021). Advances in sustainable approaches utilizing orange peel waste to produce highly value-added bioproducts. *Critical Reviews in Biotechnology*, pp. 1–20.

Morton, J. F. (1987). *Fruits of warm climates.*

Morton, J. F. (2007). *Cashew apple, Anacardium occidentale L. Fruits of warm climates.* Center for New Crops and Plant Products, Department of Horticulture and Landscape Architecture, Purdue University. Lafayette, W. Archived from the original (pp. 03–15).

Nayar, N. M. (2016). *The coconut: Phylogeny, origins, and spread*. Academic Press.

Nazari, M. M., San, C. P., & Atan, N. A. (2019). Combustion performance of biomass composite briquette from rice husk and banana residue. *International Journal on Advanced Science, Engineering and Information Technology, 9*(2), 455–460. https://doi.org/10.18517/ijaseit.9.2.2408

Nuriana, W., Anisa, N., & Martana. (2014). Synthesis preliminary studies durian peel bio briquettes as an alternative fuels. *Energy Procedia, 47*, 295–302. https://doi.org/10.1016/j.egypro.2014.01.228

Nyakuma, B. B., Ahmad, A., Johari, A., Abdullah, T. A. T., & Oladokun, O. (2015). *Torrefaction of pelletized oil palm empty fruit bunches*. arXiv preprint arXiv:1505.05469.

Okello, C., Pindozzi, S., Faugno, S., & Boccia, L. (2013). Bioenergy potential of agricultural and forest residues in Uganda. *Biomass and Bioenergy, 56*, 515–525. https://doi.org/10.1016/j.biombioe.2013.06.003

Okot, D. K., Bilsborrow, P. E., & Phan, A. N. (2018). Effects of operating parameters on maize COB briquette quality. *Biomass and Bioenergy, 112*, 61–72. https://doi.org/10.1016/j.biombioe.2018.02.015

Okot, D. K., Bilsborrow, P. E., & Phan, A. N. (2019). Briquetting characteristics of bean straw-maize cob blend. *Biomass and Bioenergy, 126*, 150–158. https://doi.org/10.1016/j.biombioe.2019.05.009

Oladeji, J. T., & Enweremadu, C. C. (2012). The effects of some processing parameters on physical and densification characteristics of corncob briquettes. *International Journal of Energy Engineering, 2*(1), 22–27. https://doi.org/10.5923/j.ijee.20120201.04

Olorunnisola, A. (2007). *Production of fuel briquettes from waste paper and coconut husk admixtures*.

Orhevba, B. A., & Olatunji, O. (2021). Development of improved briquettes from three agricultural residues. In *5th International conference of the international commission of agricultural and biosystems engineering* (CIGR), Hosted by the Canadian Society for Bioengineering (CSBE/SCGAB).

Oyibo, M. C., Elinge, C. M., Birnin-Yauri, A. U., Ige, A. R., Ogundele, O. A., & Dabai, M. D. (2020). Combustion profiles of fuel briquettes produced from alkali treated and untreated sugarcane bagasse. *American Journal of Engineering Research, 9*(3), 268–274.

Papini-Terzi, F. S., Rocha, F. R., Vêncio, R. Z., Felix, J. M., Branco, D. S., Waclawovsky, A. J., Del Bem, L. E., Lembke, C. G., Costa, M. D., Nishiyama, M. Y., & Vicentini, R. (2009). Sugarcane genes associated with sucrose content. *BMC Genomics, 10*(1), 1–21.

Papuangan, N., & Jabid, A. W. (2019). Predesign of bio-briquette production using kenari shell. In *IOP conference series: Earth and environmental science* (Vol. 276, No. 1, p. 012051). IOP Publishing.

Patil, R. A., Deshannavar, U. B., Ramasamy, M., Emani, S., Issakhov, A., & Khalilpoor, N. (2021). Briquetting of dry sugarcane leaves by using press mud, cow dung, and buffalo dung as binders. *International Journal of Chemical Engineering, 2021*, 1–12. https://doi.org/10.1155/2021/8608215

Penfound, W. T., & Earle, T. T. (1948). The biology of the water hyacinth. *Ecological Monographs, 18*(4), 447–472. https://doi.org/10.2307/1948585

Pham, L. J., & Dumandan, N. G. (2015). Philippine Pili: Composition of the lipid molecular species. *Journal of Ethnic Foods, 2*(4), 147–153. https://doi.org/10.1016/j.jef.2015.11.001

Pimchuai, A., Dutta, A., & Basu, P. (2010). Torrefaction of agriculture residue to enhance combustible properties. *Energy and Fuels, 24*(9), 4638–4645. https://doi.org/10.1021/ef901168f

Plíštil, D., Brožek, M., Malaták, J., Roy, A., & Hutla, P. (2005). Mechanical characteristics of standard fuel briquettes on biomass basis. *Research in Agricultural Engineering, 51*(2), 66–72.

Potip, S., & Wongwuttanasatian, T. (2018). Combustion characteristics of spent coffee ground mixed with crude glycerol briquette fuel. *Combustion Science and Technology, 190*(11), 2030–2043. https://doi.org/10.1080/00102202.2018.1482888

Rao, Y. S., & Mathew, K. M. (2012). Tamarind. In *Handbook of herbs and spices* (pp. 512–533). Woodhead Publishing.

Rezania, S., Md Din, M. F. M., Kamaruddin, S. F., Taib, S. M., Singh, L., Yong, E. L., & Dahalan, F. A. (2016). Evaluation of water hyacinth (*Eichhornia crassipes*) as a potential raw material source for briquette production. *Energy, 111*, 768–773. https://doi.org/10.1016/j.energy.2016.06.026

Rhén, C., Öhman, M., Gref, R., & Wästerlund, I. (2007). Effect of raw material composition in woody biomass pellets on combustion characteristics. *Biomass and Bioenergy, 31*(1), 66–72. https://doi.org/10.1016/j.biombioe.2006.06.016

Rhofita, E. I., Hutardo, P., & Miraux, F. (2018). The characterization of rice straw briquette as an alternative fuel in Indonesia. In *Proceedings of the built environment, science and technology international conference* (pp. 304–309). https://doi.org/10.5220/0008908203040309

Riaz, M. N. (2005). *Soy applications in food*. CRC Press.

Rodiah, S., Al Jabbar, J. L., Ramadhan, A., & Hastati, E. (2021). Investigation of mango (*Mangifera odorate*) sap and starch as organic adhesive of bio-briquette. In *Journal of physics: Conference series* (Vol. 1943, No. 1, p. 012185). IOP Publishing. https://doi.org/10.1088/1742-6596/1943/1/012185

Rybak, W. (2006). *Combustion and co-combustion of solid biofuels* (in Polish). Wroclaw University of Technology. ISBN 83-7085-938-0.

Sánchez, E. A., Pasache, M. B., & García, M. E. (2014). Development of briquettes from waste wood (sawdust) for use in low-income households in Piura, Peru. In *Proceedings of the world congress on engineering* (Vol. 2).

Santos, C., Fonseca, J., Aires, A., Coutinho, J., & Trindade, H. (2017). Effect of different rates of spent coffee grounds (SCG) on composting process, gaseous emissions and quality of end-product. *Waste Management, 59*, 37–47. https://doi.org/10.1016/j.wasman.2016.10.020

Sawadogo, M., Tchini Tanoh, S. T., Sidibé, S., Kpai, N., & Tankoano, I. (2018). Cleaner production in Burkina Faso: Case study of fuel briquettes made from cashew industry waste. *Journal of Cleaner Production, 195*, 1047–1056. https://doi.org/10.1016/j.jclepro.2018.05.261

Shah, A., Darr, M. J., Dalluge, D., Medic, D., Webster, K., & Brown, R. C. (2012). Physicochemical properties of bio-oil and biochar produced by fast pyrolysis of stored single-pass corn stover and cobs. *Bioresource Technology, 125*, 348–352. https://doi.org/10.1016/j.biortech.2012.09.061

Shewry, P. R. (2009). Wheat. *Journal of Experimental Botany, 60*(6), 1537–1553. https://doi.org/10.1093/jxb/erp058

Smith, I. E., Probert, S. D., Stokes, R. E., & Hansford, R. J. (1977). The briquetting of wheat straw. *Journal of Agricultural Engineering Research, 22*(2), 105–111. https://doi.org/10.1016/0021-8634(77)90054-3

Smith, B. D. (1995). *The emergency of agriculture.*

Song, A., Zha, F., Tang, X., & Chang, Y. (2019). Effect of the additives on combustion characteristics and desulfurization performance of cow dung briquette. *Chemical Engineering and Processing—Process Intensification, 143*, 107585. https://doi.org/10.1016/j.cep.2019.107585

Song, X., Zhang, S., Wu, Y., & Cao, Z. (2020). Investigation on the properties of the bio-briquette fuel prepared from hydrothermal pretreated cotton stalk and wood sawdust. *Renewable Energy, 151*, 184–191. https://doi.org/10.1016/j.renene.2019.11.003

Sriram, N., Sikdar, D. C., Sunil, H., & Shetty, M. K. (2014). Study on briquetting of cotton waste. *International Journal of Technical Research and Applications, 2*(6), 102–106.

Sukarta, I. N., Sastrawidana, I. D. K., Ayuni, N. P. S. (2018). Proximate analysis and calorific value of pellets in biosolid combined with wood waste biomass. *Journal of Ecological Engineering, 19*(3), 185–190. https://doi.org/10.12911/22998993/86153

Suprianto, F. D., Kasrun, A. W., & Siahaan, I. H. (2017). the effects of particle size and pressure on the combustioan characteristics of carbera mangas leaf briquettes. *ARPN Journal of Engineering and Applied Sciences, 12*(4), 1–6.

Tan, S. H., Mailer, R. J., Blanchard, C. L., & Agboola, S. O. (2011). Canola proteins for human consumption: Extraction, profile, and functional properties. *Journal of Food Science, 76*(1), R16–R28. https://doi.org/10.1111/j.1750-3841.2010.01930.x

Taulbee, D., Patil, D. P., Honaker, R. Q., & Parekh, B. K. (2009). Briquetting of coal fines and sawdust Part I: Binder and briquetting-parameters evaluations. *International Journal of Coal Preparation and Utilization, 29*(1), 1–22. https://doi.org/10.1080/19392690802628705

Tekulu, K., Taye, G., & Assefa, D. (2020). Effect of starter nitrogen and phosphorus fertilizer rates on yield and yield components, grain protein content of groundnut (*Arachis hypogaea L.*) and residual soil nitrogen content in a semiarid north Ethiopia. *Heliyon, 6*(10), e05101. https://doi.org/10.1016/j.heliyon.2020.e05101

Thabuot, M., Pagketanang, T., Panyacharoen, K., Mongkut, P., & Wongwicha, P. (2015). Effect of applied pressure and binder proportion on the fuel properties of holey bio-briquettes. *Energy Procedia, 79*, 890–895. https://doi.org/10.1016/j.egypro.2015.11.583

Tumuluru, J. S., Tabil, L. G., Song, Y., Iroba, K. L., & Meda, V. (2015). Impact of process conditions on the density and durability of wheat, oat, canola, and barley straw briquettes. *Bioenergy Research, 8*(1), 388–401. https://doi.org/10.1007/s12155-014-9527-4

Uemura, Y., Omar, W., Othman, N. A., Yusup, S., & Tsutsui, T. (2013). Torrefaction of oil palm EFB in the presence of oxygen. *Fuel, 103*, 156–160. https://doi.org/10.1016/j.fuel.2011.11.018

Ugwu, K. E., & Agbo, K. E. (2011). Briquetting of palm kernel shell. *Journal of Applied Sciences and Environmental Management, 15*(3), 447–450.

Ugwu, K., & Agbo, K. (2013). Evaluation of binders in the production of briquettes from empty fruit bunches of Elais Guinensis. *International Journal of Renewable and Sustainable Energy, 2*(4), 176–179. https://doi.org/10.11648/j.ijrse.20130204.17

Ujjinappa, S., & Sreepathi, L. K. (2018). Production and quality testing of fuel briquettes made from Pongamia and tamarind shell. *Sādhanā, 43*(4), 1–7. https://doi.org/10.1007/s12046-018-0836-8

Umogbai, V. I., & Iorter, H. A. (2013). Experimental evaluation of the performance of a prototype solar dryer box. *African Journal of Food Science and Technology, 4*, 110–115.

Velasco, R., & Licciardello, C. (2014). A genealogy of the citrus family. *Nature Biotechnology, 32*(7), 640–642. https://doi.org/10.1038/nbt.2954

Velusamy, S., Subbaiyan, A., Kandasamy, S., Shanmugamoorthi, M., & Thirumoorthy, P. (2022). Combustion characteristics of biomass fuel briquettes from onion peels and tamarind shells. *Archives of Environmental and Occupational Health, 77*(3), 251–262. https://doi.org/10.1080/19338244.2021.1936437

Vijayaraghavan, P., Kalaiyarasi, M., & Vincent, S. G. P. (2015). Cow dung is an ideal fermentation medium for amylase production in solid-state fermentation by Bacillus cereus. *Journal of Genetic Engineering and Biotechnology, 13*(2), 111–117. https://doi.org/10.1016/j.jgeb.2015.09.004

Waluyo, J., & Pratiwi, Y. (2018). Analysis proximate, ultimate, and thermal gravimetric based on variations dimensions of briquettes from waste jackfruit crust. *International Journal of Scientific Engineering and Science, 2*(10), 36–39.

Wan, D., Wu, L., Liu, Y., Zhao, H., Fu, J., & Xiao, S. (2018). Adsorption of low concentration perchlorate from aqueous solution onto modified cow dung biochar: Effective utilization of cow dung, an agricultural waste. *Science of the Total Environment, 636*, 1396–1407. https://doi.org/10.1016/j.scitotenv.2018.04.431

Wang, J. W., Sun, Y. S., Jin, Z. R., Feng, Z. Y., & Dong, F. G. (2013). Effect of densification conditions on physical quality of densified sorghum straw briquette fuel. In *Applied mechanics and materials* (Vol. 339). Trans Tech Publications Ltd. https://doi.org/10.4028/www.scientific.net/AMM.339.651

Wang, Q., Han, K., Gao, J., Li, H., & Lu, C. (2017). The pyrolysis of biomass briquettes: Effect of pyrolysis temperature and phosphorus additives on the quality and combustion of bio-char briquettes. *Fuel, 199*, 488–496. https://doi.org/10.1016/j.fuel.2017.03.011

Waweru, J., & Chirchir, D. K. (2017). Effect of the briquette sizes and moisture contents on combustion characteristics of composite briquettes. *International Journal of Innovative Science Engineering and Technology, 4*, 102–111.

Widodo, S., Astriani, A., & Asmiani, N. (2019). Utilising of canary shell as the material of bio-briquette. *International Journal of Engineering and Science Applications, 6*(1), 31–38.

Widodo, S., Asmiani, N., Jafar, N., Artiningsih, A., Nurhawaisyah, S. R., Chalik, C. A., Thamrin, M., Husain, J. R., Husain, J. R., Husain, J. R., et al. (2021). The effect of raw material composition of mixed carbonized canary shell and coal bio briquettes on caloric value. In *IOP conference series. IOP conference series: Earth and environmental science* (Vol. 921, No. 1, p. 012027). IOP Publishing. https://doi.org/10.1088/1755-1315/921/1/012027

Wilaipon, P. (2008). Density equation of bio-coal briquettes and quantity of maize cob in Phitsanulok, Thailand. *American Journal of Applied Sciences, 5*(12), 1808–1811. https://doi.org/10.3844/ajassp.2008.1808.1811

Xu, Q., Chen, L. L., Ruan, X., Chen, D., Zhu, A., Chen, C., Bertrand, D., Jiao, W. B., Hao, B. H., Lyon, M. P., Chen, J., Gao, S., Xing, F., Lan, H., Chang, J. W., Ge, X., Lei, Y., Hu, Q., Miao, Y., Ruan, Y., et al. (2013). The draft genome of sweet orange (*Citrus sinensis*). *Nature Genetics, 45*(1), 59–66. https://doi.org/10.1038/ng.2472

Yadav, R. D., Jain, S. K., Alok, S., Prajapati, S. K., & Verma, A. (2011). Pongamia pinnata: An overview. *International Journal of Pharmaceutical Sciences and Research, 2*(3), 494.

Yank, A., Ngadi, M., & Kok, R. (2016). Physical properties of rice husk and bran briquettes under low pressure densification for rural applications. *Biomass and Bioenergy, 84*, 22–30. https://doi.org/10.1016/j.biombioe.2015.09.015

Yuhazri, M. Y., Sihombing, H., Umar, N., Saijod, L., & Phongsakorn, P. T. (2012). Solid fuel from empty fruit bunch fiber and waste papers Part 1: Heat released from combustion test. *Global Engineers and Technologists Review, 2*(1), 7–13.

Yuliah, Y., Kartawidjaja, M., Suryaningsih, S., & Ulfi, K. (2017). Fabrication and characterization of rice husk and coconut shell charcoal based bio-briquettes as alternative energy source. In *IOP conference series. IOP conference series: Earth and environmental science* (Vol. 65, No. 1, p. 012021). IOP Publishing. https://doi.org/10.1088/1755-1315/65/1/012021

Yuliansyah, A. T., Hirajima, T., Kumagai, S., & Sasaki, K. (2010). Production of solid biofuel from agricultural wastes of the palm oil industry by hydrothermal treatment. *Waste and Biomass Valorization, 1*(4), 395–405. https://doi.org/10.1007/s12649-010-9045-3

Zhang, X., Cai, Z., Chen, L., Chen, Y., & Zhang, Y. (2014). Research on the compressing mechanism and quality evaluation parameters of densified biomass fuel. *Kezaisheng Nengyuan/renewable Energy Resources, 32*(12), 1917–1921.

Zhang, G., Sun, Y., & Xu, Y. (2018). Review of briquette binders and briquetting mechanism. *Renewable and Sustainable Energy Reviews, 82*, 477–487. https://doi.org/10.1016/j.rser.2017.09.072

Zhao, Y., Chang, H., Ji, D., & Liu, Y. (2001). The research progress on the briquetting mechanism of fine coal. *Journal of Coastal Conservation, 24*, 12–14.

Zhou, X., Jellen, E. N., & Murphy, J. P. (1999). Progenitor germplasm of domisticated hexaploid oat. *Crop Science, 39*(4), 1208–1214. https://doi.org/10.2135/cropsci1999.0011183X003900040042x

Zohary, D., & Hopf, M. (2000). *Domestication of plants in the Old World: The origin and spread of cultivated plants in West Asia, Europe and the Nile valley* (3rd ed.). Oxford University Press.

Chapter 13
Using Project-Based Collective Action Theory to Identify Key Success Factors and Key Difficulties for Circular Economy Projects: A Case Study of Pays de la Loire Region, France

Chaymaa Rabih and Nicolas Antheaume

13.1 Introduction and Research Question

Recent developments in the field of circular economy (CE) have led to a proliferation of studies focused on explaining CE as a paradigm, its relationship with sustainable development (Geissdoerfer et al., 2017; Prieto-Sandoval et al., 2018), and the concepts that define it (Kirchherr et al., 2017). Prieto-Sandoval et al. (2018) proposed that the CE can be understood through four specific components. The first one is the recirculation of resources and energy, the minimization of resource demand, and waste recovery. This cyclical flow has two advantages: It avoids depleting a finite environmental stock of materials, ensures an "infinite" supply, and allows for an independent collection of materials within nature's capacity (Tibbs, 2006). The second component is a multi-level approach due to its implementation at the micro (firms and consumers) (Park et al., 2010), meso (integrated economic agents) (Geng et al., 2012), and macro (city, regions, and governments) levels (Yuan et al., 2006). The third component is its importance as a pathway to sustainable development (Prieto-Sandoval et al., 2017; Velenturf & Purnell, 2017). The fourth component is its close relationship with how society innovates (Prieto-Sandoval et al., 2018).

In simple terms, resource recirculation in the CE stems from a transformation process through using and returning materials. Industries take resources from the environment and transform them into goods and services. Then, they are distributed

C. Rabih (✉) · N. Antheaume
LEMNA/IAE Nantes—Economie & Management, Nantes Université, 44000 Nantes, France
e-mail: Chaymaa.rabih@univ-nantes.fr

N. Antheaume
e-mail: Nicolas.antheaume@univ-nantes.fr

S. A. Bandh and F. A. Malla (eds.), *Waste Management in the Circular Economy*,
https://doi.org/10.1007/978-3-031-42426-7_13

and used by consumers, or other businesses to provide other goods and services. Finally, these goods are returned as materials and energy in other value chains (MacArthur Foundation, 2013; Park et al., 2010). As for waste, depending on whether it is a biological or a technical resource, it can be redirected and returned either to the biosphere or an industrial cycle (McDonough & Braungart, 2002), a process that can be operated at different levels (micro, meso, and macro).

This new approach to business is becoming increasingly important, attracting interest from both academics and practitioners. The number of projects on this topic, both from industry and academia, has grown exponentially in recent years (D'Amato et al., 2017). However, in reality, this approach is challenging, and therefore, companies fail to address the transition to the CE. To enable managers to successfully adopt CE approaches, a clear understanding of the challenges of circular business models is necessary. This chapter provides an overview of the drivers and barriers of CE projects, based on the project-based theory of collective action, allowing for a multi-dimensional reading and analysis of the case studies. The majority of existing studies have focused on the circular side of the concept and mainly on engineering processes (Nikolaou & Stefanakis, 2022). Korhonen et al. (2004) pointed out that despite the potential of engineering and natural science to support and facilitate organizations in introducing CE principles into their daily operations, they failed to make these circular models understandable by learning from management science. To this end, the authors acknowledge that CE does not only depend on nature science and engineering but also on economic and management aspects, which are very important for the implementation of CE principles in organizations.

The collective action project-based theory attracted our attention because it allows us to explicitly ask the question of the existence and emergence of an enterprise and, more broadly, of collective action using simple questions during the "subject framing" phase. It guides the researcher in investigating these questions: the question of why (why are enterprises created? what is their fundamental purpose?), the question of what (what do they do?), the question of how (how do they work), and the question of which (which methods they follow?). By enriching the current knowledge with a different analytical view through the project approach regarding the actions of CE organizations, this paper tries to answer the following questions: What is a circular economy project from a managerial perspective, which dimension(s) can influence its sustainability, and how do the different project dimensions occur? It discusses the application of the multi-dimensional analysis framework to three case studies from the reuse and recycling sector in the Pays de la Loire region in France.

13.2 Context and Utility of Project-Based Collective Action Theory

The perspective of CE adopted in this article goes beyond material, energy, and natural resource flows. We conceptualize it as an economy that includes a societal project with multi-dimensional collective progress by drawing on project-based collective action theory (Desreumaux & Bréchet, 2009, 2018). As CE involves environmental sustainability, economic prosperity, and social equity (Kirchherr et al., 2018), implementing circularity is not trivial. It is a complex process that requires multi-level and multi-stakeholder engagement (Milios, 2021). While the CE requires essential changes in current production and consumption patterns (Korhonen et al., 2018), it also requires changes along the value chain, adapted market strategies and business models (Rosa et al., 2019), management strategies, and consideration of the vision of different stakeholders. Furthermore, stakeholder expectations and consumer perceptions are key determinants that can accelerate the transition process to a strong CE (Hartley et al., 2020; Milios & Matsumoto, 2019) and ensure its sustainability.

Moving to a new business model requires defining new value propositions for customers and indicates how companies design their value chain with partners to create and capture new value (Zott & Amit, 2010, 2013; Malik et al., 2018). In other words, the CE reinforces the need for collaboration among multiple actors and calls for collective action. A CE organization can never act alone on the entire life cycle of a product or service. It must at least consider its stakeholders and the regulations in force. Indeed, observing a CE organization means watching a collective project, regulations (practices, tools, devices), and multiple considerations and values held by the stakeholders involved in the project, explicitly asking the question of the existence and emergence of the enterprise and more broadly of collective action.

13.2.1 The Foundations of Project-Based Collective Action Theory

We have chosen to rely on the project-based collective action theory (henceforth, PCAT) with a regulationist perspective. According to this vision (Reynaud, 1991), collective action arises from a project in which a stakeholder takes shape as a social actor by adopting individual or singular rules in their conception and practice. Agreeing on rules, that is, the constitutive and organizational regulation of collective action, consecrates the possibility of joint action and tends to organize a social world via other processes such as negotiation, commitment, and coordination. These processes are based on management devices and tools designed, arranged, and transformed to achieve goals (Aggeri & Labatut, 2010). The PCAT with a regulationist perspective is characterized by the following key concepts (Desreumaux & Bréchet, 2019):

(a) The project has to be viewed from a non-technical perspective. Any enterprise, by extension any form of collective action—if we want to adapt it to the contemporary world (company, association, hospital)—is understood as a collective progress project. The project concept we are talking about here should not be confused with a business management project, considered from a technical or instrumental point of view, but rather a phenomenological project vision of collective action, from a more fundamental point of view that of Jean-Pierre Boutinet (2012). He defines it as an operative anticipation of a partially determined character or a fuzzy type of the desired future from an anthropological perspective. According to this author, this means that the fuzzy mode of determination of the project implies continuous back-and-forth iterations between a desired orientation and its implementation. According to this perspective, a project is composed of three dimensions. The ethico-political dimension which means that any project is based, consciously or not, on a given vision of the world. The technical–economic dimension which describes the missions the company intends to accomplish through its operations and the skills which it uses. Finally, an organizational dimension describes the methods and the means of action.
(b) The phenomenological and ontogenetic singularity of the enterprise. This is a question of investigating how collective action emerges and takes shape. It can be described as the subjectivist effort of an actor (individual or collective) committing his inventiveness and creativity through the use of his imagination and his judgment. This commitment follows an attempt to make the project understood, followed by action and construction phases based on how the stakeholders anticipate a desired future and meaning for the project. From a semantic point of view, the term "anticipate" does not involve forecasting or predicting, but rather it involves the idea of evoking the "realities" of a future and then assuming it in words (narration), images, or figures. The perspective that this theory gives to the project is multi-dimensional; for the project to be viable, it involves existential and instrumental considerations. In other words, instrumental choices (management devices, materials, practices, business model) and axiological choices (ethical, social, environmental, political) are intertwined and merged.
(c) The processual or developmental character of the action. PCAT is interested in the action because it means implementing a value creation project based on perspectives a and b described above.

As mentioned in the introduction, PCAT incorporates a simple questioning method during the "topic framing" phase. It guides the researcher who embraces this theory through the "what, why and how" questions (Desreumaux & Bréchet, 2019). For our research field, we build on this theory to develop and make a multi-dimensional analysis and understand how circular economy projects actually work. This framework formalizes and explains the characteristics of a project. It allows us to emphasize the

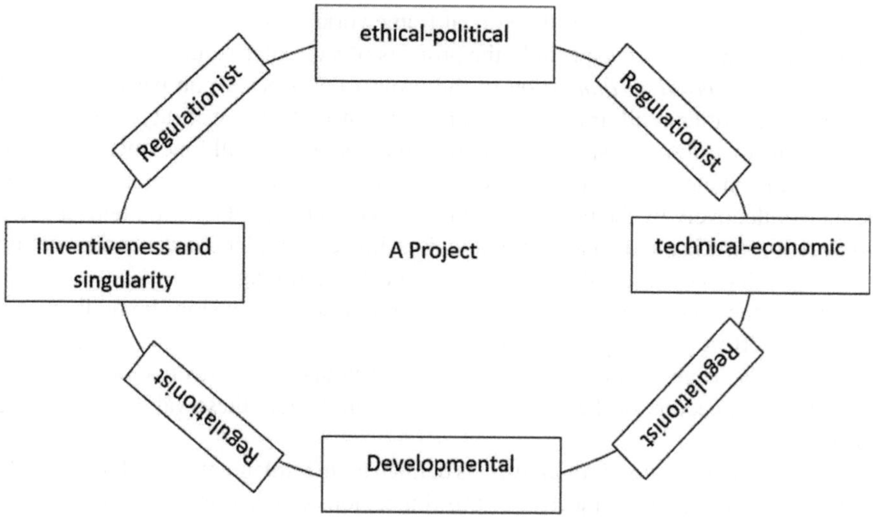

Fig. 13.1 Multi-dimensional analysis framework

multi-dimensional and regulationist nature of the CE context. Based on the theoretical framework, we have identified the following five dimensions (Fig. 13.1): ethico-political, techno-economic, singularity, developmental, and regulationist dimensions (each dimension is explained in detail below).

13.2.2 Toward a Multi-dimensional Analysis Framework for Circular Economy Projects

The circular economy concept and its characteristics are often described in terms of a business model concept. Geissdoerfer et al. (2020) define it as a value proposition and then an organizational form that cycles, extends, intensifies, and/or dematerializes material and energy loops in order to reduce resource inputs and waste and emission leakages. It is therefore rare to find implementations of circular economy organizations that focus on the concepts of project and collective action. Doing so, however, would allow for a different and more constructive dialogue on its characteristics and specificities. Consequently, there is a considerable lack of clarity about its theoretical conceptualization and position in the management literature.

We, therefore, develop a framework that would be (a) deductive (i.e., emerging from theory rather than empirical research) and (b) flexible (i.e., applicable to a wide range of forms of collective action) in the review of specific and explicit dimensions of CE projects. Therefore, its primary focus is analytical, not prescriptive.

Figure 13.1 shows our multi-dimensional analysis framework. It includes the ethical–political, techno-economic, singularity, developmental, and regulationist dimensions. According to our theoretical framework, these are the dimensions that characterize any project, especially the projects of a circular nature.

The ethical–political dimension of the project describes a non-partisan field of activity, accessible to all, inviting awareness, a sense of responsibility and commitment in an environmental situation. These values, which we call fundamental values, support and invite the creation of even more project-specific values that must be defined collectively by the project community. The ecological ethic remains a guide for decision-making and collective action. It is, however, under constant construction because each environmental situation is unique and evolving.

Beyond the fundamental values, the ethical–political dimension should therefore call for an effort to clarify the utility sought on the social, economic, and environmental levels. Then, it should determine the individual and collective objectives of the authors and actors of the project in order to implement these values in concrete terms.

The technical–economic dimension is defined as an implementation of the ethical–political dimension. The objective of this dimension is to determine a solution to an environmental situation that is technically feasible and reliable, as well as economically acceptable, in collaboration with the project stakeholders. It also translates the missions that a collective action project intends to accomplish through the activity it chooses to carry out and the skills it uses. The technical–economic aspect can be observed or measured through attributes such as the nature of the technology, resources, jobs, sustainability metrics.

The developmental dimension describes the evolution of the project in terms of its content, form, and potential for development. It can be translated, for example, by the extent of its geographical presence, the evolution of its sales, the change of legal status, the evolution of its performance, the categories of stakeholders involved from one period to another, etc.

The singularity and inventiveness dimension denotes a subjectivist perspective, meaning that any enterprise is the result of a stakeholder (individual and/or collective) committing his inventiveness and creativity through the use of his imagination and his judgment to address an environmental situation. An effort in terms of intelligibility (Desreumaux & Bréchet, 2009) is necessary to make the inventiveness of the project explicit and to connect stakeholders, the latter being a crucial point in the implementation of a circular project.

The regulationist dimension describes how collective action is formed by adopting rules, and how they are designed and put into practice. In other words, it shows how the previous dimensions interact with each other and the project stakeholders. This dimension tends to organize a social world through other processes such as negotiation and coordination. These processes are based on management devices and tools designed, arranged, and transformed to fulfill their goals.

13.3 Methods

The empirical material for this chapter comes from three exploratory case studies of three local projects in the recycling and reuse sector, focusing on using the conceptual framework explained above to understand how the project is run. In the following subsections, we describe the data collection and analysis and the case studies.

Data collection and analysis

The empirical field was the Pays de la Loire region in western France. To explore the challenges that CE projects face in adopting CE practices from a management perspective, we adopted a qualitative case study research design. As we were at an early stage in understanding CE as a phenomenon, an exploratory research approach was appropriate (Barratt et al., 2011; Eisenhardt & Graebner, 2007; Makadok et al., 2018). The case study method was chosen because of its accuracy in observing how activities take place in practice and its potential to generate new knowledge from a single study context (Benbasat et al., 1987; Meredith, 1998). The projects studied were: a return and deposit scheme for glass bottles, a project to extend the life span of out-of-use recreational boats, and a bio-waste management and networking initiative project. These projects were selected because the University the authors belong to is involved in several research projects on circular economy and sustainability issues and is located in the region where the projects are taking place. A close relationship between several stakeholders in this field and the University has been ongoing over the years, which was expected to improve prospects for participation.

Data were collected between June 2021 and June 2022 using several sources, including interviews, documents, and direct observations (see Table 13.1). The data were triangulated to propose different perspectives and insights (Benbasat et al., 1987; Eisenhardt, 1989; Gibbert & Ruigrok, 2010). The first author conducted all of the semi-structured interviews and site visits to enrich the data and increase the validity of the results (Eisenhardt, 1989). The data were discussed and analyzed to avoid bias and misinterpretation, and the interpretations were validated jointly with the second author. Subsequently, presentations to challenge the results took place in meetings with fellow researchers/practitioners (Goffin et al., 2019) who were familiar with the reuse and recycling sector but did not participate in the primary data collection and analysis. Due to requests for anonymity, the names of the organizations and individuals interviewed are not disclosed.

The materials collected are of text type and have been read twice, guided by the terminology of project-based collective action theory, in particular by the multi-dimensional analysis framework, as unifying lexicons of analysis to identify the success factors and the main difficulties of circular economy projects.

Table 13.1 Data collection method

Data collection method		Data characteristics
Semi-structured interviews	Project holders	Fifteen interviews (15 interviewees, 6.75 h total)
	The project's co-founders	
	Internal and external stakeholders	
Document analysis	Reports	Two reports
	Presentations	Five presentations (5.0 h total)
	Websites	Three websites of three cases and other websites of the projects' co-founders
Direct observation	Site visits Participation in meetings	Five visits (12 h)
Other	Informal interactions	Multiple interactions in various contexts with, for example, project founders

Case Studies

First Case

The ReBottle organization is developing a regional project to implement a deposit system for glass bottles as a system to improve the return rates of glass bottles. After being phased out progressively in the 70's and replaced by recycling schemes the deposit system is currently making a comeback in France, not as a country-wide system but as a multitude of local initiatives. Local governments, as stakeholders, are aware of the environmental benefits of such a system. Indeed, glass alone represents half the tonnage of household packaging put on the French market each year (2.5 million tons per year). Recycling glass limits the use of new raw materials, but its energy balance is heavy because it involves melting the raw material at 1500 °C for 24 h. In the Pays de la Loire region, used glass must also be transported over long distances because there are no glass manufacturers. However, a glass bottle can be washed and reused up to 50 times without going through the waste stage. Managed locally, this allows a major ecological gain compared to recycling: −76% energy, −33% water, −79% greenhouse gas emissions. Therefore, to minimize the impact of transporting the bottles to the local washer, ReBottle mutualizes the collections with the usual routes of partner companies rather than using additional trucks.

From an economic point of view, the deposit scheme is also advantageous for local governments because when glass is collected and recycled, they pay for the bottles deposited in the glass containers to be recovered. As the glass is crushed, the communities sell it to the glass companies in charge of melting it at a lower price than the collection price. They pass on this cost to the taxpayer through the household waste tax because even if local governments sell the crushed glass to the glass manufacturers, they still have to pay the extra cost of collection. Reuse, washing, and collection are included in the price paid by producers for used bottles. This system gives local businesses an advantage because it is run and managed locally. ReBottle works with local producers and retailers.

Moreover, this system can create jobs locally in washing and logistics, while there is no recycling site in the Pays de la Loire region. Finally, the reuse of bottles can create many jobs in the area, especially in bottle washing plants. Because of all these advantages, ReBottle is trying to bring together consumers, local producers, retail stores, a collection transporter, and local authorities to develop a deposit system for glass packaging. ReBottle is housed in an incubator and its founders are working on transitioning from a non-profit organization to a for-profit business and seeking investors.

Second Case

The BioRez organization is developing a bio-resource (bio-waste) management project. Its members are legal entities and other specialized actors whose purpose is to structure local composting and anti-food waste initiatives. BioRez is a local collective and cooperative project that brings together stakeholders from the metropolitan area of the city of Nantes to build a complete pathway to develop bio-resources locally in priority areas. Through the stakeholders gathered around BioRez, the project offers customized solutions adapted to the specific situation of the applicants. The program includes awareness-raising, anti-waste actions, local collection and composting, and on-the-job training.

BioRez targets local authorities by offering a systemic approach to the recovery of organic matter in their territory, such as collecting and managing unsold food from the farmers' market. It also targets social housing providers and private individuals by proposing to combine shared composting at the bottom of buildings with gardening, as the compost returns directly to the land through a very short loop. It also transforms dedicated areas into real meeting places that can build connections between residents, encourage exchanges, and foster openness to other neighborhood residents. BioRez is a well-established collective project on the territory of Grand Bellevue, an area that stands across the neighboring cities of Nantes and Saint-Herblain, drawing on different complementary actors–stakeholders and working as a network. The actions carried out by Biorez are part of the public policy framework supported by the ANRU + program of the Agence Nationale de la Rénovation Urbaine (ANRU).

Third Case

The NeoBoat Company gives a new life to out-of-service leisure boats. NeoBoat belongs to the social and solidarity economy. It is a Social Utility Solidarity Company (Entreprise Solidaire 'd'Utilité Sociale, ESUS) that received the national social and solidarity economy award in December 2019. NeoBoat's project is to reuse end-of-life speedboats and sailboats to transform them into unusual habitats on dry land for tourism professionals, lodges for individuals, meeting or co-working spaces, guest room extensions to houses, etc. The transformation of boats by NeoBoat is a possible solution to the problem of the end of life of the thousands of sailboats and speedboats which are to be destroyed in the years to come. About 90% of the leisure boats at the end of their life are built in polyester, a solid material but both complicated and very expensive to recycle. This project requires using various skills such as carpentry, painting, interior design, composite materials, and metal fabrication, which are likely to attract young apprentices and unemployed people. It takes more than 400 h of work to transform a boat into a NeoBoat.

In two years, the project has hosted more than 50 people and has provided more than 5000 h of training and apprenticeship for these people using old boats. The life cycle of the boats is optimized; NeoBoat and old boats are sourced locally, within a radius of less than 80 km on average from the shipyard, to reduce the impacts of collection and transport. This "circular" business model of extending the life of boats implies, to achieve it, the co-creation of an ecosystem of public, private, and associative partners around an approach intended to create "sustainable" economic, environmental, and social values shared by all stakeholders.

13.4 Results and Discussion

Based on the data collected and presented, we can now use the dimensions of the theoretical framework to describe and analyze the projects studied (Table 13.2).

First of all, it appears that in their ethical–political dimension, while having a common goal, which is the development of the CE, the projects have different dominant logics. The ReBottle project aims to develop the circular economy by fitting into a classical economic logic and proposing a service offer likely to convince investors. The BioRez project aims to preserve a dominant environmental and social logic by addressing local authorities whose mission is to finance actions that respond to particular missions of general interest. BioRez recognizes that its services are part of the general interest services and are something they can sell to local authorities rather than having to ask for grants and subsidies. By showing what the NeoBoat project could bring to the actors, the founder wanted to keep the social and environmental logics present, by appealing to the responsibility of the actors concerned by the professional insertion and the management of boats at the end of their life.

In their technical–economic dimension, what characterizes the projects is the diversity of the technical–economic means implemented. If the theoretical framework is helpful, as a checklist to describe and not forget essential elements, observation does not allow us to propose elements common to the three projects. Regarding the regulatory system, the three projects have in common that they have legal texts that regulate their activity and can constitute either a powerful driver or a potential barrier that calls for the negotiation of new rules. Whether or not the actors involved share common values, the negotiation of rules will be more or less complicated.

For the ReBottle project, the French government deposit decree and all the accounting and tax rules related to the deposit are strong and restrictive rules. In addition, there are project-level rules concerning the size of the bottles and the type of glue to be used for the labels, and also the rules that have been developed, following

Table 13.2 Observed dimensions

Dimension to observe	Sub-attributes	ReBottle	BioRez	NeoBoat
Ethico-political	Concern	The disappearance of glass bottle recycling, in favor of reuse, which consumes less energy	The waste of organic matter and the reinforcement of the social link in a popular district	Professional integration, and the problem of the end of life of leisure boats
	Dominant logic	Economic and environmental	Environmental and social	Social and environmental
	Desired utility	Development of a virtuous local economic sector, through the reuse of glass bottles rather than their recycling	Structuring of the local composting and anti-waste food sector by bringing together a diversity of expertise within a single structure to be able to address solutions around organic matter	Professional integration of disadvantaged people, awareness of environmental issues
	Speech			Need to create an ecosystem of actors
Technico-economic	Job/mission	Organize local collection and washing of glass bottles	Collect and return organic matter to the soil by organizing the shortest possible collection routes	Extend the life of out-of-use pleasure boats by giving them a new function, while training unemployed people for jobs required by companies. To set an example so that this activity can be reproduced elsewhere
	Target/clients	Customers: retailers and producers, Other targets: committed consumers, local authorities	Customers: local authorities, private companies Other targets: social landlords, fruit and vegetable sellers on markets, farmers, inhabitants	Customers: camping sites, co-working spaces, homeowners, banks Other targets: boat builders, local authorities

(continued)

Table 13.2 (continued)

Dimension to observe	Sub-attributes	ReBottle	BioRez	NeoBoat
	Resources/ uses	A washing unit A semi-trailer truck A stock of bottles and racks A coordination team	A team of operators who come twice a week to set up composting paloxes in the markets A processing for the transformation of unsold fruit and vegetables into processed food (such as canned soup)	A singular shipyard for the transformation of boats and a coordination team
	Value to create	Create local jobs for bottle collection and washing, better use natural resources and energy	Create social links, Reduce food waste, Create local jobs for the collection of bio-waste	Professional integration of people in difficulty, extension of the life of an asset through a change of function
Regulationist	Rules	Rules for collecting bottles and organizing washing, common bottle standards Rules concerning the glue used on the labels Deposit decree	Legal obligation of sorting at the source, for all producers of bio-waste by 2023	Rules for determining the good condition of a boat at the end of its life. Rules for maintaining the status of boats after their transformation. National rules on extended Producer responsibility
	Technical devices and management tools	Experiments with voluntary producers and distributors, then feedback before generalization Accounting systems, including financial accounting	Metrics collection and management software	Business model developed according to the BM3C2 template[1]

(continued)

Table 13.2 (continued)

Dimension to observe	Sub-attributes	ReBottle	BioRez	NeoBoat
	Cognitive capabilities	Potentially strong because of the values shared by the project's actors: producers and distributors with environmental values. However, these capacities are limited by a lack of accounting, legal and fiscal skills on the deposit	Potentially strong due to the homogeneous nature of the BioRez founders. However, these capabilities are limited by a lack of expertise in the development of indicators appropriate to the desired utility	Weak because of the disparity in the values of the actors. Difficult to share information and develop common knowledge
Inventiveness/ singularity	Levels of membership	Average. The actors involved have common values and membership is constrained by the objective that the project can be economically viable, without subsidies and the necessity for all actors to not lose out financially	Strong: great homogeneity of the values of the founding members of BioRez	Weak (great heterogeneity of values of the actors involved, no spirit of cooperation). For example, boat producers perceive this project as a competitor to their efforts to manage the end of life of boats, especially in terms of access to financial resources
	Intelligibility	Strong	Strong	Average
	Singularity	The most advanced bottle recycling project in France	There are many organic matter management projects in France However, BioRez is one of the first organic matter management associations that wants to be paid for a service rendered and not receive a subsidy	The only project to extend the life of leisure boats in France. An award-winning project (social and solidarity economy prize)
Developmental	Presence on a territory	Pays de la Loire Region	Metropolitan area of Nantes	Pays de la Loire Region

(continued)

Table 13.2 (continued)

Dimension to observe	Sub-attributes	ReBottle	BioRez	NeoBoat
	Category of actors	Wine, Beer, juice producers, retailers, consumers, logistics and washing companies, local governments, and the national ecological transition agency	Nantes and St Herblain cities, founding members of BioRez, social housing providers, fruit and vegetable retailers in "farmers" markets, the local population	Local governments, Chamber of Commerce and Industry, government administrations, boat manufacturers, camping grounds, co-working spaces
	Sustainability	Reconciliation of the ecological interest of the project with a more traditional economic dimension. Search for investors	Transition to a logic of selling services of general interest to local authorities. The social and economic dimensions remain dominant by addressing clients whose mission is the general interest	Resale of the NeoBoat company by its founder to ensure more stable financing and a buyer sharing more values with the investment world and the boat builders
	Change of status	Transition from non-profit to corporate status		

[1]Business Model Multi-acteurs Multi-niveaux Circulaire et Collaboratif (For more details see Boldrini & Antheaume, 2021)

experiments, for the organization of the logistic flows (collection, washing, and provision of the washed bottles). The fact that the founders of the ReBottle project chose first to solicit local producers and retailers with strong environmental values certainly facilitated the search for common rules that would allow the project to get started.

> I would say that we joined the project because I found it interesting from a sustainable development point of view. That was one of the first elements; it was new. We were asked for our logistical expertise. The goal was not to make money on logistics; it was mainly to help launch the project, at least initially, with a view to the social and solidarity economy.... (Verbatim of a logistics actor in the ReBottle project)

In the case of the BioRez project, the legal obligation that producers of bio-waste will have to sort their waste at the source in 2023 is a strong driver to get stakeholders to adhere. In addition, the founders of BioRez see themselves as complementary and see the creation of BioRez as a tool created together to facilitate cooperation. In such a context, the possibility of negotiating common rules is facilitated.

In the case of the NeoBoat project, the law on extended producer responsibility is potentially an important driver. It stipulates that any producer of a good is responsible for its end of life and the waste management it generates. With such an obligation, NeoBoat potentially offers an interesting service to the producers of leisure boats, when these products reach the end of their life. Indeed, they can be transformed into leisure housing units or workplaces. Their life is extended by at least ten years, and they keep the legal status of a boat. This means that you do not need planning permission to install these housing units or workplace units on a plot of land. It also provides more time to identify the best waste management solutions at the end of their life. Yet, financial resources for end-of-life management are limited, and boat builders and a company like NeoBoat do not share many standard references and values on social inclusion and the end-of-life of boats. Moreover, boat producers see the NeoBoat project as a competitor, not as a complement to their projects, especially to obtain financial resources to set up waste treatment channels for used boats. Such a context does not allow for the discussion of common rules.

As for the regulationist dimension, it seems to be linked to cognitive capacities and management tools. The first aspect is the sharing of common values by the actors and the existence of skills that allow the development of cognitive capacities. Thus, in the case of the NeoBoat project, the disparity in the actors' values, which was omnipresent from the start, made it difficult to share information and develop a common knowledge among the stakeholders for a management solution for end-of-life recreational boats. On the contrary, in the case of the ReBottle project, the producers and retailers involved strongly shared environmental values. However, the potential for these shared values to create information-sharing capabilities was limited by a lack of knowledge on accounting, legal, and tax rules which apply to deposit operations. Very few certified professional accountants still have the professional experience required to advise on these rules. Finally, the BioRez project has potentially strong cognitive capacities due to the homogeneous and complementary nature of the BioRez founders. However, these capacities are limited by a lack of existing benchmarks to facilitate the development of indicators appropriate to measure the desired outcomes. More research is needed on this subject.

Concerning the singularity dimensions, it has been observed that too much singularity in the project can lead to difficulties in terms of intelligibility and, therefore of, adhesion. If it is appropriate to be singular, it is not advantageous to be too singular for a project to be explained and understood clearly, and thus to progress rapidly. Therefore, the NeoBoat project, the first of its kind in France, is perhaps the most difficult to explain, and gaining support for it is more complicated. In contrast, we observe that the management of organic matter and the reuse of bottles are practices that are already known and well identified, even if they are not as widespread as we would like. Because of the existence of similar projects, it is no longer necessary for the founders of ReBottle and BioRez to spend a lot of energy defending the interest of reusing bottles or returning organic matter to the land. These projects are easier to explain and, therefore, more intelligible. Moreover, it seems to us that adhesion and intelligibility are also favored by the homogeneity of the stakeholders and their common values. This increases their potential for cooperation. In this perspective, the ReBottle and BioRez projects are to be contrasted with the NeoBoat project.

> .... To launch the project, I had in my accounts and in my speech a modern policy of waste management, and a social policy of training and integration of people far from work by strengthening their independence on the labor market later ... I wanted an accounting tool in order to be able to present the project and its potential gains by creating a dynamic around all that. (Verbatim of the founder of NeoBoat)

Finally, the developmental dimension is very present in all three projects. The founders of the projects are constantly questioning what will ensure their sustainability. The answers are different according to the projects and, by feedback effect, influence the dominant logic in the ethical–political dimension. The three projects studied aim to find sources of financing, but they do not follow the same paths. The ReBottle project relies on the reconciliation of economic and ecological logic to convince investors and thus ensure its sustainability. At the same time, the BioRez project aims to preserve the predominance of the environmental and social dimension by targeting local authorities as clients rather than as subsidizers. Indeed, since local authorities seek the common interest, they will be willing to finance actions that fall within the framework of services of public interest. As for the NeoBoat project, its resale to three buyers (a non-profit organization, an investment fund, and an entrepreneur) aims to bring the project closer to the values of boat builders, to ensure sustainable financing, and the possibility to preserve its social and environmental objectives.

13.5 Conclusions

Based on project-based collective action theory, we have proposed a multidimensional analysis framework using the project approach. The aim is to facilitate the analysis of how EC projects emerge, how they are managed, and how they develop by focusing on the emergence of collective action around a project with its multiple dimensions. We provide a platform for practitioners and researchers that goes beyond a purely technical or engineering focus in the CE field. The proposed framework highlights the critical dimensions of CE projects and facilitates strategic discussion on their development and improvement potential. The framework could help stakeholders answer questions such as: how does my project define itself differently and what are its specific dimensions and indicators? Which dimensions (and possibly indicators) are likely to change soon? How will this affect the sustainability of the project and the potential for joint action? Which dimensions or indicators of our project could be improved for better effect?

Our research makes essential contributions to research on CE projects. The results clearly show that the five dimensions are present, define the project, and justify the existence of collective action. We can also say that the dimensions are dynamically linked to the regulatory dimension. The latter determines the continuity of the project. The project participants were voluntarily enthusiastic (ethical–political dimension) about implementing CE despite several obstacles to their activities, such as supply and demand problems, regulatory clarity, cash flow uncertainties, and lack of management tools (technical–economic and regulatory dimensions).

The current framework could be further developed, and specific indicators (qualitative and quantitative) could be proposed through empirical research for each dimension. Ultimately, we hope that the framework will inspire new thinking about CE projects and enterprise organization even if further improvements and applications are required to improve it.

Funding This research is funded by the PARC 2021 doctoral grant from the Association Francophone de Comptabilité and the Conseil Supérieur de l'Ordre des Experts Comptables.

References

Aggeri, F., & Labatut, J. (2010). La gestion au prisme de ses instruments. *Finance Contrôle Stratégie, 13*(3), 5–37.

Barratt, M., Choi, T. Y., & Li, M. (2011). Qualitative case studies in operations management: Trends, research outcomes, and future research implications. *Journal of Operations Management, 29*(4), 329–342. https://doi.org/10.1016/j.jom.2010.06.002

Benbasat, I., Goldstein, D. K., & Mead, M. (1987). The case research strategy in studies of information systems. *MIS Quarterly, 11*(3), 369–386. https://doi.org/10.2307/248684

Boldrini, J. C., & Antheaume, N. (2021). Designing and testing a new sustainable business model tool for multi-actor, multi-level, circular, and collaborative contexts [Circular]. *Journal of Cleaner Production, 309*, 127209. https://doi.org/10.1016/j.jclepro.2021.127209

Boutinet, J. P. (2012). La figure du projet comme forme hybride de créativité. *Spécificités, N*(1), 7–20. https://doi.org/10.3917/spec.005.0007

D'Amato, D., Droste, N., Allen, B., Kettunen, M., Lähtinen, K., Korhonen, J., Leskinen, P., Matthies, B. D., & Toppinen, A. (2017). Green, circular, bio economy: A comparative analysis of sustainability avenues [Circular]. *Journal of Cleaner Production, 168*, 716–734. https://doi.org/10.1016/j.jclepro.2017.09.053

Desreumaux, A., & Bréchet, J. P. (2009). 3. Quels fondements pour les théories de la firme? Plaidoyer pour une théorie artificialiste de l'action collective fondée sur le projet. In *Analyses et transformations de la firme* (pp. 61–89). La découverte.

Desreumaux, A., & Bréchet, J. P. (2018). Repenser l'entreprise: Pour une théorie de l'entreprise fondée sur le Projet. Presses Univ. Septentrion.

Desreumaux, A., & Bréchet, J. P. (2019). *Repenser l'entreprise: Une théorie de l'entreprise fondée sur le Projet*. Presses Universitaires du Septentrion.

Eisenhardt, K. M. (1989). Building theories from case study research. *Academy of Management Review, 14*(4), 532–550. https://doi.org/10.2307/258557

Eisenhardt, K. M., & Graebner, M. E. (2007). Theory building from cases: Opportunities and challenges. *Academy of Management Journal, 50*(1), 25–32. https://doi.org/10.5465/amj.2007.24160888

Geissdoerfer, M., Pieroni, M. P. P., Pigosso, D. C. A., & Soufani, K. (2020). Circular business models: A review. *Journal of Cleaner Production, 277*, 123741. https://doi.org/10.1016/j.jclepro.2020.123741

Geissdoerfer, M., Savaget, P., Bocken, N. M. P., & Hultink, E. J. (2017). The circular economy–a new sustainability paradigm? *Journal of Cleaner Production, 143*, 757–768. https://doi.org/10.1016/j.jclepro.2016.12.048

Geng, Y., Fu, J., Sarkis, J., & Xue, B. (2012). Towards a national circular economy indicator system in China: An evaluation and critical analysis. *Journal of Cleaner Production, 23*(1), 216–224. https://doi.org/10.1016/j.jclepro.2011.07.005

Gibbert, M., & Ruigrok, W. (2010). The '"What"' and '"How"' of case study rigor: Three strategies based on published work. *Organizational Research Methods, 13*(4), 710–737. https://doi.org/10.1177/1094428109351319

Goffin, K., Åhlström, P., Bianchi, M., & Richtnér, A. (2019). Perspective: State-of-the-art: The quality of case study research in innovation management. *Journal of Product Innovation Management, 36*(5), 586–615. https://doi.org/10.1111/jpim.12492

Hartley, K., van Santen, R., & Kirchherr, J. (2020). Policies for transitioning towards a circular economy: Expectations from the European Union (EU). *Resources, Conservation and Recycling, 155*, 104634. https://doi.org/10.1016/j.resconrec.2019.104634

Kirchherr, J., Piscicelli, L., Bour, R., Kostense-Smit, E., Muller, J., Huibrechtse-Truijens, A., & Hekkert, M. (2018). Barriers to the circular economy: Evidence from the European Union (EU). *Ecological Economics, 150*, 264–272. https://doi.org/10.1016/j.ecolecon.2018.04.028

Kirchherr, J., Reike, D., & Hekkert, M. (2017). Conceptualizing the circular economy: An analysis of 114 definitions. *Resources, Conservation and Recycling, 127*, 221–232. https://doi.org/10.1016/j.resconrec.2017.09.005

Korhonen, J., Nuur, C., Feldmann, A., & Birkie, S. E. (2018). Circular economy as an essentially contested concept. *Journal of Cleaner Production, 175*, 544–552. https://doi.org/10.1016/j.jclepro.2017.12.111

Korhonen, J., Von Malmborg, F., Strachan, P. A., & Ehrenfeld, J. R. (2004). Management and policy aspects of industrial ecology: An emerging research agenda. *Business Strategy and the Environment, 13*(5), 289–305.

MacArthur, E. (2013). Towards the circular economy. *Journal of Industrial Ecology, 2*(1), 23–44.

Makadok, R., Burton, R., & Barney, J. (2018). A practical guide for making theory contributions in strategic management. *Strategic Management Journal, 39*(6), 1530–1545. https://doi.org/10.1002/smj.2789

Malik, A., Pereira, V., & Budhwar, P. (2018). Création et capture de valeur à travers les pratiques de gestion des ressources humaines: Regarder à travers le prisme du modèle d'entreprise. *Dynamique Organisationnelle, 47*(3), 180–188.

McDonough, W., & Braungart, M. (2002). Design for the triple top line: New tools for sustainable commerce. *Corporate Environmental Strategy, 9*(3), 251–258. https://doi.org/10.1016/S1066-7938(02)00069-6

Meredith, J. (1998). Building operations management theory through case and field research. *Journal of Operations Management, 16*(4), 441–454. https://doi.org/10.1016/S0272-6963(98)00023-0

Milios, L., & Matsumoto, M. (2019). *Consumer perception of remanufactured automotive parts and policy implications for transition.*

Milios, L. (2021). Towards a circular economy taxation framework: Expectations and challenges of implementation. *Circular Economy and Sustainability, 1*(2), 477–498.

Nikolaou, I. E., & Stefanakis, A. I. (2022). A review of circular economy literature through a 3-fold level framework and engineering-management approach. In *Circular economy and sustainability* (pp. 1–19).

Park, J., Sarkis, J., & Wu, Z. (2010). Creating integrated business and environmental value within the context of China's circular economy and ecological modernization. *Journal of Cleaner Production, 18*(15), 1494–1501. https://doi.org/10.1016/j.jclepro.2010.06.001

Prieto-Sandoval, V., Jaca-García, C., & Ormazabal-Goenaga, M. (2017). *Economía circular: Relación con la evolución del concepto de sostenibilidad y estrategias para su implementación.*

Prieto-Sandoval, V., Jaca, C., & Ormazabal, M. (2018). Towards a consensus on the circular economy. *Journal of Cleaner Production, 179*, 605–615. https://doi.org/10.1016/j.jclepro.2017.12.224

Reynaud, J. D. (1991). Pour une sociologie de la régulation sociale. *Sociologie Et Sociétés, 23*(2), 13–26. https://doi.org/10.7202/001632ar

Rosa, P., Sassanelli, C., & Terzi, S. (2019). Towards circular business models: A systematic literature review on classification frameworks and archetypes. *Journal of Cleaner Production, 236*, 117696. https://doi.org/10.1016/j.jclepro.2019.117696

Tibbs, H. (2006). 35 La boucle de valeur—Un nouveau cadre pour la pensée commerciale. In *Le manuel international sur la gestion des technologies environnementales* (p. 534).

Velenturf, A. P., & Purnell, P. (2017). Valorisation des ressources à partir des déchets: Rétablir l'équilibre entre la rareté des ressources et la surcharge des déchets. *Durabilité, 9*(9), 1603.

Yuan, Z., Bi, J., & Moriguichi, Y. (2006). The circular economy: A new development strategy in China. *Journal of Industrial Ecology, 10*(1–2), 4–8. https://doi.org/10.1162/108819806775545321

Zott, C., & Amit, R. (2010). Business model design: An activity system perspective. *Long Range Planning, 43*(2–3), 216–226.

Zott, C., & Amit, R. (2013). The business model: A theoretically anchored robust construct for strategic analysis. *Strategic Organization, 11*(4), 403–411.

Chapter 14
Green Manure Management as a Sustainable and Economical Alternative for Intensive Crop Fertilization in the Framework of the Circular Economy

Francisco José Castillo-Díaz, Luis J. Belmonte-Ureña, Francisco Camacho-Ferre, and Julio César Tello-Marquina

14.1 Introduction

Agriculture is a strategic sector within a society. In Spain, the agri-food system generated more than 1.1 million jobs and accounted for 5.8% of the country's Gross Value Added in 2020, which was 34.6% higher than the European average (Maudos et al., 2020). Thus, we can visualize a dependent relationship between the socioeconomic development of some Spanish territories and the presence of agriculture (Maudos et al., 2019). This relationship has resulted in the development of high-yielding agricultural systems, such as greenhouse agriculture in the province of Almeria (Almeria Model), one of the most critical at the international level (Camacho-Ferre, 2004). In the 2020/2021 campaign, just over 3.5 million tons of fruit and vegetable products were obtained from 32,554 ha of greenhouse crops. Economically speaking, the

F. J. Castillo-Díaz · F. Camacho-Ferre · J. C. Tello-Marquina
Department of Agronomy, Research Centre CIAIMBITAL, University of Almería, 04120 Almería, Spain
e-mail: fcastillo@ual.es

F. Camacho-Ferre
e-mail: fcamacho@ual.es

J. C. Tello-Marquina
e-mail: jtello@ual.es

F. J. Castillo-Díaz · L. J. Belmonte-Ureña (✉)
Department of Economy and Business, Research Centre CIAIMBITAL, University of Almería, 04120 Almería, Spain
e-mail: lbelmont@ual.es

S. A. Bandh and F. A. Malla (eds.), *Waste Management in the Circular Economy*,
https://doi.org/10.1007/978-3-031-42426-7_14

sector generated 2937.6 million euros from the sale of its production, of which almost 80% was exported to E.U. member states. Production comprises eight vegetable species: tomatoes, bell peppers, watermelon, zucchini, cucumbers, melons, and beans (Cajamar, 2021).

However, agricultural production has also caused negative externalities that have compromised ecosystem well-being, fauna, and flora. Among these are the loss of genetic diversity (European Union, 2019), the degradation of agricultural soil (Gómez-Tenorio et al., 2021b; Jacobs et al., 2022), and the overexploitation and loss of quality in some water bodies (Dhaoui et al., 2022). The impacts of agriculture have intensified since the Green Revolution, which caused agricultural productivity to increase (e.g., 240% increase in cereal crops from 1961 to 2017) (IPCC, 2019). This revolution was preceded by the application of aggressive cropping practices involving higher agrochemical consumption, increased mechanization needs, higher energy consumption, and a change in crop varieties (FAO, 2015, 2017). The result has been a tripling of agricultural systems' demand for natural resources over the last fifty years (FAO, 2019a), especially for cropland and water (FAO, 2019b).

In the Almeria Model, some environmental impacts are caused by the poor management of abundant agricultural biomass generated at the end of the production cycle. Its seasonal nature means that more than 70% of the residual material (about 1.3 million tons) must be managed in only three months (Duque-Acevedo et al., 2020; Junta de Andalucía, 2016).

Negative externalities are inconsistent with the principles of sustainable development (Fundación Encuentro, 1992). In 2015, members of the United Nations (UN) signed the 2030 Agenda, where 17 Sustainable Development Goals (SDGs) can be identified. These aim to address the most crucial social, economic, and environmental problems globally and achieve sustainability in its three aspects. SDG 2—Zero Hunger seeks to implement sustainability in agricultural systems. To this end, agroecology should be established in food production models (UN, 2015). This approach is based on the theory of agro-systems, i.e., the conception of agriculture as an "organized whole" comprised modifiable and interdependent factors (Navarro-Garza et al., 1993). These factors are to be transformed by applying national agricultural and environmental policies in search of the desired balance, i.e., sustainability (European Commission, 2020a).

The European Union has incorporated the latest international agreements into its policy. The European Green Pact lays out the future of member states until 2050 and aims to transform the E.U. into a resilient territory whose growth is decoupled from the consumption of natural resources. This is done by restricting net greenhouse gas emissions, which requires systemic change in all sectors (European Commission, 2019). In addition, the E.U. requires that this goal can be reached by applying the circular economy principles (European Commission, 2020b). Doing so leads to reducing inputs, reusing by-products, and recycling waste so that it does not negatively impact the biodiversity of ecosystems (European Commission, 2020c).

Sustainable agriculture has been boosted in Europe through the Farm to Fork strategy, one of the most important initiatives of the European Green Pact (European

Commission, 2019, 2020a). This strategy aims to expand the sustainability of agricultural systems by reducing the use of pesticides by 50%, curtailing the use of fertilizers by 20%, which could reduce nutrient losses by 50%, and increasing the amount of land under European organic certification to 25%. This certification prohibits the use of inorganic fertilizers and synthetic phytosanitary products. In addition, cultivation practices that maintain or increase soil fertility must be applied. Green manure is a necessary measure for this purpose (European Union, 2018). This situation reduces the demand for energy in agricultural systems during food production (Basavalingaiah et al., 2022). However, it also calls for the elimination of some practices that have been in place for more than sixty years in production systems (FAO, 2015, 2017). Therefore, it is relevant to the search for new farming models based on the circular economy that reduce the demand for inputs without reducing crop productivity. Previous research has justified the use of agricultural biomass incorporated through the biosolarization technique as a methodology capable of obtaining production and net profit before taxes, similar to conventional crops fertilized with inorganic fertilizers in greenhouse agriculture in Almeria (Castillo Díaz et al., 2021, 2022; Gómez-Tenorio et al., 2016). The biosolarization technique should be applied to thermally inactivate pathogens associated with agricultural biomass to reduce the risk of expression during the following crop (Gómez-Tenorio et al., 2018). Therefore, value could be obtained by analyzing its control effect on different pathogens associated with agricultural biomass.

Soil biosolarization is included in biodisinfection techniques. It is a methodology that merges into the same procedure as the protocols of solarization (Katan et al., 1976) and biofumigation (Kirkegaard et al., 1993). Traditionally, it has been used to mitigate the infection of soil-borne diseases that affect plant organisms (Chamorro et al., 2015; Gómez-Tenorio et al., 2018; Núñez-Zofio et al., 2012), although during its use it has also achieved an improvement in soil fertility (Castillo Díaz et al., 2021, 2022). In this technique, organic amendments of diverse origins, such as plants or commercial compounds obtained from the *Brassicaceae* genus (Castillo-Díaz et al., 2022; Kirkegaard & Sarwar, 1998; Masahiko et al., 2014), manure (Chamorro et al., 2015), agricultural biomass (Gómez-Tenorio et al., 2018), agroindustry residues (Achmon et al., 2016), or any organic amendment possessing a carbon/nitrogen ratio higher than 8 (Bello et al., 1997) can be used. Previous research has focused primarily on evaluating the disinfection power of the technique and not on the fertilization capacity offered by the organic amendments that are added to this technique. Crops have continued to be managed under commercial practices involving fertilizer consumption (Chamorro et al., 2015; Guerrero et al., 2010; Marín-Guirao et al., 2016; Mauromicale et al., 2010, 2011). The dual functionality of the biosolarization technique (i.e., pathogen control and fertilization capacity) makes it a suitable strategy in the transition toward sustainable and circular agriculture (Castillo Díaz et al., 2021, 2022; Gómez-Tenorio et al., 2018).

On the other hand, radish and mustard plants, both included in the *Brassicaceae* genus, can be grown as green manure to be incorporated into the soil once it has finished its growth stage before serving as an organic amendment for biosolarization processes (Bhuiyan & Jannat, 2018; Masahiko et al., 2014). The necessary use of this technique in organic farming (European Union, 2018) suggests the need to evaluate its fertilization power.

In this scenario, the **objective** of this research was to evaluate the fertilizing power of different green manures (mustard and radish) and the remains of tomato plants affected by *Fusarium oxysporum f.sp. radicis-lycopersici* (FORL) versus conventional inorganic fertilization and no fertilization in short cycles of greenhouse tomato production. The parameters analyzed were yield, quality, water consumption, and net profit before taxes.

14.2 Materials and Methods

14.2.1 Localization and Greenhouse Characteristics

The experiment was conducted at the "Catedrático Eduardo Fernández" farm (UAL-ANECOOP Foundation) located in the province of Almeria (Spain). The experimental greenhouse was the "Raspa y Amagado" type, with a minimum and maximum height of 3.40 and 4.70 m, respectively. The greenhouse was covered with a 200 μm thick plastic sheet and had lateral and zenithal windows protected by anti-strip nets. The orientation of the greenhouse and its crop rows was northwest–southwest, with a total area of 1784 m^2. The irrigation system consisted of two independent subsectors, which allowed the application of different amounts of water and inorganic fertilizers. The nominal flow rate of the emitters used was 3 L h^{-1}. The previous tomato crop did not show any disease of edaphic origin.

14.2.2 Crop, Experimental Design, and Description of Treatments

A tomato cycle of 169 days after transplanting (DAT) was carried out. Tomato seedlings were transplanted during the first week of September and extended until the third week of February of the following year. A tomato variety (*Solanum lycopersicum* Mill.) cultivar "Pitenza F1" (Enza Zaden, Enkhuizen, The Netherlands) was used with a transplanting density of 2 plants/m^2. The cultural practices applied to the tomato crop were those recommended by Camacho-Ferre (2004). However, the plants were guided only by the trellising raffia without using trellising clips to facilitate the self-management of plant debris. Pests and diseases that affected the crop were controlled according to integrated production (IP) regulations and did not cause any loss of production during cultivation.

The treatments applied depended on the fertilization plan and their location. The experimental area was 20 m^2 for each repetition (one crop line with 40 tomato plants).

Table 14.1 Treatments

Code	Composition
IF[a]	Conventional cultivation with inorganic fertilization
Test[b]	Without fertilization
PD	Exclusive fertilization with 3.5 kg m^{-2} of tomato plant debris affected by *Fusarium oxysporum* f.sp. *radicis-lycopersici* (FORL)
PR[c]	Exclusive fertilization with 2.0 kg m^{-2} of radish plants through their incorporation as green manure
PM[c]	Exclusive fertilization with 1.6 kg m^{-2} of mustard plants through their incorporation as green manure

[a] Inorganic blanket fertilization was based on the equilibrium reported by Steiner in his ideal solution up to an electrical conductivity of 3 dS m^{-1} (water + nutrient solution)
[b] A commercial tomato crop was grown in the experimental area, irrigated only with water and biosolarized with 2 kg m^{-2} of poultry manure during the previous season
[c] Biomass generated during green manure cultivation

The rows at the extremes were excluded to avoid contamination between experimental plots (i.e., edge effect). The experimental design corresponds to a single-factor design with four replications ($n = 4$).

The treatments that were fertilized only with organic amendments and those not fertilized were located in a different sector from the ones that received inorganic fertilization (Table 14.1).

14.2.3 Radish and Mustard Planting

In the experimental area dedicated to the PR and PM treatments, green manure of radish and mustard plants was sown at a planting density of 15 kg ha^{-1} and 75 kg ha^{-1}, respectively. The plant species were sown in the last week of April, and cultivation lasted 41 days.

14.2.4 Biosolarization Procedure

After growing the green manures, all experimental plots underwent a disinfection treatment (solarization or biosolarization). The FORL-affected tomato plants were placed on the central concrete aisle of the greenhouse and crushed with a hammer chopper once the trunk branches were collected from the growing surface. After placing them in their respective experimental plots and homogenizing the organic amendments in the upper soil, tillage was done using a rotovator at a depth of 20–30 cm. The tilling was carried out independently in each experimental plot to avoid contamination between treatments. It incorporated green manure (i.e., radish and mustard plants), which had been grown in the area assigned for the PR and PM treatments. Afterward, the irrigation system was reinstalled, its correct functioning was verified, and the soil was covered with a 50 μm transparent plastic. The plastic cover was sealed with staples using tape when located between posts with a rectangular trench measuring 20 cm at the base and 30 cm in height between edges. This procedure prevents moisture and gases generated during the biodecomposition of organic amendments from escaping. Finally, irrigation was applied until the field capacity of the soil was reached (49.8 L m^{-2}). The disinfection treatment lasted three months and extended from June to September (Fig. 14.1).

Fig. 14.1 Incorporation procedure of green manures and biosolarization. **a**: growth of radish and mustard plants; **b**: biomass generated by mustard plants during their cultivation; **c**: green manure on the experimental plots (PR and PM) soil; **d**: tomato plant debris on the central aisle of the greenhouse; **e**: shredding of tomato plant remains; **f**: plant remains distributed over the greenhouse soil; **g**: plant remains mixed in the upper soil profile; **h**: unsealed plastic solarization cover and irrigation check; **i**: sealed plastic cover and irrigation at field capacity

14.2.5 Variables Analyzed

Incidence FORL

A symptomatologic assessment of the plants in the PD treatment's experimental plots was carried out at 170 DAT, to detect plants affected by tomato vascular fusarium foot rot (causal agent FORL). The clues recommended by Tello-Marquina and del Moral (2001) were used for this purpose.

Crop yield

The cumulative production per unit area and the fruit mass of the crop were determined after the first harvest at 98 DAT. To this end, the crops obtained from each experimental plot were measured separately during the seven harvests carried out throughout the experiment. Fruit mass was determined independently on 25 randomly selected fruits from each experimental plot and harvest. A scale with a sensitivity of 0.1 g (Mettler Toledo, Columbus, OH, USA) was used for the measurements. The fruits were harvested according to the maturity indexes required by the marketing entity.

Harvest quality

Tomato fruit quality was analyzed three times during the experiment. For this, ten fruits were selected from each experimental plot on each analysis day. The following parameters were analyzed for each fruit: equatorial diameter calculated with a digital caliper with a sensitivity of 0.01 mm (Mitutoyo; Kanagawa, Japan); average firmness of the fruit pulp after eliminating the epidermis obtained from three points separated at a distance of 120° and on a surface of 0.5 cm^2 using a durometer with a sensitivity of 0.001 kg cm^{-2} (Penefel DFT14, Agrosta, Compainville, France); the pH of the fruit pulp with a pH meter at a sensitivity of 0.01 pH units (pH-25, Crison, Barcelona, Spain); and the content of soluble solids in the fruit pulp via refractometer with a sensitivity of sensitivity 0.1°Brix (PAL-1, Atago, Tokyo, Japan).

Water

The amount of water applied to the treatments during tomato cultivation and the volume of water required by the green manure (i.e., radish and mustard plants) during cultivation were calculated. The measurement was obtained from the irrigation frequency, irrigation time, some emitters, and nominal flow rate of the emitters. The results were finally expressed in L m^{-2}.

Economic performance

An economic analysis was carried out to evaluate the crop viability of the five field treatments (i.e., TEST, IF, PD, PR, and PM) to maximize farmers' profit. Based on the results obtained in the field, a reduction of some items in the expense account (fertilizers, water, chemical disinfectants, etc.) was proposed. The indicator selected was the pre-tax profit under the recommendations made by other authors (Castillo-Díaz et al., 2022; Honoré et al., 2019). The variable was calculated under the following mathematical expression:

$$NPb_t = TNR - TC \tag{14.1}$$

NPb_t: profit before taxes; TNR: total annual revenue; TC: total annual costs.

- **Expense structure**

The structure of income and expenses used in this analysis is based on the recommendations offered by Castillo-Díaz et al. (2022) and Honoré et al. (2019), which fits the structure used by the "Catedrático Eduardo Fernández" farm in the UAL-ANECOOP Foundation. The values were adapted to the economic conditions of the 2014/2015 campaign through the national European Classification of Individual Consumption by Purpose (ECOICOP) index.

- **Revenue structure**

Revenues were obtained by multiplying the production obtained from each experimental plot by the original price of smooth tomatoes reported by the Observatory of Prices and Markets of the Junta de Andalucía in the 2014/2015 campaign (i.e., between December and February).

14.2.6 Statistical Treatment

The statistical treatment consisted of applying an analysis of variance (one-way ANOVA). The effect of the applied treatments was observed (i.e., TEST, IF, PD, PR, and PM) on each of the parameters evaluated (i.e., final commercial production, harvest quality, equatorial diameter, fruit mass and firmness, pH and total soluble solids concentration of the fruit pulp, and pre-tax economic profit). Subsequently, a post hoc least significant difference (LSD) test was applied at a 95% confidence level. Previously, the assumptions of normality and homoscedasticity were verified through the Shapiro–Wilk and Bartlett tests, respectively. The statistical analyses were carried out with the STATGRAPHIC CENTURION XVIII statistical software (Manugistics Incorporate, Rockville, Maryland) for Windows.

14.3 Results and Discussion

14.3.1 **Fusarium oxysporum** ***f.sp.*** **radicis-lycopersici** ***Expression***

Pre-biosolarization inoculation with FORL-infected tomato plants did not affect any plant in the experimental plots of the PD treatment despite being susceptible to tomato vascular fusarium foot rot. Inoculation was also performed on the same experimental plots during the previous growing season and FORL did not affect any plant (unpublished data). The biosolarization technique could thermally inactivate the pathogen found on the diseased plant material. Gómez-Tenorio et al. (2018) postulated tomato agricultural biomass affected by nematodes of the genus *Meloidogyne* as a sustainable organic amendment to be used in biosolarization protocols due to the thermal inactivation capacity of the biodisinfection methodology. Thus, it is necessary to apply the biosolarization technique as a preventive measure when self-managing the agricultural biomass generated by crops due to pests and diseases that may affect the plant by-product (Castillo-Díaz et al., 2019). By doing so, they are thermally inactivated, and the probability of expression in the following crop and the allocation of phytosanitary products are reduced. This strategy follows the principles proposed by the European Union in its Farm to Table strategy to achieve the proposed goal of reducing agrochemicals while expanding the sustainability of agricultural systems in compliance with the UN SDGs (European Commission, 2020a; UN, 2015) while also favoring the use of by-products generated within the framework of the circular economy.

14.3.2 Cumulative Commercial Production

At the end of the production cycle (169 DAT), the conventional treatment, which received inorganic fertilization during the whole crop cycle (i.e., conventional crop), obtained a final cumulative yield significantly lower than the TEST, PD, PR, and PM treatments (Fig. 14.2). The production of the IF treatment started to decrease with the fifth harvest (i.e., 144 DAT). This result differs from that of Castillo-Díaz et al. (2021) in their three tomato production cycles conducted consecutively in the same greenhouse following this experiment. At this time of cultivation, the treatments fertilized with tomato plant debris or inorganic fertilization showed higher production than those without fertilization and were equal to each other. The IF treatment did not register any production loss due to pests and diseases, so the behavior obtained in our trial could be registered by a difference in production quality.

The different plant remains evaluated (i.e., FORL-infected tomato plants, radish plants, and mustard plants) showed adequate viability to be used as fertilizers. Two of them were applied as green manure (i.e., radish plants and mustard plants) (Fig. 14.3). This result increases the number of plant remains used as fertilizer by Almeria producers in their greenhouse agriculture cycles of up to 169 DAT. Until now, only the tomato plant remains incorporated through the biosolarization technique has shown this performance (Castillo Díaz et al., 2021, 2022; Gómez-Tenorio et al., 2016). The latter has also proven to be a suitable fertilizer in production cycles of more than seven and a half months (i.e., known as long production cycles). This fact increases the tools at the disposal of farmers to meet the new demands made by the European Union in terms of fertilizer reduction (European Commission, 2020a). This strategy is included in the circular economy framework (European Commission, 2020b) as it allows a 100% reduction of inorganic fertilizers routinely required

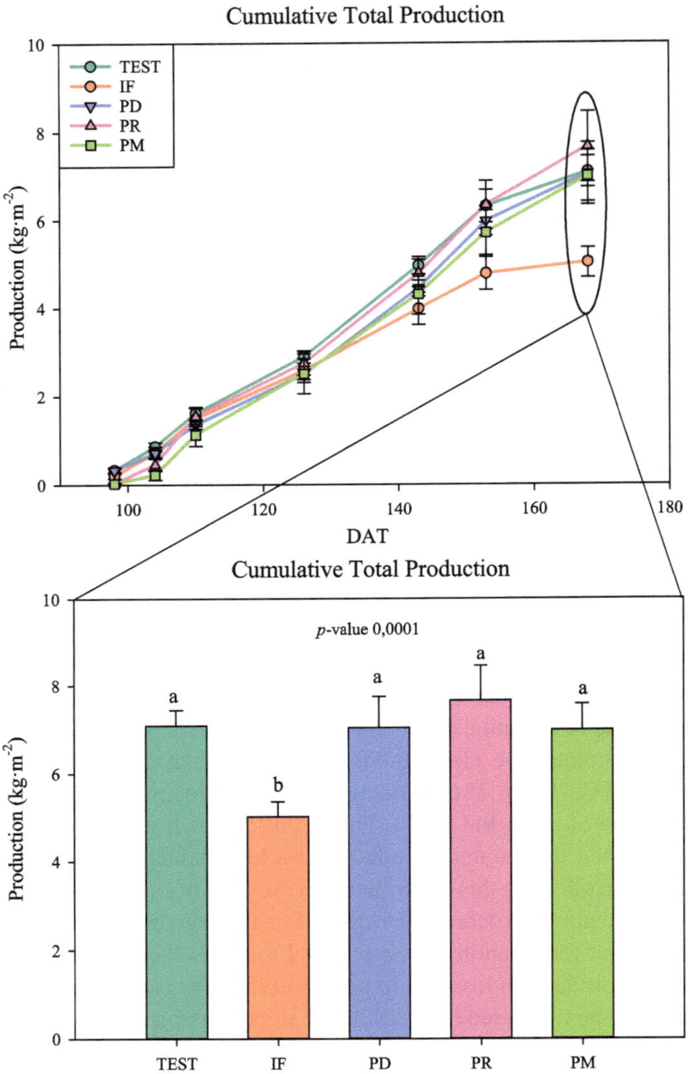

Fig. 14.2 Cumulative production at the end of the study season (September–February cycle). TEST: no fertilization; IF: conventional cultivation with inorganic fertilization; PD: exclusive fertilization with 3.5 kg m^{-2} of tomato plant debris affected by FORL; PR: exclusive fertilization with 2.0 kg m^{-2} of radish plants; PM: exclusive fertilization with 1.6 kg m^{-2} of mustard plants. Different letters indicate significant differences ($p \leq 0.05$, LSD test). DAT: days after transplanting

by crops, which expands the sustainability of agricultural systems while reducing external dependence on inputs and energy (Basavalingaiah et al., 2022). In addition, the use of green manure, or the self-management of crop plant remains, can be used by producers under the European organic certification, an area with high input restrictions (European Union, 2018). The incorporation of plant remains, with or without the biosolarization technique, has also been described as a methodology capable of improving the fertility of agricultural soils (Castillo Díaz et al., 2021, 2022; Salinas et al., 2020). The fertility of agricultural soil is one of the requirements of Regulation 2018/848, of May 30, 2018, on organic production, which stipulated that the addition of organic amendments or the application of green manures is necessary measures to maintain or increase soil quality (European Union, 2018). Green manure of the genus *Brachypodium* and other species autochthonous to the province of Almeria and Granada (Spain) can improve the fertility of almond tree soil. Some of them belong to the *Fabaceae* family and can establish symbiosis with the *Rhizobium* genus, which can fix atmospheric nitrogen through this relationship. This genus of bacteria is present in the greenhouse soil of the agricultural system of Almeria. Gómez-Tenorio et al. (2021a) identified the presence of *Rhizobium* in 91.6% of 12 soil samples collected from greenhouses in Almeria. Therefore, applying green manure formed by leguminous plants should be favored to continue expanding soil fertility.

The nutrient reserves in the soil of the experimental plots of the TEST treatment could have allowed this treatment to be one of the most productive at the end of the production season. Monge et al. (2007), Salas et al. (2003), and Yepis et al. (1999) stated that the supply of high fertilizer content could lead to both the accumulation of these substances in the soil and their percolation into groundwater, with over-fertilization being a common practice in agriculture (Cánovas et al., 2003; Murcia Region, 2019). The response of the experimental plots to this treatment during

Fig. 14.3 Tomato crop fertilized with radish and mustard plants at 150 DAT

the three subsequent seasons was consistent with what was expected: a progressive reduction in the production of 46.3% (Castillo Díaz et al., 2021).

14.3.3 Harvest Quality

Fruit mass and equatorial diameter showed the same characteristics previously reported for final cumulative yield (Fig. 14.4). Thus, the differences manifested in final cumulative production were induced by these physical quality parameters. Mauromicale et al., (2010, 2011) communicated an increase in fruit mass with the addition of different doses of organic amendments versus their control treatment. However, these authors applied inorganic fertilizers to all their treatments. Bilalis et al. (2018) reported an increase in the fruit mass harvested from plots fertilized with inorganic fertilizers versus their unfertilized treatment in tomato production cycles. These results are contrary to that obtained in this experiment. Castillo-Díaz et al., (2021, 2022) reported that the mass and equatorial diameter of tomatoes harvested from plots fertilized with inorganic fertilizers or organic amendments were similar, and both were superior to unfertilized tomato crops, although their crop cycles were up to 217 DAT.

The rest of the parameters evaluated (i.e., firmness, pH, and total soluble solids content of the fruit pulp) were very close. However, there were significant differences in favor of the TEST treatment regarding fruit pulp firmness and the PR treatment regarding fruit pulp acidity (Fig. 14.2). Polat et al. (2010) obtained similar firmness, pH, and soluble solid content between fertilizations with inorganic and organic fertilizers of a commercial tomato crop, as did Pieper and Barrett (2009) for pH and total soluble solids content. Castillo-Díaz et al., (2021, 2022) reported similar results to ours when they compared exclusive fertilization with organic amendments versus inorganic fertilization applied through the irrigation system with no fertilization.

14.3.4 Water Consumption

Incorporating green manure (i.e., PM and PR) and tomato plant debris (i.e., PD) into the soil through the biosolarization technique created a 43.9% reduction in the amount of water applied during tomato cultivation of 169 DAT when compared to conventional cultivation. However, water applied to green manure during its cultivation (i.e., PM and PR) reduced this amount to 22.4% (Fig. 14.5 and Table 14.2). The experimental greenhouse soil reduced the hydraulic conductivity of the soil in the plots of the PD, PR, and PM treatments during the following three years with the self-management of tomato plant debris versus conventional cultivation (Castillo Díaz et al. 2021), which implied a reduction of the water endowment by 37.2% (Castillo-Díaz et al., 2022). The biosolarization technique has been described as a methodology capable of modifying soil hydraulic conductivity in greenhouse crops

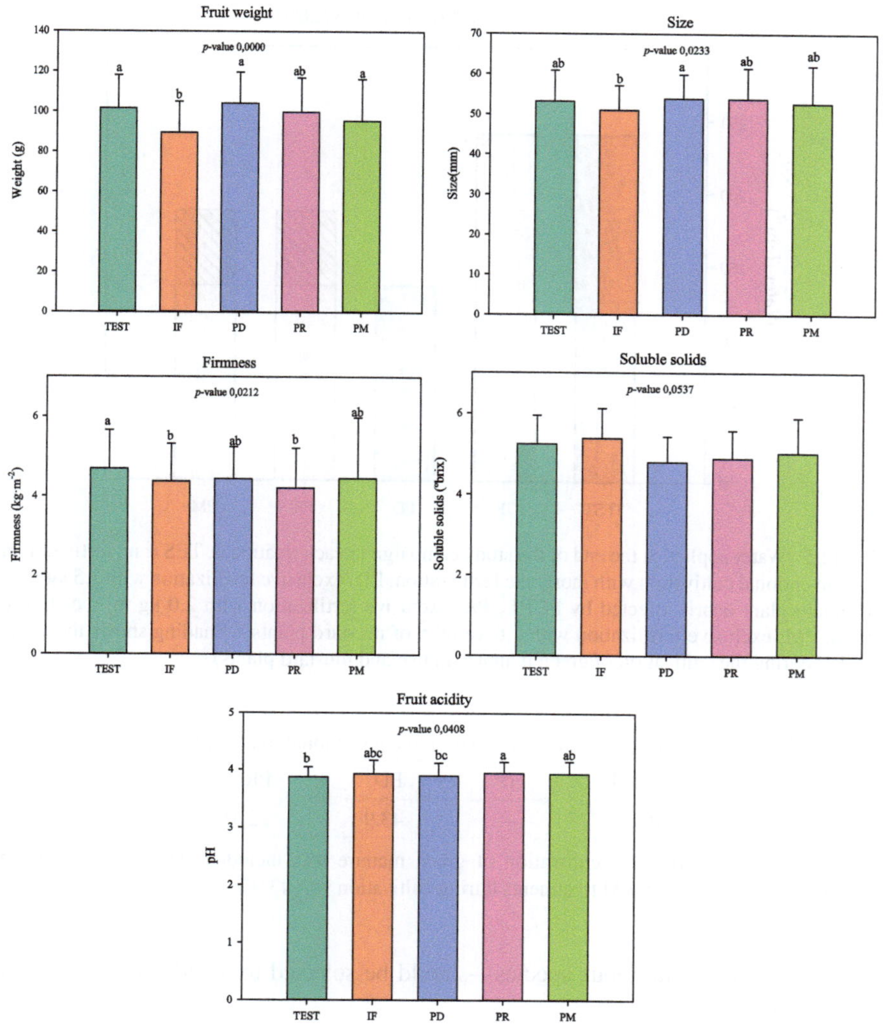

Fig. 14.4 Fruit quality parameters in the study season (September–February cycle). TEST: no fertilization; IF: conventional cultivation with inorganic fertilization; PD: exclusive fertilization with 3.5 kg m^{-2} of tomato plant debris affected by FORL; PR: exclusive fertilization with 2.0 kg m^{-2} of radish plants; PM: exclusive fertilization with 1.6 kg m^{-2} of mustard plants. Different letters indicate significant differences ($p \leq 0.05$, LSD test)

in Murcia during the first year after its implementation (Fernández et al., 2018). The infiltration rate of the PD, PR, and PM treatment plots experienced a reduction in our trial. Soil fertility has also improved, which allowed the assignment of a different irrigation endowment and frequency for each treatment.

Given the need to establish green manure in organic farming (European Union, 2018), those species adapted to the edaphoclimatic conditions of the Almeria

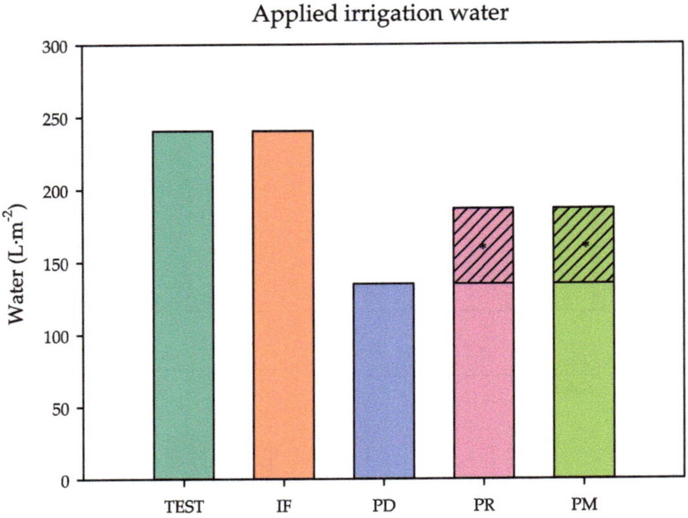

Fig. 14.5 Water applied at the end of the study campaign in each treatment. TEST: no fertilization; IF: conventional cultivation with inorganic fertilization; PD: exclusive fertilization with 3.5 kg m^{-2} of tomato plant debris affected by FORL; PR: exclusive fertilization with 2.0 kg m^{-2} of radish plants; PM: exclusive fertilization with 1.6 kg m^{-2} of mustard plants * Shading shows the water applied during the cultivation of green manure (radish and mustard plants)

Table 14.2 Reduced irrigation water with respect to conventional treatment

	TEST	IF	PD	PR	PM
Variation (%)	0.0	–	43.9	22.4[a]	22.4[a]

[a] The water used during the cultivation of green manure was included. The water reduction experienced by the PR and PM treatments during cultivation was 43.9%

Model—mainly leguminous species—should be selected to avoid excessive water consumption.

Adding organic amendments through the biosolarization technique is a strategy postulated within the circular economy framework to reduce greenhouse crop water footprint (Castillo Díaz et al., 2021, 2022; Fernández et al., 2018). This reduction is achieved by expanding agricultural systems' sustainability while reducing their environmental footprint. In Spain, irrigated crops consume approximately 80.5% each year (Ministry of the Presidency, 2022). Even though water consumed by Spanish irrigation has been reduced by 10% during the last decade (INE, 2020), it is necessary to apply all strategies that favor a lower consumption of such a basic resource as water.

14.3.5 Economic Analysis

Production costs ranging from 54,446.4 to 57,041.4 € were obtained in the crop cycle, similar to those offered by Cajamar (2015) for a typical farm for the 2014/2015 season (Table 14.3).

The difference between the production costs of each treatment was due to the reduction in some of the items in the expense account of each treatment (i.e., TEST,

Table 14.3 Production costs obtained at the end of the study season (September–February cycle)

	TEST	IF	PD	PR	PM
Total variable cost (€)	45,598.6	47,549.3	44,954.2	45,659.2	45,579.2
	149.3	149.3	149.3	149.3	149.3
Removal of plant debris	971.0	971.0	0.0	971.0	971.0
Incorporation of plant residues	0.0	0.0	869.5	0.0	0.0
Green manure seeds	0.0	0.0	0.0	160.0	80.0
Green manure irrigation	0.0	0.0	0.0	443.5	443.5
Veneers	0.0	0.0	0.0	0.0	0.0
Water for solarization	223.6	223.6	125.5	125.5	125.5
Chemical disinfectant	0.0	0.0	0.0	0.0	0.0
Covering and structure (€)	2292.1	2292.1	2292.1	2292.1	2292.1
Seeds and seedling production (€)	5530.7	5530.7	5530.7	5530.7	5530.7
Labor, supplies, etc.	35,417.6	35,417.6	35,417.6	35,417.6	35,417.6
Water	1014.3	1014.3	569.4	569.4	569.4
Fertilizers	0.0	1950.6	0.0	0.0	0.0
Total fixed costs (€)	9492.2	9492.2	9492.2	9492.2	9492.2
Soil maintenance (€)	994.5	994.5	994.5	994.5	994.5
Covering and structure (€)	1991.0	1991.0	1991.0	1991.0	1991.0
Energy and fixed supplies (€)	786.6	786.6	786.6	786.6	786.6
Insurance, management, and financial services (€)	1731.1	1731.1	1731.1	1731.1	1731.1
Equipment and irrigation system (€)	3988.9	3988.9	3988.9	3988.9	3988.9
Total expenses (€/ha)	55,090.8	57,041.4	54,446.4	55,151.4	55,071.4
Decrease in production costs (%)	−3.4	0.0	−4.5	−3.3	−3.5

TEST: no fertilization; IF: conventional cultivation with inorganic fertilization; PD: exclusive fertilization with 3.5 kg m^{-2} of tomato plant debris affected by FORL; PR: exclusive fertilization with 2.0 kg m^{-2} of radish plants; PM: exclusive fertilization with 1.6 kg m^{-2} of mustard plants. Source: own elaboration based on other authors (Castillo-Díaz et al., 2022; Honoré et al., 2019).

PD, PR, and PM) compared to the conventional management applied in the Almeria Model (i.e., IF) (Table 14.3). The decrease occurs in the "variable costs" due to the replacement of some habitual inputs by agroecological techniques (i.e., green manure and biosolarization) or by-products (i.e., agricultural biomass) in line with the principles of the circular economy and sustainability criteria required by the European Union (European Commission, 2020a, 2019) and the UN through the SDGs (UN, 2015). The benefits of using plant residues or green manures as the sole and exclusive fertilizer resulted in a decrease in the cost of water and inorganic synthetic fertilizers. The drop in production costs was 3.4%, 4.5%, 3.3%, and 3.5% for TEST, PD, PR, and PM, respectively. The treatment where agricultural biomass was utilized experienced a lower reduction in production costs than that reported by other authors for greenhouse tomato cultivation in Almeria. However, the differences could be explained by the different lengths of the production cycles, which extended up to 217 DAT, while in our trial, it was 169 DAT (Castillo-Díaz et al., 2022).

The conventional crop fertilized with inorganic synthetic fertilizers (i.e., IF) significantly reduced the net profit before taxes compared to treatments fertilized with green manure, FORL-affected plant debris, or non-fertilized treatments. This inequality was due to a difference in yield quality, mainly in fruit mass and equatorial diameter. However, despite these results, all treatments obtained economic losses at the end of the production season (Fig. 14.6).

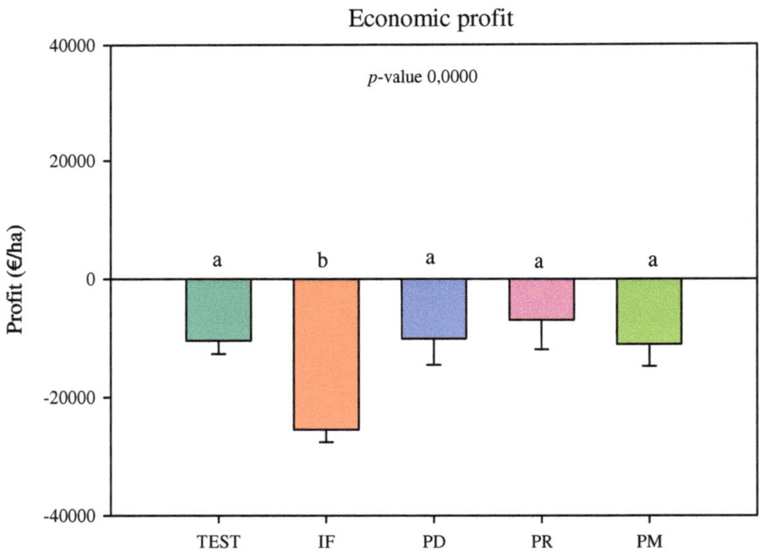

Fig. 14.6 Pre-tax profit obtained at the end of the study campaign (September–February cycle). TEST: no fertilization; IF: conventional cultivation with inorganic fertilization; PD: exclusive fertilization with 3.5 kg m^{-2} of tomato plant debris affected by FORL; PR: exclusive fertilization with 2.0 kg m^{-2} of radish plants; PM: exclusive fertilization with 1.6 kg m^{-2} of mustard plants. Different letters indicate significant differences ($p \leq 0.05$, LSD test)

The decrease in production costs could not alleviate the profitability problem suffered by some producers of the greenhouse agricultural system in the province of Almeria due to the stability of prices at origin and the increase in production costs in recent years (Castillo-Díaz et al., 2022; Honoré et al., 2019). However, this analysis has not contemplated decreases in variable expenses that can help mitigate economic losses due to their crucial influence on the producer expense accounts (Castillo-Díaz et al., 2022; Honoré et al., 2019). The first of these expenses is the labor demanded by the crops grown in the Almeria Model. The figure can vary from 21.2 to 86.4% of annual campaign expenditures, depending on the cultivated plant species (Honoré et al., 2019). In periods of low profitability, this labor can be substituted by family labor or assumed by the farm owner himself to reduce the number of wage earners by being a regular member of the workforce (Valera-Martínez et al., 2014) despite increasing his working hours. In the 2014/2015 season, de Andalucia (2015) estimated family labor for the crop at 8800 €/ha. This would reduce production costs by 16%. The second variable expense item is the net profit before taxes that the farm owner renounces in the productive campaigns where economic losses are generated. The minimum inter-professional wage of the owner is contemplated in the salaried labor expenses item as he is one more member of the workforce and the economic benefit obtained complements this wage. The third variable is the indebtedness ratio of farms, which may represent a significant part of the income. The use of agricultural biomass or green fertilizer through the biosolarization technique could help mitigate the impact of economic losses in combination of these three expense items (Castillo-Díaz et al., 2022).

The economic situation experienced by some of the traditional crops in the Almeria Model pushes the search for new crops to diversify the range of products offered by protected agriculture in Almeria. In this line, some authors have proposed papaya as a viable alternative to high economic interest for use in greenhouse agriculture in Almeria (Honoré et al., 2019).

14.4 Conclusions

The results of this research postulate agricultural biomass affected by limiting diseases in tomato crops (i.e., *Fusarium oxysporum* f.sp. *radicis-lycopersici*) and green manuring with radish and mustard plants as an alternative to conventional inorganic fertilization. This expands the number of tools available to greenhouse growers in the circular economy framework to reduce the use of fertilizers and increase the amount of organic land as some of the goals proposed by the European Union for 2030.

Regarding economic and environmental footprint, biomass is more suitable than green manure because it allows for a higher reduction in production and water costs, the latter due to the amount of water demanded by green manure during its growth. Therefore, the use of agricultural biomass on farms should be expanded to increase the sustainability of production models.

The profitability of the tomato crop analyzed was nil due to the stability of prices at origin and the rise in production costs. To maintain farm profitability, the application of measures and cultivation tasks that allow lowering production costs should be favored. In this sense, using agricultural biomass can help mitigate economic losses in combination with family labor, the indebtedness ratio of the farm owner, and the financial benefit that the farm owner renounces in seasons when economic losses occur.

The fertilizing power of different types of agricultural waste biomass generated in agrarian systems should be evaluated in future research. It should also be determined which native leguminous species adapted to the soil and climatic conditions of greenhouse crops could serve as green manure. These would demand less water, and their water footprint would be reduced as proposed by the regulatory requirements of the European Union's Organic Production Regulation for their establishment.

Acknowledgements The authors wish to thank the dedication of the members of the AGR-200 Research Group of the University of Almeria, especially César Antonio Ruiz Olmos, and the UAL-ANECOOP Experimental Farm.

References

Achmon, Y., Harrold, D. R., Claypool, J. T., Stapleton, J. J., VanderGheynst, J. S., & Simmons, C. W. (2016). Assessment of tomato and wine processing solid wastes as soil amendments for biosolarization. *Waste Management, 48*, 156–164. https://doi.org/10.1016/j.wasman.2015.10.022

Basavalingaiah, K., Paramesh, V., Parajuli, R., Girisha, H. C., Shivaprasad, M., Vidyashree, G. V., Thoma, G., Hanumanthappa, M., Yogesh, G. S., Misra, S. D., Bhat, S., Irfan, M. M., & Rajanna, G. A. (2022). Energy flow and life cycle impact assessment of coffee-pepper production systems: An evaluation of conventional, integrated and organic farms in India. *Environmental Impact Assessment Review, 92*, 106687. https://doi.org/10.1016/j.eiar.2021.106687

Bello, A., González, J. A., & Tello-Marquina, J. C. (1997). La biofumigación como alternativa a la desinfección de suelos. *Horticultura Internacional, 17*, 41–43.

Bhuiyan, K., & Jannat, R. (2018). Effect of biofumigation and soil solarization on stem canker and black scurf diseases of potato (Solanum tuberosum L.) caused by Rhizoctonia solani solate PR2. *Advan in Agricul Sci., 6*, 33–48.

Bilalis, D., Krokida, M., Roussis, I., Papastylianou, P., Travlos, I., Cheimona, N., & Dede, A. (2018). Effects of organic and inorganic fertilization on yield and quality of processing tomato (Lycopersicon esculentum Mill.). *Folia Horticulturae, 30*(2), 321–332. https://doi.org/10.2478/fhort-2018-0027

Cajamar (2015) *Análisis de la campaña hortofrutícola de Almería*. Campaña. (2014/2015). Cajamar-Caja Rural. Almería, España.

Cajamar (2021) *Análisis de la campaña hortofrutícola de Almería*. Campaña. (2020/21). Cajamar-Caja Rural. Almería, España.

Camacho-Ferre, F. (2004). *Técnicas de producción en cultivos protegidos*. Caja Rural Intermediterránea.

Cánovas, J., Alcatraz, N., Gómez, M. C., & Vicente, F. E. (2003). *Contaminación por nitratos en pimiento de invernadero en el campo de Cartagena (Campañas 1998–99, 1999–00 y 2000–01).* Comunidad Autónoma de la Región de Murcia. Consejería de agricultura, agua y Medio ambiente, región de Murcia, España.

Castillo-Díaz, F. J., Ruiz-Olmos, C. A., Marín-Guirao, J. I., & Tello-Marquina, J. C. (2019). Biosolarización y su relación con enfermedades del cultivo de tomate bajo invernadero. In I. I. Congreso de Jóvenes, & I. en Ciencias Agroalimentarias (Eds.) (p. 246). Almería, España.

Castillo-Díaz, F. J., Belmonte-Ureña, L. J., Camacho-Ferre, F., & Tello Marquina, J. C. (2022). Biodisinfection as a profitable fertilization method for horticultural crops in the framework of the circular economy. *Agronomy, 12*(2), 521. https://doi.org/10.3390/agronomy12020521

Castillo-Díaz, F. J., Marín-Guirao, J. I., Belmonte-Ureña, L. J., & Tello-Marquina, J. C. (2021). Effect of repeated plant debris reutilization as organic amendment on greenhouse soil fertility. *International Journal of Environmental Research and Public Health, 18*(21), 11544. https://doi.org/10.3390/ijerph182111544

Chamorro, M., Miranda, L., Domínguez, P., Medina, J. J., Soria, C., Romero, F., López Aranda, J. M., & De los Santos, B. (2015). Evaluation of biosolarization for the control of charcoal rot disease (Macrophomina phaseolina) in strawberry. *Crop Protection, 67*, 279–286. https://doi.org/10.1016/j.cropro.2014.10.021

de Andalucia, J. (2015). *Costes de producción: Campaña 2014/15 Tomate larga vida.* 2014. Consejería de Agricultura, Ganadería Pesca y Desarro Sostenible la Junta Andalucía, Sevilla, España.

de Región de Murcia, C. G. (2019) *Decreto-Ley 2/2019 de 26 de diciembre, de protección integral del Mar Menor.* Boletín of la Región Murcia 36008–36089.

Dhaoui, O., Antunes, I., Agoubi, B., & Kharroubi, A. (2022). Integration of water contamination indicators and vulnerability indices on groundwater management in Menzel Habib area, south-eastern Tunisia. *Environmental Research, 205*, 112491. https://doi.org/10.1016/j.envres.2021.112491

Duque-Acevedo, M., Belmonte-Ureña, L. J., Plaza-Úbeda, J. A., & Camacho-Ferre, F. (2020). The management of agricultural waste biomass in the framework of circular economy and bioeconomy: An opportunity for greenhouse agriculture in Southeast Spain. *Agronomy, 10*(4), 489. https://doi.org/10.3390/agronomy10040489

European Commission. (2019). *The European green deal.* Publications Office of the European Union.

European Commission. (2020a). *A farm to fork strategy for a fair, healthy and environmentally friendly food system.* Publications Office of the European Union.

European Commission. (2020b). A new circular economy action plan. In *For a cleaner and more competitive Europe.* Publications Office of the European Union.

European Commission. (2020c). *EU biodiversity strategy for 2030.* Bringing nature back into our lives. Publications Office of the European Union.

European Union. (2018). Regulation (EU) 2018/848 of the European Parliament and of the Council of 30 May 2018 on organic production and labelling of organic products and repealing Council Regulation (EC), 834/2007. *O Journal of the europ union L, 150*, 1–92.

European Union. (2019). *Audit preview.* Biodiversity in farming. Publications Office of the European Union.

Fernández, P., Lacasa, A., Guirao, P., & Larregla, S. (2018). Effects of Biosolarization with fresh sheep manure on soil physical properties of pepper greenhouses in Campo de Cartagena. In F. Artés-Hernández, J. E. Cos, J. A. Fernández-Hernández et al. (Eds.), *Proceedings of the 6th Workshop on Agri-Food Research* (pp. 97–100). Universida. Cartagena.

Food and Agriculture Organization. (2015). *World's soil resources.* Food and Agriculture Organization of the United Nations.

Food and Agriculture Organization. (2017). *Towards zero hunger 1945–2030.* Food and Agriculture Organization of the United Nations.

Food and Agriculture Organization. (2019a). Background notes on sustainable, productive and resilient agro-food systems. Food and Agriculture Organization of the United Nations.

Food and Agriculture Organization. (2019b). *The international for the sustainable use Code of Conduct and management of fertilizers*. Food and Agriculture Organization of the United Nations.

Fundación Encuentro. (1992). *Conferencia de Rio sobre el Medio Ambiente y el Desarrollo*. Artydis, SA.

Gómez-Tenorio, M. A., Lupión-Rodríguez, B., Boix-Ruiz, A., Ruiz-Olmos, C., Marín-Guirao, J. I., Tello-Marquina, J. C., Camacho-Ferre, F., & de Cara-García, M. (2018). Meloidogyne-infested tomato crop residues are a suitable material for biodisinfestation to manage Meloidogyne sp. in greenhouses in Almería (south-east Spain). *Acta Horticulturae*, (1207), 217–222. https://doi.org/10.17660/ActaHortic.2018.1207.29

Gómez-Tenorio, M. Á., Magdaleno-González, J., Castillo-Díaz, F. J. et al. (2021a). Presencia de rizobios en suelos agrícolas de las provincias de Granada y Almería. In *Córdoba 2020 congreso nacional de ciencias hortícolas* (Vol. XVI, p. 143). Córdoba, España.

Gómez-Tenorio, M. Á., Magdaleno-González, J., & Tello-Marquina, J. C. (2021b). *Evaluación e implementación de técnicas regenerativas para la mejora de la fertilidad en el cultivo del almendro en las provincias de Almería y Granada*. Portal TecnoAgrícola Madrid.

Gómez-Tenorio, M. A., Ruiz-Olmos, C. A., Marín-Guirao, J. I., et al. (2016). La biodesinfección de un suelo arenado y relación con el lixiviado de nitratos. *Acta Horticulturae, 327*, 16–21.

Guerrero, M. M., Ros, C., Lacasa, C. M., Martínez, V., Lacasa, A., Fernández, P., Martínez, M. A., Núñez-Zofío, M., Larregla, S., Díez-Rojo, M. A., & Bello, A. (2010). Effect of biosolarization using pellets of brassica carinata on soil-borne pathogens in protected pepper crops. *Acta Horticulturae*, (883), 337–344. https://doi.org/10.17660/ActaHortic.2010.883.42

Honoré, M. N., Belmonte-Ureña, L. J., Navarro-Velasco, A., & Camacho-Ferre, F. (2019). Profit analysis of papaya crops under greenhouses as an alternative to traditional intensive horticulture in Southeast Spain. *International Journal of Environmental Research and Public Health, 16*(16), 2908. https://doi.org/10.3390/ijerph16162908

INE. (2020). *Encuesta sobre el uso del agua en el sector agrario*. Año 2018. https://www.ine.es/dyngs/INEbase/es/operacion.htm?c=estadistica_C&cid=1254736176839&menu=ultiDatos&idp=1254735976602. Retrieved May 2, 2022.

IPCC. (2019). *An IPCC Special Report on climate change, desertification, land degradation, sustainable land management, food security, and greenhouse gas fluxes in terrestrial ecosystems*. Food and Agriculture Organization of the United Nations.

Jacobs, A. A., Evans, R. S., Allison, J. K., Garner, E. R., Kingery, W. L., & McCulley, R. L. (2022). Soil and Tillage Research Cover crops and no-tillage reduce crop production costs and soil loss, compensating for lack of short-term soil quality improvement in a maize and soybean production system. *Soil and Tillage Research, 218*, 105310. https://doi.org/10.1016/j.still.2021.105310

Junta de andalucía. (2016) *Líneas de actuación en materia de gestión de restos vegetales en la horticultura de Andalucía*. Consejería de Agricultura, Ganadería Pesca y Desarro Sostenible la Junta Andalucía, Sevilla, España.

Katan, J., Greenberger, A., Alon, H., & Grinstein, A. (1976). Solar Heating by polyethylene mulching for the control of diseases caused by soil-borne pathogens. *Phytopathology, 66*(5), 683–688. https://doi.org/10.1094/Phyto-66-683

Kirkegaard, J. A., Gardner, J., Desmarcherlier, J. M., & Angus, J. F. (1993). Biofumigation using Brassica species to control pest and diseases in horticulture and agriculture. In N. Wrather & R. J. Mailes (Eds.), *Proceedings of 9th Australian Research Assembly on Brassicas* (pp. 77–82). Waga Wagga, Australia, October 5–7.

Kirkegaard, J. A., & Sarwar, M. (1998). Biofumigation potential of brassicas: I. Variation in glucosinolate profiles of diverse field-grown brassicas. *Plant and Soil, 201*(1), 71–89. https://doi.org/10.1023/A:1004364713152

Marín-Guirao, J. I., Tello, J. C., Díaz, M., Boix, A., Ruiz, C. A., & Camacho, F. (2016). Effect of greenhouse soil bio-disinfection on soil nitrate content and tomato fruit yield and quality. *Soil Research, 54*(2), 200–206. https://doi.org/10.1071/SR15106

Masahiko, I., Masakazu, H., Nobuko, F., Tomohiro, K., & Yasujiro, M. (2014). Glucosinolate metabolism, functionality and breeding for the improvement of Brassicaceae vegetables. *Breeding Science, 64*, 48–59.

Maudos, J., Salamanca, J., Ballesteros, M., & Miravalles, B. (2019). *Observatorio sobre el sector agroalimentario de las regiones españolas. Informe 2019.* Cajamar-Caja Rural, Almería, España.

Maudos, J., Salamanca, J., Raya, A., & Miravalles, B. (2020). *Observatorio sobre el sector agroalimentario español en el contexto europeo Informe 2020.* Cajamar-Caja Rural. Almería, España.

Mauromicale, G., Lo Monaco, A. L., & Longo, A. M. G. (2010). Improved efficiency of soil solarization for growth and yield of greenhouse tomatoes. *Agronomy for Sustainable Development, 30*(4), 753–761. https://doi.org/10.1051/agro/2010015

Mauromicale, G., Longo, A. M. G., & Lo Monaco, A. (2011). The effect of organic supplementation of solarized soil on the quality of tomato fruit. *Scientia Horticulturae (amsterdam), 129*, 189–196. https://doi.org/10.1016/j.scienta.2011.03.024

Ministry of the Presidency. (2022). *Council of Ministers*. https://www.lamoncloa.gob.es/consejodeministros/referencias/Paginas/2022/refc20220322v02%20(2).aspx. Retrieved May 2, 2022

Monge, E., Espada, L., Blanco, A., & Val, J. (2007). Efecto de la sobrefertilización en la calidad de la manzanas. *Acta Horticulturae, 49*, 195–201.

Navarro-Garza, H., Colin, J., & Milleville, P. (1993). *Sistemas de producción y desarrollo agrícola.* Colegio de Postgraduados. de México y ORSTOM. Ciudad, México, México.

Núñez-Zofio, M., Larregla del Palacio, S., & Garbisu, C. (2012). Repeated biodisinfection controls the incidence of Phytophthora root and crown rot of pepper while improving soil quality. *Spanish Journal of Agricultural Research, 10*(3), 794–805. https://doi.org/10.5424/sjar/2012103-571-11

Pieper, J. R., & Barrett, D. M. (2009). Effects of organic and conventional production systems on quality and nutritional parameters of processing tomatoes. *Journal of the Science of Food and Agriculture, 89*(2), 177–194. https://doi.org/10.1002/jsfa.3437

Polat, E., Demir, H., & Erler, F. (2010). Yield and quality criteria in organically and conventionally grown tomatoes in Turkey. *Scientia Agricola, 67*(4), 424–429. https://doi.org/10.1590/S0103-90162010000400008

Salas, M., Urrestarazu, M., & Valera-Martínez, D. L. (2003). La contaminación por nitratos en los sistemas agrícolas. *Vida Rural, 179*, 38–41.

Salinas, J., Meca, D., & del Moral, F. (2020). Short-term effects of changing soil management practices on soil quality indicators and crop yields in greenhouses. *Agronomy, 10*(4), 582. https://doi.org/10.3390/agronomy10040582

Tello-Marquina, J. C., & del Moral, F. (2001). Las enfermedades no víricas del tomate. In F. Nuez (Ed.), *El cultivo del tomate, Mundi-Pren* (pp. 385–468). Almería.

UN. (2015). *Draft outcome document of the United Nations summit for the adoption of the post-2015 development agenda.* United Nations, New York, USA.

Valera-Martínez, D. L., Belmonte-Ureña, L. J., Molina Aiz, F. D., & López Martínez, A. (2014). *Los invernaderos de Almería. Análisis de su tecnología y rentabilidad.* Cajamar-Caja Rural, Almería, España.

Yepis, O., Fundora, O., Pereira, C., & Cresco, T. (1999). La contaminación ambiental por el uso excesivo de fertilizantes nitrogenados en el cultivo del tomate. *Scientia Gerundensis, 24*, 5–12.

MIX
Papier aus verantwortungsvollen Quellen
Paper from responsible sources
FSC® C105338

FSC
www.fsc.org

If you have any concerns about our products,
you can contact us on
ProductSafety@springernature.com

In case Publisher is established outside the EU,
the EU authorized representative is:
Springer Nature Customer Service Center GmbH
Europaplatz 3, 69115 Heidelberg, Germany

Printed by Libri Plureos GmbH
in Hamburg, Germany